ID0787423

# SEDITION HUNTERS

# SEDITION HUNTERS

## HOW JANUARY 6TH BROKE THE JUSTICE SYSTEM

# RYAN J. REILLY

PUBLICAFFAIRS

New York

Copyright © 2023 by Ryan J. Reilly
Cover design by Pete Garceau
Cover photograph © Shay Horse/Nurphoto/Shutterstock
Cover copyright © 2023 by Hachette Book Group, Inc.

Hachette Book Group supports the right to free expression and the value of copyright. The purpose of copyright is to encourage writers and artists to produce the creative works that enrich our culture.

The scanning, uploading, and distribution of this book without permission is a theft of the author's intellectual property. If you would like permission to use material from the book (other than for review purposes), please contact permissions@hbgusa.com. Thank you for your support of the author's rights.

PublicAffairs
Hachette Book Group
1290 Avenue of the Americas, New York, NY 10104
www.publicaffairsbooks.com
@Public_Affairs

Printed in Canada

First Edition: October 2023

Published by PublicAffairs, an imprint of Hachette Book Group, Inc. The PublicAffairs name and logo is a trademark of the Hachette Book Group.

The Hachette Speakers Bureau provides a wide range of authors for speaking events. To find out more, go to www.hachettespeakersbureau.com or email HachetteSpeakers@ hbgusa.com.

PublicAffairs books may be purchased in bulk for business, educational, or promotional use. For more information, please contact your local bookseller or the Hachette Book Group Special Markets Department at special.markets@hbgusa.com.

The publisher is not responsible for websites (or their content) that are not owned by the publisher.

Library of Congress Control Number: 2023940796

ISBNs: 9781541701809 (hardcover), 9781541701823 (ebook)

MRQ-C

10 9 8 7 6 5 4 3 2 1

*For my family and democracy.*

# Contents

# Preface

## Early 2023
## Undisclosed Location

The beating heart of the largest FBI investigation in American history isn't in the J. Edgar Hoover Building, or in Quantico, or in the type of drab federal office building a screenplay writer would refer to as nondescript. It's in places like Alex's garage.

When the US Capitol came under attack two years earlier, on Jan. 6, 2021, Alex was hundreds of miles away, and he was pissed. He'd watched the day unfold on Fox News, and he couldn't believe what he saw. That was *his* Capitol. That was *our* Capitol. One of his daughters had a basketball game that night, and when she joined him on the bleachers after her game ended, he couldn't quite find the words to articulate to her the significance of the day. She'd been in school, so she wasn't aware of all that took place in Washington. She probably wouldn't appreciate it until later, he told her, but Jan. 6 was a very sad day for the country.

"I was still almost teary at that point after watching it," Alex said. "I was trying to talk to her and just trying to explain to her what a big deal it was."

Some of his friends didn't seem to think Jan. 6 was all that bad, so Alex set out to prove them wrong, spending part of Jan. 7 surfacing videos that showed the extent of the violence and chaos at the Capitol.

"There was debate over whether there was antifa in the crowd, you know, all of the idiot arguments," Alex said. "I was basically looking for videos to send to some of my right-wing friends just to show them this was not a peaceful rally."

Alex stumbled upon some of the budding efforts to identify the rioters, watching as social media users began collaborating online. He's a software guy, and he thought his skills might be useful. "Mass collection and standardization, and even web scraping of data, is something that is in my wheelhouse," he said. His brain started churning.

*How do you organize all this material? How do you archive all that content? How the hell do you quickly find the piece of evidence that you need? How do you enable others?*

A couple of months later, Alex made the tough decision to skip out on a family vacation. He had an app to launch.

Flash forward to 2023. A community of online sleuths—"Sedition Hunters," as they'd dubbed themselves—had aided in hundreds of cases against Jan. 6 defendants. What's more, the sleuths had identified more than 700 Jan. 6 participants who had not yet been arrested.

There was a funeral home director who'd sprayed cops with a wasp and hornet spray. A celebrity photo collector who'd had his photo taken with Rihanna, Selena Gomez, and Kim Kardashian. An ex-NFL player. A neurosurgeon. A stand-up comic named Kevin Downey Jr., who'd been on *America's Got Talent* a decade earlier.[1] A Trump enthusiast who'd flashed a gun at the Capitol and then fatally stabbed a nineteen-year-old at a park a few months later.[2] Another male model, another corrections officer, another police officer, another real estate agent. A fan of anthropomorphized animals, seen on the Senate floor on Jan. 6, who was identified because a sleuth did a deep dive into the world of furries and found the man's name (and his pseudonym, or "fursona") because the man hosted a furry Thanksgiving party at his "den." A guy who'd previously been arrested for playing a musical instrument naked in public, and a guy who'd since been arrested for walking around his neighborhood without pants.[5] A man associated with the Proud Boys who'd been at the Capitol with his son and was subsequently arrested—with the help of DNA—for the decades-old murder of a seventeen-year-old girl. They'd soon find a husband and wife, both of them Trump administration officials, who climbed through a Capitol window.

The new identifications weren't stopping. One day in early 2023, an online investigator—after combing through what he described as "an avalanche of dicks"—identified a former gay porn actor who appeared to assault a police officer. That wasn't a first.

"The things I do for this country," the sleuth, who I'll call Josh, jokingly messaged me when he got a match. "The number of dicks I've had to see in the name of preserving our democracy, someone should give me a goddamn medal."

The sleuths accomplished all this, in no small part, thanks to an app Alex had whipped up in his garage. Alex's app was one of the main tools of the manhunt: driving investigations, organizing information, generating new leads.

Oh! One other thing: Alex voted for Trump. Twice.

That one came as a bit of a shocker to Alex's fellow Sedition Hunters.

"There are moments in life that you never forget where you are when they happened. Sept. 11. Michael Jackson dying. Finding out Alex voted for Trump in 2020," Josh joked. "It was less of a record scratch moment and more of a record getting blown up by a tomahawk missile moment."

————— ◆ —————

In the two years since the Capitol attack, the feds had racked up impressive numbers: nearly one thousand cases, about five hundred guilty pleas, dozens of significant prison sentences, and the largest number of seditious conspiracy verdicts in modern American history. Federal prosecutors around the country had stepped up to assist the US Attorney's Office for the District of Columbia, which was overseeing a breathtaking number of cases. Assistant US attorneys from across the country were dialing in for hearings and flying in for trials. Now some of those offices were wondering how long this was going to last, how much longer they'd have to dedicate their personnel to the Capitol probe.

"Our work is far from over," Attorney General Merrick Garland assured the American public at the two-year mark. "We remain committed to ensuring accountability for those criminally responsible for the January 6 assault on our democracy. And we remain committed to doing everything in our power to prevent this from ever happening again."

Alex wasn't so sure. He's a data and numbers guy, and the math just didn't work out. The clock on the statute of limitations was ticking, and the feds were quickly approaching the halfway mark. The sleuths had long given up on the notion that everyone who stormed the Capitol or assaulted law enforcement that day— more than three thousand people, they'd determined—would be held accountable. Now they were concerned that the feds wouldn't even charge all the violent offenders in time.

There are many ways to describe the federal investigation in the Jan. 6 attack on the US Capitol. Unprecedented. The largest FBI investigation in American history (by number of defendants, Sept. 11 investigation alums will remind you). Garland called the investigation into Jan. 6 and efforts to interfere with the peaceful transfer of power "the most wide-ranging investigation, and the most important investigation, that the Justice Department has ever entered into."[4]

Another way to describe it? A clusterfuck.

FBI special agents and federal prosecutors were overwhelmed. Hundreds of thousands of tips had rolled in, which would have been a tremendous logistical challenge even if the

FBI weren't a massive, sprawling bureaucracy operating with antiquated communication and organization technology. Critical evidence against key violent defendants was buried within the bureaucracy, siloed into bunkers, and not readily searchable or organized.

And there was an *enormous* amount of digital evidence. The FBI had nearly four million files, thirty thousand videos from body-worn cameras, surveillance video, and the footage rioters had captured on their own devices. You'd be able to watch all the video in just under a year, provided you didn't sleep. The FBI's information technology (IT) offerings—*yesterday's technology tomorrow*, went the unofficial motto—were less than agile, burdened by onerous security settings that made simple operations like sharing large files and reviewing publicly available evidence burdensome.

Some sleuths had formalized relationships with the FBI at this point, and the community knew how much their work was appreciated and thought the bureau special agents they dealt with were dedicated. But they also knew how bad the bureaucracy could be. Sedition Hunters were building entire cases for the FBI from soup to nuts, but the bureau's rules made it difficult for special agents to offer even basic updates on the status of investigations. The one-way street was frustrating. The best written feedback they could hope for was maybe one word: "received." One FBI informant was thrilled just to get thumbs-up emojis from her FBI handler.

Politics was also clearly playing a role, although it was tough to say to what extent. There had long been rumblings about certain FBI field offices being less than enthusiastic about bringing Capitol attack cases, and now some of those internal battles had spilled into public view.

For someone like Alex—who, in the private sector, was used to being able to pivot and adapt quickly, to make changes when things weren't working—it was maddening.

"There's something that's so fundamentally broken," Alex told me. "We're going to figure it out one day, and we're all going to be pissed that we allowed it to happen."

———————◆———————

Since a mob of Donald Trump supporters whipped up by his lies about the 2020 election stormed the US Capitol on Jan. 6, I've immersed myself in the communities of online sleuths who are driving the FBI investigation. I've learned the names of hundreds of not-yet-arrested Jan. 6 participants, including the names of over one hundred individuals on the FBI's Capitol Violence page who are not yet facing charges.

I've spent months getting to know many of the sleuths. I've talked to sleuths all over the country about their techniques, their motivations, and their biggest finds. I know some by their real names, others only by their handles and their investigative track records. We've bonded over child-rearing, attention deficit hyperactivity disorder (ADHD, a common diagnosis among sleuths, it turns out!), sports rivalries, bingeable shows, and memes. Lots of memes.

Many I've even met IRL ("in real life") on their trips to DC. I took one sleuth to the restaurant in the Mayflower Hotel where former FBI director J. Edgar Hoover ate lunch every day for twenty years, which now bears the name Edgar. I took others to the restaurant where, months later, federal prosecutors on the first successful seditious conspiracy trial in modern history would celebrate. One weekend, I even took my daughter on a playdate with a sleuth's kid, walking around the Capitol grounds and then watching as they wolfed down ice cream on the National Mall and entered into a heated debate on whether "chitchat" was a real term. Another sleuth, who'd made the jump back into journalism as a result of his Sedition Hunters work, proved himself a very adept aide when my kid brought out the Where's Waldo? books during his visit.

I've covered the Justice Department for more than a decade, and I'd talked with FBI informants before. But none of them hold a candle to the impact the Sedition Hunters have had. Working out of their home offices, from their couches, kitchen tables, bedrooms, garages, and—in one case—from the sleeper cab of their semitruck, this group of anonymous Americans has been working to hold the FBI's feet to the fire to make sure these cases don't get buried.

Jan. 6 was a pivot point for American democracy. It was also a pivot point for the FBI and law enforcement, which were caught flat-footed despite all the warning signs flashing online ahead of Jan. 6 and were left playing catch-up with open-source researchers moving at internet speed.

"How was it that this mass of civilian sleuths were able to compile all this information, and so rapidly?" DC Metropolitan Police Department (MPD) officer Michael Fanone, whose assailants were identified by online sleuths, later asked me. "Maybe we need to be changing what it is that we're looking for in our hiring, because I sure as shit don't know how to do any of that crap. I still fucking turn off my computer by pulling the plug."

———————•◆•———————

The sleuths aren't a monolith. They're from across the country, from the east coast to California, from the deep south to the Midwest. Especially early on, there were several sleuths living outside the country, including a meme-making grandma from the Hague who made TikToks with her granddaughter. There were sleuths from right around DC too, including the spouse of a police officer who served on Jan. 6.

There were a handful of different sleuth groups that organized under different banners, with some members who belonged to multiple groups. A lot of the initial buzz was on Twitter, but soon moved to the messaging applications that prospered during the pandemic. Soon sleuths were bonding and forming real friendships, even when they were talking to people they only knew by their screennames and profile images. Early on, after one sleuth mentioned that they were expecting a child, another sleuth offered up breastfeeding advice, only to learn they were talking to a soon-to-be dad. Now they exchange Christmas cards.

Some sleuths kept at it a few weeks, others a few months. Others popped in-and-out, as their life and schedule allowed. Some were already plotting what was going to happen after Jan. 6, 2026, when the statute of limitations expired, and the justification for withholding the names of identified rioters from the public—namely, that making those names known would negatively impact the cases they hoped the feds were building—was no longer a concern.

They'd scour Facebook, Instagram, YouTube, Twitter, TikTok, Rumble, Gab, and Telegram. Venmo, the payment app that defaulted to lax security settings, proved particularly useful in making connections. Facial recognition played a big role in creating initial leads, but the biggest dopamine hits came from the finds that took some work. One rioter was identified because of the "dumb ass way he folds the brim of his hat"—a sleuth was scouring Parler profiles and the hat fold caught their eye.

Fresh video was the lifeblood of the investigation, and it would help in ways that were tough to imagine. A lot of rioters did a frustratingly solid job of keeping their face covered the entire time they were at the Capitol. But in one instance, in a moment caught on video, a rioter whose face was fully covered pulled out their phone. The video caught a glimpse of the screen, revealing a background image that showed the name of the rioter's small business. That's all it took.

———◆———

There's a person playing an important role in all this: a lightning rod, a "middle man," or "middle woman"—the FBI has kept their gender a secret, and I'll do so here too. They're a confidential human source, or CHS, for the FBI. I'll call them Chris. Read through the thousand or so FBI affidavits in Jan. 6 cases, and you'd eventually stumble upon a paragraph from a case in mid-2022 that describes Chris like this:[5]

> *This CHS was an established source who led a team of open-source researchers who collaborated shortly after January 6, 2021, to identify United States Capitol rioters. This CHS group was motivated by a desire to assist law enforcement regarding the events of January 6, 2021, at the United States Capitol. The CHS did not have personal knowledge of the subjects of reporting, but rather derived the information solely through open-source research.*

In other FBI affidavits, you'd learn that Chris wasn't working off any charges, that "his/her sense of outrage regarding the attacks on law enforcement during the events of January 6, 2021" had motivated them, and that they were willing to testify if necessary.[6] You'd learn that Chris had never steered them wrong. Two years after Chris and other sleuths had dedicated countless hours to aiding the FBI investigation, an affidavit against a Montana man who had brought his minor child inside the Capitol on Jan. 6 revealed that Chris had, by that point, "been compensated for his/her time."[7]

Chris knew the sleuths were having a big impact. It wasn't until they visited the FBI sometime in 2021 that they realized the extent of it, when they were able to meet with members of law enforcement who had served on Jan. 6 and appreciated the work of the online sleuths identifying those who had assaulted them.

Chris had done a lot of things in their life. They'd raised tens of thousands of dollars for charity, volunteered thousands of hours, and even saved a guy's life once. But when a Capitol Police officer handed them a challenge coin, they felt as if this was one of the most important things they'd ever done.

The nation's premier law enforcement organization was relying on a ragtag group of so-called amateurs from the internet to piece together the story of Jan. 6. Chris became increasingly tight-lipped about their work with the FBI as time went on, and I'd heard that the bureau would regularly question them about their interactions with me. Soon after that first meeting with the bureau, though, Chris was clearly floored by just how critical the sleuths had been to the investigation.

"The impact we had is huge," Chris said, a fedora festooning their head, like a detective in a film noir. In an almost Trumpian style, they motioned with their hands and widened their eyes. "*Huge!*" they mouthed.

# PART I

# CHAPTER 1

# "Locker Room Talk"

The presidential election was approaching, and a small group of Donald Trump supporters was plotting.

One thought Trump was "the man."[1] They all posted right-wing memes on Facebook. They said they were prepared to kill their perceived enemies, the people Trump said hated America. They were militia members, and they planned to provide security at Trump campaign events. There was talk of overthrowing the government if Trump was declared the loser, talk of placing a bomb at the US Capitol, talk of dragging politicians from the building and lynching them in the streets.

Trump had warned them that there'd be fraud, that the election was "rigged" against him. He asked his supporters to go to "certain areas" to watch voters at the polling place.[2] "It's one big fix," Trump told his supporters a few weeks before the election. "This whole election is being rigged."[3] Perhaps, Trump suggested, "the Second Amendment people" would do something if he lost.[4]

Curtis Allen had posted a Trump campaign sign outside his trailer in a mobile home park in Liberal, Kansas, a town of nineteen thousand just north of the border with the Oklahoma Panhandle. He was a military veteran who enlisted in the US Marine

Corps at the age of seventeen, getting his parents, Jamie and Fred, to sign off on a contract just a few days before he began his senior year of high school in Ashland, Kansas. His twin brother, Kevin, enlisted too. Just after graduation, one week after the twins turned eighteen, Curtis arrived in San Diego. He'd spend four years in the Marines, and he later enlisted in the US Army National Guard. When foreign terrorists attacked the United States on Sept. 11, 2001, Allen was mobilized as an airport security officer in Garden City and Dodge City, Kansas, where he supervised twelve soldiers and worked as a liaison between airport officials, law enforcement, and his command structure.[5]

As the war on terror ramped up, so did Allen's responsibilities. He went to Germany in support of the war in Afghanistan and then served in Iraq during one of the bloodiest stretches of the conflict. He was staying at a camp nicknamed Rocket City, which had been described as the most attacked military base in Iraq. He'd been knocked down by the force of rockets that struck the runway. He saw the dead bodies of Iraqi children, a memory that now moved him to tears. When a car bomb went off nearby, he "observed a large pail of water with a man's face floating in it."

He talked all about it when he walked into the Veterans' Affairs (VA) Medical Center in Wichita. He had trouble sleeping and felt "restless, keyed up, and on edge almost every minute of every day." They diagnosed him with major depressive disorder and then posttraumatic stress disorder. He eventually began to receive a partial disability payout.

Now Allen was almost fifty, working as an alarm salesman, and he took on a second job at a mobile home dealership. He was thinking about starting his own ammunition business.[6] He only sporadically talked to his daughter and brother. Much of his time was spent on Facebook, where he'd digest and repost right-wing memes.

He had an on-again, off-again girlfriend named Lula Harris, who'd sometimes stay at the trailer home he'd surrounded with stones painted red, white, and blue. They'd met at a bar in

Wichita a decade earlier, not far from the VA. She was trying to get an intoxicated friend to the car, but she gave Curtis her phone number, and they met up at the same bar again a few days later. Eventually Lula moved in. But Lula said Curtis was sleeping around, including with an old flame, the wife of a police officer from Washington state.

"Short, tall, skinny, fat, he just loved them all," Lula told me. "He just had a fetish for women."

Allen could be abusive. He wasn't supposed to own a gun because of his record of domestic violence. But there were some good times. The couple were preppers who talked about moving up to Alaska and living off the grid with solar panels. She made quilts; Allen made a tepee.

There were also, of course, the carnal benefits. Dan Day, a fellow gun enthusiast who spent time with Lula and Curtis, said Lula used to joke about how she was only with Curtis because of the size of his penis.

"That's what she was saying, 'He's got a nine-inch cock, and that's why I stayed around,'" Day said. "She would say that to everybody."

She certainly wasn't shy about it. "The sex was good," Lula told me.[7] "Hate to say that, but, yup."

In the leadup to the election, Allen had been a commander of the Kansas Security Force, or KSF, and had cards with his name and phone number on them. "Recruiting Now!" the card read, featuring the logo of the far-right Three Percenters movement. He went by CO, for "commanding officer." Day served as the "vetting" and "intelligence officer" for the KSF division that covered the western part of the state, and was known as Minuteman. Patrick Stein and Gavin Wright were members too, with Stein going by XO, for "executive officer," and Wright going with the nickname Sparky, for his knowledge of electrical work.

Lula, who had just gotten out of cancer treatment, would sometimes listen in to their calls over a walkie-talkie application called Zello. It wasn't too difficult to hear what was up since Curtis had terrible hearing and used the speakerphone. He was also watching videos about explosives, and he played a CD of *The Anarchist Cookbook* on repeat so many times that Lula knew the bomb recipes.[8]

Indeed, a plan was forming. Stein, a former drug addict and current farmer, laid out his thinking in a message to a like-minded man set to help them acquire bomb materials.

"We are dedicated patriots with love of country that doesn't end and we are all willing to die for this country if that is what it takes to get it back," he wrote.

They'd wait until after voters had cast their ballots. A preelection attack would harm Trump's campaign, he reasoned, giving Democrats a talking point.

"I don't want to give them any ammunition at all if we don't have to," Stein wrote. "My thoughts are that it will not be until right after the election. That obstacle will be out-of-the-way at that point."

The stakes were just far too high, and the group couldn't do anything that would hurt Trump's chances, Stein said. After all, they thought Trump was going to lose, or at least that's what the media would say.

"We cannot in any way let Hillary back in to the White House," he texted.[9]

———◆———

The plot out of Kansas unfolded during the 2016 presidential election, before Trump's surprise electoral college victory and chaotic presidency. Before Charlottesville, when Trump said there were "very fine people" among the neo-Nazis and white supremacists, one of whom ran over demonstrators with his Dodge Challenger, killing a young woman named Heather Heyer. Before a Trump "superfan" who lived in a van in Florida mailed pipe bombs to Democratic politicians, media figures, and people Trump had attacked. Before El Paso, where a gunman shot and killed twenty-three people in a Walmart and wrote a manifesto that echoed Trump's language about the "invasion" on the Mexican border.

Allen and the three others—Stein, Wright, and Day—spent hours and hours discussing what would have been one of the worst domestic terrorist attacks in American history. They also became associated with the Liberty Restoration Committee, an effort led by a truck driver with a YouTube channel who wanted to abolish the federal government.[10]

"We can clear out the government, get a new constitution, and if they don't do it, we'll take this up to them, and if they won't voluntarily leave, we will kill them and force them out," Day later said, describing the plan. Stein was even more blunt.

"The only fucking way this country's ever going to get turned around is it will be a bloodbath and it will be a nasty, messy motherfucker," Stein said. It was time that "people in this country wake up and smell the fucking coffee and decide they want this country back," he said.[11]

They talked about a lot of potential targets to send a political message, to wake up the rest of the country. There was even talk of attacking Washington.

"I've said it for fucking years that if we, the American people knew what is going on in D.C. we would storm that place— fucking place, we would grab every motherfucker up there, we'd lynch them downtown and burn the whole motherfucker to the ground," said Stein.

Wright chimed in. "Park the bomb next to the fucking Senate and blow the whole fucking building up," he said.[12]

Maybe they could team up with the Liberty Restoration Committee and other militias and "unite and fucking just go up there and grab the motherfuckers," Stein said. "Or Trump comes in. I mean, there's a whole scenario."

As I scrolled through their Facebook pages during the last few weeks of the 2016 campaign, Stein stuck out as the biggest Trump enthusiast.

"I WOULD VOTE FOR TRUMP AT HIS WORST ANY DAY OF THE WEEK OVER HILLARY AT HER BEST," read one meme Stein posted on Facebook. Another meme—a Photoshopped image of Trump shaking Obama's hand—featured the caption "DONALD TRUMP GRABBING A PUSSY."[13]

Given their admiration for Trump, it wasn't surprising that one of Trump's top targets on the campaign trail was also in the crosshairs of the group that dubbed themselves "the Crusaders."

"I think Islam hates us," Trump had said, as though the world's nearly two billion Muslims were a single inseparable and evil being. Trump called for a Muslim ban, a "total and complete shutdown" of Muslims entering the United States. He mocked the Gold Star mother of Muslim soldier Humayun Khan, who died in a 2004 car bombing in Iraq. He claimed, without evidence, to have seen footage of "thousands" of Muslims celebrating the Sept. 11 attack in Jersey City.[14]

The Crusaders took things even further. Stein called himself "Orkinman," after the pest control company, and referred to Muslims as "cockroaches" he wanted to exterminate. They zeroed in on an apartment complex in Garden City that was home to refugees from Somalia, many of whom worked at the Tyson Foods beef slaughterhouse.

"There were Muslims all over the place," Lula told me. One time, she said, Curtis told her he went to Walmart to buy a bacon pan for the microwave, and a Muslim associate had to get someone else to help him find the pan because they did not want to be associated with a product used to cook pork. He started boycotting Walmart after that. Lula had her concerns too, which seem ripped right from a right-wing Facebook algorithm.

"In schools, we've had change," Lula later complained to me. "It used to be Christmas break; now it's winter break. It used to be 'Merry Christmas'; now it's 'Happy holidays.'"

She listened to Curtis rant about Sharia law, a possible Muslim takeover. So when she heard Curtis and his militia buddies talking about harassing Muslims with dead hogs, she didn't have any strong objections.

"If they cut a pig's throat and ran it into a mosque, I'd probably have laughed. If it hadn't affected anybody, I'd have probably laughed. That would've been something funny, because just the beliefs of them and bacon. I would've laughed about something like that if it didn't hurt anybody," Lula said. "But this was not something that was, you know, I guess it was way more serious."

More serious, indeed. Stein was the most vocal of the bunch, acting in a way that even embarrassed someone like Lula, whom nobody would describe as shy. It was one thing to rant and rave in close company, but Stein took it beyond that. She remembered him using the N-word to refer to then president Barack Obama when they stood in line waiting for barbeque at a county fair.

"We've got Black people standing a few people away from us," Lula said. "He was loud, obnoxious, I mean, just wanting a confrontation with somebody."

That wasn't even the worst of Stein's public outbursts. Once, when scouting out buildings in Garden City with Day, Stein pulled his gun from his truck console and threatened to kill two Muslim women—"raghead bitches," he called them—on the street.

What neither Stein nor the other men knew—and wouldn't learn until after the FBI arrests and their first court appearance—was that Day was an FBI confidential human source, or CHS. "Minuteman" was what his FBI handler called him. And when Stein pulled his gun, Day had his hand on his weapon too.

"I was not going to let him kill these innocent women because they were who they were, because they were Somalis or Muslims," Day said.[15]

Tony Mattivi, a Kansas-based federal prosecutor with an affinity for American-flag socks, wasn't so sure about Day at first. Mattivi had been an assistant US attorney since 1998, working drug-trafficking cases along with bank robberies and gang cases.

He did a stint in Iraq, advising Iraqi prosecutors who were bringing cases against members of Saddam Hussein's regime, including the man known as "Chemical Ali," who was hung for gassing more than five thousand Kurds.[16] Mattivi also led the prosecution of Abd al-Rahim al-Nashiri, the man accused of organizing the al Qaeda bombing of the USS *Cole* in 2000, which killed seventeen soldiers. He supervised the trial team for nearly five years, until 2013, arguing motion after motion at a pop-up courtroom set near an old, dilapidated airplane hangar in Guantanamo Bay, Cuba.[17] The military commissions process at Guantanamo is notoriously messy, and just like the Sept. 11 tribunals, the USS *Cole* case still hasn't begun more than two decades after both attacks.

Back home in Kansas, Mattivi took on other terrorism cases. One was against Terry Lee Loewen, a fiftysomething al Qaeda sympathizer who came onto the FBI's radar after he befriended someone who regularly posted about violent jihad. Loewen worked at the Wichita airport and planned to use his access card to enter a secure area at the airport. But the explosives he was provided by undercover FBI employees were inert, and Loewen was arrested when he tried to use his key card.[18]

Another of Mattivi's cases was against a twenty-year-old US citizen named John T. Booker who had been preparing to join the US Army. Three weeks before he was supposed to report to basic training, he fired up his Facebook page. "I will soon be leaving you forever so goodbye! I'm going to wage jihad and hopes that i die," he wrote in a public post. Someone tipped off the FBI, the FBI interviewed him, and Booker admitted he was planning to commit an insider attack, as Nidal Hassan had done in Fort Hood. Booker was denied entry into the military, and the FBI stayed on his tail. Months later, under the eyes of two FBI informants, he was filming a jihad video and purchasing bomb components. The FBI arrested him as he prepared to set off an inoperable bomb at Fort Riley, an attack he thought he was carrying out on behalf of the Islamic State.[19]

These types of FBI stings against supporters of groups like al Qaeda and ISIL were relatively common. There was an FBI playbook for these cases, one that played out in the hundreds of stings the FBI had run since Sept. 11.[20] Such cases often came under heavy scrutiny from defense attorneys and civil liberties advocates, who said that the defendants often lacked the financial, and mental, capacity to actually pull off the attacks. But courts routinely rejected claims of entrapment.

The FBI takes a much different approach to investigating "foreign" and "domestic" terrorism. The distinction has nothing to do with the location of the attack but rather the origin of the underlying ideology motivating it. Under US law, groups like al Qaeda and ISIL were designated foreign terrorist organizations, meaning they were subject to a different set of laws. Those laws don't apply to "domestic" terrorist organizations, like the Ku Klux Klan or your local right-wing militia. Terrorist attacks motivated by racist hate or by anti-Muslim bigotry? American as apple pie.

Comparatively, there's a relatively low bar for foreign terrorism charges. Just retweeting a message supporting one of those groups could be the basis of a case alleging material support for a designated terrorist organization.[21] There was also a very real threat: Islamic extremists had pulled off several successful terrorist attacks in the years since Sept. 11, including the 2013 Boston Marathon bombing, the 2015 San Bernardino attack, and the 2016 Pulse nightclub shooting, which would kick the Crusaders' plot into overdrive. But the threat had evolved. Between 1994 and 2020, right-wing terrorists perpetrated the majority of the attacks and terrorist plots in the US, according to a report from the Center for Strategic and International Studies in June 2020, which predicted future terrorist attacks, particularly around the upcoming presidential election.[22]

Mattivi had a well-established track record in the foreign terrorism space. But domestic terrorism cases, with their higher standards and potential for political blowback, were a new challenge.

"Man, it gets really murky, really fast," Mattivi told me.

Domestic organizations, no matter how odious their views, enjoy the protection of the First Amendment, an essential element of American democracy that guarantees people the right to express even the most morally repugnant views without fear of government prosecution. And there's reason to be cautious with federal investigations into political organizations: the FBI has a sordid history of targeting disfavored political interests, including Martin Luther King Jr.

There's not a stand-alone crime of domestic terrorism on the federal level, which means that even domestic terrorists like Dylann Roof—the white supremacist who slaughtered nine people inside a church in Charleston in 2015—were not charged with a federal crime of terrorism.

Roof may have been charged with hate crimes, but his attack was clearly domestic terrorism under federal law. The lack of a federal terrorism charge, however, can shape the entire public discussion around an event, with enormous societal and legislative consequences. It's something I pressed former FBI director James Comey about when he sat down with reporters a few weeks after the attack. By that time, Roof's racist manifesto had made his intent clear.

"Given the nature of my business, I only operate in a legal framework," Comey told me. "The only world I live in is when you bring charges against someone and charge them with something under a particular provision that is a terrorism statute, and so that's the framework through which I look at it."[23]

There were a "whole lot" of FBI analysts and agents who worked domestic terrorism cases, Comey said, and he was confident the bureau was putting in the resources where they made sense.

Even when the feds did bring a case against domestic terrorists, it wasn't always clear at the outset. Without a domestic terrorism statute, federal prosecutors would bring whatever charge was applicable. Sometimes, for neo-Nazis, it was a charge under a rarely used statute that made it unlawful for a drug addict to possess a weapon. Frequently, the most readily available charge was a local crime, meaning the suspects were shuffled off to state prosecutors.

"In many instances, the government is going to be constrained, to a certain degree, from stepping in front of a podium and saying, 'Ladies and gentleman, we're revealing domestic terrorism here,'" the Justice Department's counsel for domestic terrorism, Thomas Brzozowski, explained.[24] But that doesn't mean it's not domestic terrorism.

When the feds have brought domestic terrorism cases, they've often fallen apart at trial. Not only is the legal landscape a challenge but judges and juries are just a lot more sympathetic to someone they can imagine as a family member or neighbor than, say, to an ISIL sympathizer.

So when Mattivi learned that his FBI counterparts in Kansas were investigating a militia group, he proceeded with caution.

"I was skeptical. Not just of Dan, but whether any of these groups were more than just talk," Mattivi told me.

Like Stein, Day, and the Crusaders, Mattivi wasn't a big Hillary Clinton fan. "I would rather vote for your cocker spaniel than Hillary Clinton," Mattivi joked. Yet he had a much different view of the world than did those he would end up prosecuting in what would be, at its time, one of the most significant domestic terrorism trials to go before a jury in the twenty-first century.

When Mattivi thinks of Ifrah Ahmed, he thinks of the "embodiment of the American dream." Ahmed is a community leader in Garden City who was born in a refugee camp in Kenya to Somali refugees, came to the US as a young woman, and spent time working at the Tyson plant.

"She is a refugee from a war-torn country. She comes here out of a sense of seeking safety and security. She takes a job that is difficult, gruesome, and tedious and dangerous that most Americans don't want to do," Mattivi said. She got her degree in psychology and now works to provide mental health and support services to refugee kids.

"That's the kind of person that these knuckle-draggers wanted to kill," Mattivi told me. "The person who, if you just look past her traditional dress and the color of her skin, what you see is the American dream. And yet somehow people who thought of themselves as patriots wanted to kill that person."

Mattivi was also clear minded about the role that Trump's rhetoric played in helping incite the plot against Muslim refugees like Ifrah. "These guys were whipped up by Trump himself, and a lot of his immigration rhetoric," Mattivi later told me. "And I say that as a conservative Republican."

Mattivi and the others working the case knew that the stakes were high and that it was critical to run this operation by the book, to build a strong case but also to guide the investigation to its proper end. Day, as federal appellate judges later wrote, had to be "careful not to plant ideas in defendants' heads, given the FBI's repeated instruction that he could only raise ideas that defendants themselves had already discussed."[25]

So Day kept on recording. They needed a manifesto, Allen said, one that would inspire more attacks. "We are going to try to trigger the other like-minded people across the nation to fucking stand up and start doing the same thing we are doing," Allen said.

"American people, you have to wake up while there still might be time to stop our gov't from totally selling this country out," they wrote in the manifesto. "Don't be fooled by the words 'conspiracy theory' or 'domestic terrorist,' all this is a word game, 'brainwashing' by our gov't. Standing up for the constitution is not domestic terrorism.

"This is a call to action by all Americans. Please do not just sit idle until we lose this once great nation!!"[26]

Day also introduced Stein to a black-market weapons dealer, and they all met at a remote field. Brian, the weapons dealer who was actually an undercover officer working for the FBI, soon linked up with Stein over an encrypted app. Stein saw it as divine intervention.

"I know that we have the man upstairs on our side. He works in mysterious ways and sometimes you just have to believe!" Stein wrote Brian.[27]

The FBI sting was building toward a breaking point, one that would bring the investigation to a conclusion in a safe, manageable manner.

Enter Lula. She was getting restless. She'd moved back in with Curtis as she recovered from cancer surgery, even though she thought Curtis's womanizing had given her the human papillomavirus (HPV), which she thought had caused her cancer. She was also skeptical that she was still a part of their future, the life off the grid they'd talked about, if Curtis was planning on committing a terrorist attack. "How can you sit here talking about future plans when you're going to do something this crazy, that most people don't walk away from?" Lula said.

She was also sick of taking care of his puppy, she told me. That's what ultimately sparked the fight: Curtis threatened the dog and then assaulted her, she said, when she threatened to call 911. When the cops arrived, she spilled on the whole plan, telling them that he was cooking explosives and plotting to kill Muslims.

"I really felt that if I hadn't gotten out of the house, I probably wouldn't have gotten out of the house," Lula told me. "I knew too much information." A few years later, Lula added a life event to her Facebook page. "Turned Ex in for Terrorism," she wrote.

Allen's arrest spooked Wright. But Stein moved ahead. The next day, he met with Brian, shooting fully automatic weapons provided by the FBI. They headed over to the apartment complex they planned to target. Stein told Brian he could provide three hundred pounds of fertilizer for a bomb. On Oct. 14, 2016, the duo met in a McDonald's parking lot, where Stein had the fertilizer. They went inside the restaurant to talk over the payment plan and the plot. The FBI SWAT team moved in when Stein returned to his vehicle.

"I was doing it for the country, and you know that!" he told his mom shortly after his arrest.[28]

Inside the interrogation room, FBI special agents Robin Smith and Amy Kuhn tried to gain a bit of trust, to get Stein to open up. Their attempts to build a rapport didn't go too far. Law enforcement, he believed, should be on his side, working to save America from those Democratic traitors.

"If we thought alike, this wouldn't be happening," Stein told the FBI special agents. "You'd be shaking my hand and saying thank you, my friend, my fellow citizen, for having the balls and the guts to stand for our country before we lose it."

———————◆———————

I covered the militia case when it emerged weeks before the election, and I eventually traveled out to Wichita for the trial in 2018, tag teaming the story with my colleague Chris Mathias. When we had to sell our editors on sending us out to Kansas, we told them how incredibly rare it was for domestic terrorism cases—let alone plots that had been inspired in part by the sitting president—to go to trial. Chris and I each spent long stretches in Wichita, working from a small courthouse in Kansas's biggest city, located about three and a half hours away from Garden City, Kansas, where the plot unfolded. There were precautions put in place. When Brian testified, the judge sent us to a different floor of the courthouse to watch a feed featuring only Brian's voice, as part of an effort to protect his identity.

Politics were unavoidable. Jury selection took place just over a week after former deputy FBI director Andrew McCabe was dismissed after coming under attack from Trump.[29] There was a potential juror who described himself as a "Western chauvinist," a term used by members of the Proud Boys. He didn't make the final cut.

"Locker room talk" was how one defense attorney described the defendants' words, echoing the defense Trump used when tape surfaced of him bragging about how he could assault women because he was famous.[30] They told jurors that their clients had been fooled by fake news on Facebook and that the federal government was targeting them for their conservative beliefs.

The defense tactics didn't work. The evidence was overwhelming. Day's testimony made an impact. So did that of Brody Benson, a member of the Kansas Security Forces militia, who testified that he had extremely anti-Muslim views but still was worried about what the group had planned.

"I actually thought it was not just talk—it was more of an actual action, action," Benson said. "I had a gut feeling that what was just banter back and forth, ranting and everything else, was turning into something more serious and concrete."

Mattivi emphasized to jurors that the defendants were being prosecuted not because of their beliefs but because of their actions.

"This isn't a case about the thought police," Mattivi said in his closing. "The defendants plotted to murder dozens of innocent men, women, and children. They didn't just talk. They're not here because of their words."

The jury had been deliberating for not even a full day when I got the news while sitting in a breakroom in the courthouse basement: A verdict was back. Guilty, guilty, guilty. Stein, Wright, and Allen showed little emotion as the verdicts came down.[31] Prison transport vans soon transported them from the courthouse back to jail to await their sentencing.

At a press conference across the street from the courthouse at the US Attorney's Office, I asked US Attorney Stephen McAllister—the Trump-appointed top federal prosecutor in the state—what impact he thought Trump's words had on the men's actions. He dodged.

"I can't say whether his rhetoric impacted the case or not," McAllister said. "I don't view this as a prosecution of speech at all."

Day talked to me after the trial, saying he felt a "calling from God" to step in. "I'm not an over-religious person, but I believe in God and I believe he put me there at the right time, the right place, if that makes any sense."[32] Mattivi called him a hero. Day didn't think of himself like that. He was just an average guy who did the right thing. He felt bad for the families of the extremists he helped send to prison but not as bad for the men themselves, who'd taken their political beliefs too far.

"I'd tell them fuck off," Day said. "They made their choices."

———— ◆ ————

Not long after the verdict, FBI director Chris Wray happened to be paying a visit to the FBI Kansas City Field Office as part of a tour of locations across the country. He asked Darrin Jones, the head of the field office, what sort of impact President Trump's attacks on the bureau had on the jury. Mattivi was also in Kansas City at the time, so Jones called him to the office.

"At the time it was happening, I was scared," Mattivi later explained. "I was really worried about what impact that was going to have on these jurors." He was surprised how quickly the jury came back. They'd put on four weeks of evidence, and Mattivi's rule of thumb was one day of deliberation for every week of evidence.

"I was shocked by how fast that was," Mattivi said. "This trial could have come out very differently" had Day and the FBI special agents not been such effective witnesses and maintained their credibility with the jury.

"I truly believe that had it not been for Dan's courage, and Robin and Amy's tenacity, I really believe we would have been staring over the edge of a smoking black hole in Garden City," Mattivi said. "I have no doubt that these guys were capable of it, they wanted to do it and they would have done it."

Ahead of their sentencing, prosecutors sought a terrorism enhancement.[33] Their sentencing hearing was held on Jan. 25,

2019, the day that Trump finally announced an end to the government shutdown that had left FBI employees and federal prosecutors unpaid. It was also the day that Roger Stone, the longtime Republican operative and fixture of Trump's campaign, was indicted by Robert Mueller's team on charges of obstructing an official proceeding, making false statements, and tampering with witnesses for allegedly interfering with the congressional investigation into Russian involvement in the 2016 election.[34] FBI special agents had arrived at Stone's home at dawn, and Stone later emerged from federal custody flashing peace signs à la Richard Nixon, whose face was tattooed between Stone's shoulder blades.

Inside the courtroom in Kansas, Stein apologized to his family, but not for his conduct.[35] He wanted to make sure everybody knew that he stood by his actions.

"I take full responsibility for everything I said and everything I did," Stein said at his sentencing. "Nobody forced me to do any of it."[36]

Allen, the Iraq War veteran, received twenty-five years in federal prison, plus ten years of supervised release. Wright received twenty-six. Stein was sentenced to thirty years, and would soon receive a few extra years for the dozens and dozens of child pornography images authorities found on his laptop and USB drives during the investigation.[37] His anticipated release date is in 2045, when Stein will be in his midseventies.

By the time the trio was sentenced, James Comey was long gone, and Trump had pushed out Attorney General Jeff Sessions too. The Justice Department was headed by Matthew Whitaker, who just a few years earlier was working with a company that was shut down by a federal judge and fined $26 million[38] for what the Federal Trade Commission called "an invention-promotion scam that has bilked thousands of consumers out of millions of dollars."[39] (His 2014 appointment had been touted by the company in a press release that announced the marketing launch of a "masculine toilet" for "well-endowed men."[40]) Now Whitaker—who thought that the idea of "an independent Department of Justice should concern every American"[41]—was overseeing the Mueller investigation, despite a recommendation from career staffers that he recuse himself. The legitimacy of Whitaker's appointment was on shaky constitutional grounds,[42] and he would soon be replaced by William Barr. For now, though, Whitaker was the nation's top law enforcement official.

"The Department of Justice works every day to thwart terrorist threats to the United States," Acting Attorney General Whitaker said in a statement. "The defendants in this case acted with clear premeditation in an attempt to kill innocent people on the basis of their religion and national origin. That's not just illegal—it's morally repugnant."[43]

Allen's lawyers, writing their sentencing memo a few months before the release of the Mueller report, said that their client's "misguided patriotism was inflamed by the rhetoric of the 2016 political climate and the influence of the Russian information warfare campaign against the American people" and that there was "definitive proof that Russians' false propaganda invaded Mr. Allen's Facebook account," pointing out that a group Allen liked, called "America's Freedom Fighters," had shared Russian-created content.

Wright's lawyers wrote that while a Trump fan was headed for years in federal prison for acting on hate and lies, the man who had helped boost that hate and those lies was in the White House.

It was a preview of things to come.

As long as Trump "commends and encourages violence against would-be enemies," the lawyers wrote, any sentences imposed by federal judges would do "little to deter people generally from engaging in such conduct" if they believe they are protecting their country from the commander in chief's enemies.[44]

"The speaker with the best bully pulpit in the world," they wrote, "is never sanctioned for spreading fear and advocating harm."

# CHAPTER 2
# [Distracted Boyfriend Meme]

## Oct. 19, 2020

*You have ... one minute ... remaining.*

There were just two weeks left before the 2020 election, and the female robot voice said that time was running out. I was on the line with Justin Coffman, a twenty-nine-year-old from Tennessee, a retail worker and musician who'd been named bassist of the year at the Tennessee Music Awards a couple of years back and was now trying to get an anarcho-punk band off the ground. The phone call would be exactly fifteen minutes, and it would be pricey.

*Your card was charged successfully*, the voice told me at the beginning of the call. *This call is from a corrections facility and is subject to monitoring and recording. Thank you for using GTL!*

GTL, or Global Tel Link, calls itself "the corrections industry's trusted, one-stop source for integrated technology solutions, delivering an innovative vision for the future while providing exceptional value today," providing communications services to nearly two thousand corrections facilities.[1] Critics say it's an exploitative corporation that sends the families of prison and jail inmates into debt spirals just to talk to their loved ones, whether it be over the phone or via video.[2]

For me, it was the only way to reach a man whom Attorney General William Barr's Justice Department was holding out as a poster boy for their aggressive, nationwide crackdown on antifascist protesters known as "antifa."

It had been nearly five months since George Floyd was murdered by a police officer in Minneapolis, setting off protests across the country and launching a new phase in the Black Lives Matter movement. Most protests remained peaceful, but there were plenty that sparked chaos. Violence and rioting had broken out in cities across the country. At least six people died.[3] Rioters caused millions in property damage. In response, local police frequently used excessive force.[4]

The FBI was overwhelmed, with demands flooding in from local law enforcement dealing with protests on a massive scale. David Bowdich had started off his law enforcement career as an officer with the Albuquerque Police Department before jumping to the FBI, where he'd been a SWAT team operator and sniper. During his stint at the FBI's Los Angeles Field Office, Bowdich oversaw the investigation into the San Bernardino terrorist attack[5] before ascending to the no. 2 position at the bureau in 2018 following the departure of frequent Trump target Andrew McCabe.

Nothing had quite prepared Bowdich for the crisis of the summer of 2020, with police chiefs from across the country asking for assistance on the streets and with investigations into crimes, as well as help with intelligence about what was behind the riots.

"We had violence day after day after day after day after day," said Bowdich. "We had over a thousand law enforcement officers injured. We had two shot and killed out in California, one shot in the head out in Las Vegas. We had multiples who were injured, some severely."[6]

This was a challenge the FBI wasn't prepared for. On top of everything else, the FBI had to protect their own federal buildings: Bowdich had seen the "chilling" scenes of police departments abandoning stations under siege by rioters and dispatched SWAT teams to the bureau's field offices to head off a similar fate. "We weren't going to allow that," Bowdich said. COVID-19 was raging, and vaccines weren't yet available, making it even tougher to use one of the bureau's standard tactical responses to volatile situations: bringing in overwhelming force.

"It's not like a terrorist event in one location or an active shooter in one location," Bowdich said. "The way the FBI does its business is we swamp that location. We push all sorts of assets there to help bring order to chaos, but that's one city, usually, or one region."

That's a large part of the FBI's mission. One of the bureau's four guiding principles, laid out in a jargony FBI strategy document, is all about partnerships. Teaming up with local law enforcement proved beneficial both to the FBI and to law enforcement. Historically, that's meant providing criminal background checks, maintaining the National Crime Information Center, running fingerprints, and providing forensics services that earn the bureau mentions on true crime shows like *Dateline*. Those local law enforcement partnerships greatly expanded after Sept. 11, as the federal government sought to make sure it was "left of boom," meaning before an attack took place. "Prevent, Disrupt, Defeat": that was the new mantra, and FBI-led Joint Terrorism Task Forces proliferated across the country.

Barr, who'd just turned seventy, was now a bit over a year into his second stint as the nation's attorney general, having become the first man to serve in the position twice since the Civil War. He'd been just forty-one the first time he took on the job as the nation's top cop, in 1991, under former president George H. W. Bush. The mission and technological changes that swept the bureau during the beginning of the twenty-first century were somewhat known to Barr despite his nearly three-decade-long absence from the Justice Department. That time in the private sector had been good to Barr, who was an executive vice president and general counsel for a telecommunications company that then merged with Verizon, where he was making $1.5 million a year, in 2001 dollars.[7] After retiring from Verizon in 2008, he did a bit of consulting work, served on Time Warner's board of directors, and kept a toe in politics as well.

Barr was a hard-liner. In his first round as attorney general, he had commissioned a report titled *The Case for More Incarceration* and moved hundreds of FBI counterintelligence agents onto gang cases.[8] During the 1992 riots in Los Angeles following the acquittal of four officers charged in the beating of Rodney King, when Bush invoked the Insurrection Act and sent in the military to help restore order, Barr oversaw the federal response.

Barr certainly wasn't averse to a military-adjacent role, and he relished in being referred to as "general," even though he knows the reference is based on a historical mix-up. The title of attorney general is somewhat of a misnomer that has nothing to do with the military. "General" is actually the adjective in the title, as in "general counsel." Nobody would refer to an in-house corporate lawyer as "general," but it's a relatively common way of referring to the head of the Justice Department. "It's based on a mistake, but it's a fortunate one," Barr once said. In the early Bush years, when then attorney general Dick Thornburgh named Barr acting attorney general on a temporary basis when he was headed out of town, Barr popped on an army helmet, grabbed a pair of binoculars, and went out to one of the Justice Department's balconies, joking that he was "looking for injustice."[9]

When he arrived back at the Justice Department for round two as attorney general, Barr was ready to magnify a new problem. He started kicking off weekly national security meetings with the FBI by asking what they were doing about antifa.[10]

That a septuagenarian Fox News viewer would see antifa as one of the top issues confronting the nation isn't exactly shocking.

But Barr's disdain for left-wing protesters wasn't a fresh development. Back in 1968, when Barr was a seventeen-year-old freshman at Columbia University, he was heavily involved in a group called the Majority Coalition, which sought to shut down protests in which left-wing students occupied parts of the campus in opposition to the war in Vietnam.[11] Barr joined his fellow conservative classmates and student athletes to form a blockade and prevent the Students for a Democratic Society from getting into the library. A fistfight broke out when they tried to cross the line. "Over a dozen people went to the hospital, between the two groups, when they tried to rush through," Barr later recalled. "They didn't get through," he said with a smile.[12]

In Barr's view, the rise in political violence that he believed began toward the end of the Obama administration came from the Left. When Donald Trump told his supporters to assault protesters, that was just "hyperbolic tough-guy language," and violent incidents by Trump supporters against protesters were just "low-level" scuffles. "Trump supporters weren't generating the rise in political violence," Barr wrote in his 2022 memoir, in defiance of all data showing that deadly attacks were, in fact, coming primarily from the political right. "It came from the Left."[13]

Barr believed the FBI already had a "robust program" keeping tabs on far-right threats, so he talked to FBI director Chris Wray about keeping tabs on threats "across the ideological spectrum" when he became attorney general in early 2019. The FBI had not been "as forward leaning" in monitoring left-wing extremists, Barr claimed, writing that there was "institutional reluctance" to surveil left-wing extremists because that's how it had previously "gotten into serious trouble with the political establishment."

Barr's two beliefs—that the left wing posed the most serious threat of political violence and that the FBI wasn't doing enough to stop it—are not views supported by experts or reality.

As of early 2022, antifascists had not been behind a deadly attack for a quarter century, according to data compiled by the Center for Strategic and International Studies.[14] The Anti-Defamation League found that far-right extremists were responsible for 70 percent of the 427 extremist-related killings they tracked over the course of ten years.[15] As a report by the George Washington University Program on Extremism found, when measured by human casualties and the frequency of attacks, "the violence committed by [anarchist violent extremist] actors in the past decade pales in comparison to other categories of violent extremists."[16]

Trump-appointed former US attorney Thomas T. Cullen acknowledged that reality in 2019, writing that "white supremacy and far-right extremism are among the greatest domestic-security threats facing the United States."[17] He grabbed headlines as "The Trump Appointee Who's Putting White Supremacists in Jail,"[18] prosecuting the white supremacist who rammed his Dodge Challenger into antiracist protesters in Charlottesville, Virginia, killing Heather Heyer, as well as members of the white supremacist Rise Above Movement.[19] In September 2020, when Cullen was confirmed as a federal district court judge, Barr praised his work.[20]

The stark differences between the number of deaths caused by right- and left-wing extremists are, in large part, explained by ideological principles. According to a 2022 study of extremist political views, right-wing extremists are "more often characterized by closed-mindedness and dogmatism and a heightened need for order, structure, and cognitive closure," characteristics that "have been found to increase in-group bias and lead to greater out-group hostility." Left-wingers "score higher on openness to new experiences, cognitive complexity, and tolerance of uncertainty" and are "less likely to support social dominance, which could lead to their overall lower likelihood to use violence against adversaries."[21]

When left-wing extremists engage in terrorism, they often take pains to save human lives. A radical left-wing group set off a bomb inside the US Capitol on Nov. 7, 1983, blowing a fifteen-foot-wide crater in the wall, destroying a portrait of former South Carolina senator John C. Calhoun (who'd declared slavery a "positive good"), and causing more than a million dollars in damages. What they didn't do is kill anyone: A call came into the Capitol switchboard warning that a bomb would go off in five minutes (it actually went off ten minutes after the call). In a message to National Public Radio, the group said it chose not to kill "individual members of the ruling class and government," at least not "this time."[22]

Back in the 1970s, during the heyday of the Weather Underground, left-wing bombings were extraordinarily common. They were also very rarely lethal. "During an eighteen-month period in 1971 and 1972, the FBI reported more than 2,500 bombings on US soil, nearly 5 a day," author Ryan Burrough wrote in *Days of Rage*, his book on the FBI's hunt for extremists of the era, "yet less than 1 percent of the 1970s-era bombings led to a fatality; the single deadliest radical-underground attack of the decade killed four people." The bombs "basically functioned as exploding press releases" and were "little more than a public nuisance" for much of the public.[23]

If you need a reminder of what the FBI historically did to progressive political leaders, just look at the Twitter replies that flood in every year when the bureau posts in honor of Martin Luther King Jr. Day. For years, the FBI's MLK tweets have been deluged with reminders of what the bureau did to him during his lifetime, including wiretapping his calls and writing a letter threatening him with blackmail and suggesting he kill himself. "There is but one way out for you," the letter read. "You better take it before your filthy, abnormal, fraudulent self is bared to the nation."[24]

It wasn't exactly like the federal government had gone soft on left-wing groups in the years before Barr returned to the Justice

Department. When police mass arrested hundreds of black-clad protesters on the day of Trump's inauguration in 2017 after members of the group smashed windows and set fire to a limo in front of the *Washington Post* building on K Street, federal prosecutors charged all of them with felony rioting regardless of their individual conduct. The lead detective in the case, a DC Metropolitan Police Department union official named Gregg Pemberton, spent nearly a year investigating, meticulously tracking the actions of the masked demonstrators amid tweeting about his dislike of the Black Lives Matters movement and "leftist goons."[25] When it came time for the first six defendants to go to trial, federal prosecutors conceded they had no evidence that any of the defendants actually destroyed anything or hurt anyone.[26] Jurors acquitted the first group on all charges,[27] and jurors couldn't get beyond reasonable doubt in the second trial.[28] Due to the jury verdicts, the US Attorney's Office for DC ultimately dropped more than two hundred cases, securing just one felony guilty plea (along with twenty-one misdemeanor pleas).[29]

The FBI managed to keep its hands out of that mess, leaving the investigation to DC Metropolitan Police. By the fall of 2020, the bureau thought they had a pretty good grasp on the antifascist movement, which was getting a ton of attention from Trump and in conservative media.

In gathering intel on antifa, the FBI even relied on members of the far-right Proud Boys.[30] Joe Biggs is a Proud Boys leader whom I'd met covering the unrest in Ferguson, Missouri, back in the days when he was working with Alex Jones's InfoWars and traveling with members of the far-right Oath Keepers. In July 2020, Biggs says, an FBI special agent based in Daytona Beach, Florida, called up and asked Biggs to meet at a local restaurant. Biggs sat down with two FBI special agents for two hours to hear what Biggs was "seeing on the ground," and Biggs shared what he knew about antifa in Florida and in other parts of the United States.[31]

By September 2020, FBI director Chris Wray was explaining to Congress that antifa was "more of an ideology than an organization."[32] That analysis clashed with what Barr and Trump wanted to hear. "I did not like his answers," Trump said the next day, not disputing that he was considering firing Wray over his comments on antifa. The FBI, Trump had tweeted, was letting antifa "get away with murder."[33]

By this point, Barr had been banging the drum about the threat of antifa for months, since the unrest after Floyd's death. "The violence instigated and carried out by Antifa and other similar groups in connection with the rioting is domestic terrorism and will be treated accordingly," Barr said soon after Floyd's death. Mary McCord, the former chief of the National Security Division at the Justice Department, told me it was "unusual and troubling" for Barr to suggest that a certain group was behind the attack. "It is irresponsible to cast blame on any particular group or ideology when the investigations are just beginning," she said.[34] Federal investigations would not bear out any evidence of a widespread criminal conspiracy by antifa.

Barr continued pressing. When President Trump floated a conspiracy theory about a plane "loaded with thugs" headed to DC in September, Barr demurred a bit but said that individuals had traveled to DC from cities like Portland, Washington, and Seattle "for the specific purpose of causing a riot."[35] And when members of a US Marshals task force shot and killed Michael Forest Reinoehl, a self-described antifa member charged with shooting and killing a supporter of the far-right group Patriot Prayer in Oregon,[36] Barr said that it was a "significant accomplishment" and that the "streets of our cities are safer with this violent agitator removed."[37] President Trump celebrated the killing, telling Fox News there "has to be retribution." Officers gave very different stories for why they shot thirty-seven rounds at Reinoehl, and several witnesses indicated that officers started firing as soon as they arrived.[38]

Barr's rhetoric was paired with action. In June, just after he showed up to the White House, the US Park Police operations commander said that protesters recognized Barr and began shouting at him, and the operations commander walked over to advise Barr that he was in an unsafe area and should move farther away from the crowd. That's the first time that the operations commander heard that Trump planned to come out of the White House and go into Lafayette Park.[39] Ivanka Trump had recommended that her father make a trip from the Rose Garden to St. John's Church, maybe hold a scripture reading with faith leaders. "Why don't you walk to the church, go inside, say a prayer, and show people they should not be afraid," Ivanka told her dad. "We can't tear our country apart and burn it to the ground." Simply standing in front of the church would just be awkward, Ivanka thought. But that's what Trump decided to do.[40]

Soon, while Trump held a press conference declaring himself "your president of law and order," a wall of law enforcement made their move. "They were all wearing riot gear," one journalist observed. "But there was no riot."[41] Gas canisters. Flash-bangs. Rubber pellets. Batons. Shields. Horses. It looked like a war scene.

Barr joined top officials in Trump's walk over to the photo op outside the church, although he thought it was poorly timed, and he said the president's procession to the church "had the feel of spiking the football and overdramatizing a relatively ordinary law enforcement action as a significant presidential 'victory' over the demonstrators." A couple of years later, under the Biden administration, the Justice Department would settle a lawsuit over the Lafayette Square events, and the US Park Police would update their policies.[42]

To respond to the unrest in DC, Barr brought in federal prison tactical teams from far-flung parts of the country, who typically spent their days dealing with convicted criminals with limited rights. Now, having been deputized by the US Marshals, the prison guards were overseeing American protesters near the White House. They didn't wear name badges, or even any patches indicating they were from the Bureau of Prisons (BOP). On a conference call with Justice Department beat reporters, a senior DOJ official credited Barr with coming up with the outside-the-box idea of deploying corrections officers against a civilian population. A separate source sent me a photo of Barr posing with the BOP officers in the courtyard of Justice Department headquarters.[43] "He brought those people in," the senior DOJ official said, because dealing with rioters is "exactly what they do best."[44] Days later, news emerged that members of the BOP's Special Operations Response Team had previously engaged in "inappropriate and dangerous" practice exercises that caused "significant injury" to other BOP staffers, and the Justice Department's internal watchdog recommended they suspend their training exercises.[45]

By the fall, the Justice Department was touting federal cases against more than three hundred defendants charged for their actions during demonstrations, including thirty-five charged with assaulting officers and thirty cases of civil disorder.[46] Prosecutors had already been employing rarely used federal statutes that had been written to target Black activists in the 1960s.[47] One case that went unhighlighted in the Justice Department press release was the federal charges related to the killing of a federal courthouse guard in Oakland, California, during unrest in the city, which was carried out by a pair of members of the far-right boogaloo movement. (Barr did manage to highlight the boogaloo movement alongside antifa when he formed the Task Force on Violent Anti-Government Extremists over the summer.[48])

Enter Justin Coffman.

After Floyd's death, Coffman attended two small daytime Black Lives Matter protests in Jackson, Tennessee.[49] He was a little bit of an odd man out. At one protest, he held a black banner featuring an image of a cat. "C.L.A.W.S.," it read. "Cat Lovers Against White Supremacy." At another, he rode a Onewheel, a sort of electric skateboard, and wore a black leather jacket adorned with pins that read "Punch Nazis" and with a campaign-style pin that read "Please Kill Me 2020." He sported a Guy Fawkes gaiter.

The protests Coffman attended were organized by Tracy Boyd, who was determined to keep the protests peaceful, especially in a conservative state like Tennessee. Across the country, rumors about antifa spread on local platforms like Facebook and Nextdoor, with even small towns paranoid about busloads of masked radicals invading their towns.[50] In Klamath Falls, Oregon, armed men who were worried that George Soros was busing in members of antifa stood guard, waiting to defend the city from an imaginary army that never arrived.[51] In the small resort town of Coeur d'Alene, Idaho, men carrying rifles lined the streets after unfounded fears of "ANTIFA agitators" bubbled up on Facebook pages.[52]

"Antifa members have threatened our town and said that they're going to burn everything and to kill white people, basically," a man said in a Facebook video that racked up more than one hundred thousand views.

One Trump supporter from Pennsylvania, Tammy Bronsburg, recorded a YouTube video talking about a vision she had that members of antifa were coming across a bridge in her town in military tanks. "They had, like, Biden signs on the tanks," Bronsburg said, wearing a pink "Make America Great Again" hat.[53] (Bronsburg, who said there were Illuminati symbols all over her town, would later admit to storming the Capitol on Jan. 6 and taking a shot of Jack Daniels inside a conference room.[54])

So protest leader Boyd had his guard up when the white kid dressed in black showed up to protests he helped organize in Jackson, Tennessee. *Hmm*, Boyd thought. *This may be bad.*

Boyd's worries would be calmed once he got a chance to speak with Coffman about his interest in the Black Lives Matter movement. He even realized that Coffman worked at the gas station near Boyd's house.

Coffman, it turned out, had even complimented the way that police had handled the protests. "Was out in Jackson yesterday and it's good to see the mayor and the police chief out among the people and talking to them instead of the police attacking people and inciting riots," he wrote on Facebook.

The police, meanwhile, weren't so keen on Coffman. At the time he first spotted Coffman, Boyd had been receiving messages suggesting that people were going to come down from Nashville to try to turn their peaceful protest into a violent one. He was so worried that he made a comment to two Jackson Police officers, saying they should keep an eye on the man. One of the officers, Boyd said, assured him they were on top of it, saying they'd been following Coffman on Facebook and were concerned about his posts. "He was clearly on their radar," Boyd told me.

Two days later, on June 2, 2020, President Donald Trump stood in front of St. John's Church holding a Bible following the brutal crackdown on protesters.[55] Back in Tennessee, a SWAT team; the Bureau of Alcohol, Tobacco, Firearms, and Explosives (ATF); and an FBI special agent also descended on Coffman's home, where they executed a search warrant. Major Phillip Kemper, who supervised the Jackson Police Department's Special Operations and Criminal Investigative Divisions, had told a state judge that there was probable cause to believe that Coffman was "in possession of a hoax device or otherwise actual improvised explosive incendiary device."[56]

Big, if true. But what they were referring to was just an antique bottle filled with apple cider. Working with a professional photographer, Coffman had posed for promotional photos for his band, the Gunpowder Plot. They show him standing in front of a police van, holding a fake Molotov cocktail behind his back.

"Was saving these pictures but now seems like a good time," he'd written when he posted three of the staged images on Facebook on May 28, five days before law enforcement swarmed his home. "'You will bathe in the flames born from your hatred,'" he'd written, quoting a song lyric.

The next day, a Jackson City lieutenant reached out to the owner of a tattoo parlor who knew Coffman as an acquaintance and customer. When Coffman got word, he called up the lieutenant to try to clear things up.

"The caller stated that the bottle contained apple juice and that the photograph had been taken previously to be used in support of his musical band," the search warrant application stated, without making clear that "the Gunpowder Plot" was the name of Coffman's band. They also found Coffman's use of the stage name Robert Catesby suspicious, and a Jackson Police Department criminal intelligence analyst named Sarah Webb "advised Major Kemper the 17th century historical figure known as Robert Catesby" was a co-conspirator "in a plan to blow up the king of England," according to a search warrant application.

Online, Coffman also tried to clear the air about the Molotov mocktail, which he'd used as a centerpiece when he cooked up a steak dinner for his girlfriend, a nurse named Leah Harris. "I don't know who needs to hear this but we're a band, not a terrorist organization," he'd written in one Facebook post. "Stop calling and wasting them people's time."[57]

Law enforcement went ahead with the raid anyway. Inside Coffman's home, they found some marijuana, "anarchy paraphernalia," and two guns, which—in Tennessee—isn't anything to write home about.

"Everybody in Jackson owns guns," Robert Joe Haynes, Coffman's Republican cousin, told me. "Every friend I've got has eight or ten guns at least."

Standing on Coffman's porch, an FBI special agent had some questions for Coffman. "He was asking me if I was antifa and all that bullshit, which I'm not," Coffman said.

Coffman was then taken to jail for the pot charges. He got kicked out of the home and lost his job, but he tried to piece things back together over the next few months. The FBI special agent would occasionally reach out, looking for intel on antifa. One problem.

"There ain't none of them in Jackson," Coffman said. "I didn't have anything."

Coffman said the FBI special agent expressed to him that he didn't think Coffman was a threat but noted that it was a sensitive time.

Months later, in September, Barr had a phone conference with the nation's top federal prosecutors, urging them to consider bringing sedition charges against violent demonstrators.[58] The same week, a Jackson Police Department officer on an ATF task force swore out an affidavit charging Coffman with a rare crime: being a drug addict in possession of a firearm. Coffman had admitted smoking marijuana a couple of times per day.

Coffman was arrested the next month. "Madison County Man Charged in Both Federal and State Courts for Unlawful Possession of Firearms and Possession of a Hoax Device During Civil Unrest," read the press release blasted out to Justice Department reporters. The press release, like the search warrant, never mentioned that the Gunpowder Plot was a band. I only learned that after poking around Facebook and hunting down the post myself. It wasn't just the public who were being misled: when a state grand jury indicted Coffman, they weren't even informed about the details of the alleged hoax device.[59]

In the press release, US Attorney D. Michael Dunavant praised law enforcement's "outstanding investigative work."[60] The tough-on-crime prosecutor had previously attended a Trump rally holding a "Make America Safe Again" sign and posted the image on his Twitter page. That was an odd move for a federal prosecutor, who would typically avoid overt support for political campaigns.

Coffman was working at the Circle K convenience store when two cops he knew from their stops in the store for free coffee came in and grabbed him. Now he was stuck in Madison County Jail in the middle of the COVID-19 pandemic. Breakfast was oatmeal. Lunch was a bologna sandwich, maybe a PB&J. One day they got to go outside because the shower wouldn't stop running, and another inmate told Coffman that was the first time they'd been outside in five months. The conditions were disgusting, Coffman said. There was no in-person visitation allowed.

Coffman spent weeks in custody, waiting on a detention hearing. The government wanted to keep him locked up until trial. Coffman had brought witnesses to support his case for release, but the prosecution's argument for detention was so weak that they weren't necessary. Coffman recalled that US Magistrate Judge Jon A. York was deeply skeptical of the government's request to keep him locked up.[61]

"He basically said to the [federal prosecutor], 'Your own witness doesn't think he's a threat, and the only proof you have is Facebook posts, and here in America, we don't lock people up for their First Amendment rights,'" Coffman said. Coffman was still being held on the state charge, until a reader who'd seen my coverage reached out to Coffman's girlfriend, Leah, and offered to pay his bond.

Coffman got out of jail just in time to cast his ballot in the 2020 presidential election. He walked into the polling station wearing a T-shirt mimicking the Black Flag punk band's logo, only with the words "Black Lives Matter."[62] His candidate, libertarian Jo Jorgensen, ended up with more than 29,000 votes in the state, finishing a distant third with roughly 1 percent of the total votes cast in the state. Trump received more than 1.8 million, securing over 60 percent of the vote and the state's 11 electoral college votes.

Months later, in July 2021, chief US District Court judge S. Thomas Anderson tossed the prosecution's case in the garbage pail. He wrote that the information about the band was "doubtlessly easily accessible and known by the affiant and would have been highly relevant to Judge Morgan's finding of probable cause." No "reasonably prudent person," the sexagenarian George W. Bush appointee wrote in his opinion, could believe in good faith that Coffman "was in possession of a hoax device or improvised explosive incendiary device."[63] He suppressed the results of the search, leaving the government with no evidence to present. The next month, Assistant US Attorney Hillary L. Parham moved to drop the case. The state case soon followed.

Coffman wasn't the only person swept into law enforcement's desperate search for intelligence on a movement with no hierarchical structure. Joel Feingold was questioned by a New York Police Department (NYPD) intelligence officer and an FBI special agent after he was arrested outside his Brooklyn home not long after Floyd's death.[64] According to Feingold, the "very slick" FBI special agent told him law enforcement wanted "to know who's hijacking your movement and making it violent."

In North Carolina, a man got a call from the FBI after he tweeted in jest that he was "the leader of Charlotte, NC Antifa" after Barr referred to antifa as domestic terrorists.[65] "What makes you email the FBI saying you're the leader of antifa in Charlotte?" the agent asked him, before asking if he would be interested in becoming an informant.

And in Coffman's state of Tennessee, four people who posted about Black Lives Matter had officers of an FBI Joint Terrorism Task Force show up to their homes and workplaces with questions about antifa.[66]

Barr couldn't have been surprised that FBI special agents were running scattershot investigations. In his memoir, he acknowledged that some of the tactics used by far-left groups made investigating them more difficult, including their "decentralized, cell-like structure and more sophisticated ways of operating." He wrote that left-wing groups were able to "deftly exploit the seam between rioting and legitimate First Amendment protest" by "insinuating themselves into protests and interweaving their violence with legitimate protest activity."

Chris Mathias, a journalist and former colleague of mine who is writing a book on antifa, said that as the Right started to make antifa into a boogeyman, the term really lost all meaning.

"It became a catch-all for anyone that the right disliked. Black Lives Matter was antifa, anti-Trump protesters were antifa, even

Democratic politicians were antifa, even though when you talk to actual antifascists they hate Democrats," Mathias said.

There are no real leaders or spokespeople for antifa, and while some cities have more established antifascist groups, often people just come together temporarily in support of a shared goal. Mathias pointed to Portland, where a "coalition spanning from people in black bloc to kind of normie libs" gathered to shut down the Proud Boys in what some had predicted could have been the most violent far-right rally since Charlottesville. Many of the mainstream liberals probably didn't care for the black bloc tactics, but the bigger priority was shutting down the Proud Boys.

They did outnumber the Proud Boys. But when it came time to shut down the protest, cops went after the antifascists, and Proud Boys cheered.[67]

I spent a lot of time poking around Coffman's Facebook page researching the story, and there was one post I couldn't get out of my head. Not long before his arrest, Coffman had posted a "distracted boyfriend" meme that showed conservatives, as the boyfriend, staring lustfully at "made up stuff about antifa" while ignoring "tangible proof of violent white supremacists."[68]

# CHAPTER 3

# "The Edge"

## Nov. 4, 2020

"Karen was upset," the Republican lawyer wrote.

John M. Downing Jr., an attorney from New Jersey with a focus on "motor vehicle, construction, and premises liability matters," went out to Detroit during the 2020 election as a volunteer with Lawyers for Trump. The initiative had been launched in July with a mock-up of Donald Trump as Uncle Sam, in the style of the famous US Army recruitment poster created by James Montgomery Flagg in 1917 during World War I.

"I WANT YOU TO JOIN LAWYERS FOR TRUMP," the Trump version of the recruitment poster read. "HELP PREVENT VOTER FRAUD ON ELECTION DAY."[1]

The Trump campaign was behind the effort, but it had the support of the Republican National Committee. RNC chairwoman Ronna McDaniel tweeted a copy of the image, adding on an appeal to patriotism by declaring Joe Biden's agenda "a threat to our country."[2]

For nearly four decades, since 1982, the RNC had been legally restricted from monitoring polling places. That court-enforced consent decree was put in place after the RNC, under a program

called the National Ballot Security Task Force, hired county deputy sheriffs and local police officers to target "predominantly
black and Hispanic polling places" during the gubernatorial
election in New Jersey. Officers showed up displaying revolvers
and two-way radios, wearing armbands that read "National Ballot Security Task Force."[3] The polling places were surrounded by
large signs, putting voters on notice in bright red ink:

WARNING
THIS AREA IS BEING PATROLLED BY THE
NATIONAL BALLOT
SECURITY TASK FORCE
IT IS A CRIME TO FALSIFY A BALLOT OR
TO VIOLATE ELECTION LAWS.

Thomas H. Kean, the Republican gubernatorial candidate and
future chairman of the Sept. 11 Commission, ended up squeak-
ing out a win by just 1,797 votes, the smallest margin in Garden
State history.[4] Kean owed much of his success to his chief strate-
gist, a well-dressed twenty-nine-year-old who was a partner at a
startup political consulting group named Black, Manafort &
Stone,[5] which had recently signed a thirtysomething real estate
developer named Donald Trump.[6] His name was Roger Stone.

Stone, of course, has denied having anything to do with the
Ballot Security Task Force, saying it was run by the RNC and
New Jersey Republican State Committee.[7] But Richard Richards,
an early backer of then president Ronald Reagan and then head
of the RNC, said Democrats "would have stolen the election"
without the program.[8] "Those guys would have stolen it from us,
and we protected our fannies," Richards said.[9]

Heading up the effort was a twenty-nine-year-old named John
A. Kelly, the director of the National Ballot Security Task Force,
who was also deputized as a deputy sheriff in New Jersey.[10] Kelly
was born in New York, and his parents lived in Manhattan.[11] As

an eighteen-year-old, he was the first voter under the age of twenty-one to register in New York City days after the Supreme Court upheld provisions of the Voting Rights Act of 1970 that extended the right to vote for those eighteen to twenty-one. Kelly told reporters he was a Republican and a freshman at Notre Dame.[12]

A decade later, a couple of months before the 1981 election, the chair of the Mercer County GOP laid out the plan at a diner outside of Princeton. For $15,000, he'd hire off-duty cops to become "flying squads" that targeted Trenton. Kelly liked the idea, knocked down the cost to $10,000, and ran it in three other cities with significant populations of Black voters: Newark, Camden, and Atlantic City.[13] All told, the RNC spent about $80,000 on the effort in New Jersey, and they planned to replicate it in other states.[14]

In addition to their suspicions about voters in cities, Republicans began to have questions about Kelly himself. A couple of weeks after the 1981 campaign, Kelly was suspended by the RNC after discrepancies had emerged about his résumé. The *New York Times* reported that city records showed he'd been arrested in 1976 for attempting to impersonate an officer and possessing stolen property and had been discharged as an officer in Manhattan in 1975 after complaints came in that he had twice displayed a gun in public: in a restaurant and in a traffic incident. He applied for a House select committee investigator job he hadn't received and then was ousted from the Fraternal Order of Police (FOP) for saying he had been on that congressional committee, a FOP spokesman said. His biography later claimed he was director of government affairs for the FOP, a position that did not exist.[15] He worked as a bailiff for the US Court of International Trade for six months, claiming he was a Notre Dame graduate and was attending Fordham Law School.[16] Then, while working at the RNC, he reportedly told people he worked at the White House and "was promising people U.S. judgeships," a Reagan administration official said. Despite questions about his own background, Kelly was vetting potential political appointees, making sure they were registered Republicans and Reagan supporters.

"He's just the ultimate con man," said Edward Rollins, Reagan's assistant for political affairs.[17] Carl Golden, Kean's press secretary, would later describe Kelly as "a complete psychopath," adding that he was "a short, rotund guy nobody trusted or liked" who "could strut while sitting down."[18]

Kelly resigned his post at the end of 1981, saying he didn't want to "bring unnecessary negative attention" to the RNC.[19] By the end of 1982, the RNC had entered into an agreement with the DNC that barred them from certain poll-watching activities. Among the signatories was John Barry—Trump's personal attorney and brother-in-law—who represented Kelly. A couple of decades later, in 2006, Kelly would be named cochair of Catholic outreach at the RNC[20] alongside Leonard Leo[21] (the legal powerhouse who served on Lawyers for Trump) and would pop up at a party thrown in honor of Pope Benedict during his 2008 trip to DC.[22] In 2017, he and his wife made a $1 million donation to Notre Dame,[23] the school that he'd portrayed himself as a graduate of decades earlier. A press release called him a "1974 Notre Dame alumnus." Nobody knew him as John A. Kelly at that point. He went by Jack, and he'd carved out a successful career as a DC lobbyist. The army veteran had even served as chairman of the US Advisory Committee on National Cemeteries and Memorials.

When I reached him in 2023, Kelly declined to answer detailed questions about the events that had long-spanning impacts on the RNC. "It was fifty years ago," Kelly said, speaking of events that happened forty-two years prior. Kelly's 1981 resignation from the RNC came over a year before the internet was even "born," in 1983,[24] and it would be decades before the internet would fundamentally reshape human society. In the pre-internet days, when finding a person's worst screw-ups took much more work than a quick Google search, it was possible to reinvent yourself, even when your mistakes had been thoroughly documented in the country's newspaper of record. Kelly had changed, he'd pivoted, he'd evolved. He'd become a philanthropist. Now, at seventy-one, he'd left that life behind.

Kelly may have moved on, but in the decades after the 1982 agreement, the RNC couldn't. The organization was accused of repeatedly violating the consent decree: in Louisiana in 1986, in North Carolina in 1990, in Ohio in 2004, and in New Mexico in 2008. In 2012, ahead of Obama's reelection, a federal appeals court upheld a lower court ruling that had upheld the consent decree but put an expiration date in place. Restrictions on the RNC's poll-watching activities would fall on Dec. 1, 2017, near the end of Trump's first year in office.[25]

So in 2020—for the first time since *USA Today* was founded, since Michael Jackson released *Thriller*, since Sarah Huckabee Sanders was born, and since EPCOT opened in Walt Disney World—the RNC would get to run programs targeting purported voter fraud. They had a $20 million plan and wanted to recruit fifty thousand poll watchers in fifteen states.[26]

"Democrats are deploying an 'army of poll watchers' this November, and for the first time in decades the Republican Party can too," McDaniel wrote.[27]

Downing, the Republican lawyer from New Jersey, arrived in Detroit believing a myth that had been deeply ingrained in conservative circles for decades, one that had roots in the Reconstruction era and the backlash to the Voting Rights Act: there was a lot of cheating going on.[28] It was his first time working in elections, and he told me he'd done some Zoom training before he headed out to Detroit at the suggestion of Lawyers for Trump. He spent five days in the city, including Election Day and the day after, Nov. 4.[29]

In Detroit, a record number of absentee ballots were being counted in what was then known as the TCF Center, the sixteenth-largest convention center in the United States. At 10:00 a.m. on

Monday, elections workers were finally allowed to begin process-ing absentee ballots, although they were not allowed to open the ballots, just the outer envelopes. "Absentee Ballot processing starts in Detroit today. 134 tables operating for 10 hours," tweeted Mike Roman, Trump's director of Election Day operations.[30] "Trump Challengers will have 100% coverage (unless crooked officials try to kick them out)!"[31]

Soon after the process began, a man with poll challenger cre-dentials who was wearing a full-face white mask similar to the one that Michael Myers wore in the "Halloween" series created a disturbance and was removed. It set the stage for the horror show to come.[32]

Things went relatively smoothly on Tuesday. For most of Elec-tion Day, Trump was doing better than anticipated in Michigan, and Downing went to bed thinking that the president had won the state. "I was called at 7:00 a.m.," Downing said. "It had com-pletely flipped."

From his hotel room, Downing hit the phones, calling up vol-unteer GOP poll challengers and asking them to go to the con-vention center. He wasn't the only one rounding up the troops. One message shared by a county GOP Facebook page on the morning of Nov. 4 summoned Republicans to go to TCF to "pro-tect our lead," as though Republicans were acting like goalies, blocking shots and running down the clock. "35,000 ballots were suddenly found at 3:00 am in Detroit," the message said. One message from the North Oakland Republican Club offered logis-tical advice to the suburbanites it summoned to the TCF Center in downtown Detroit: "Try parking on the roof."[33]

A Republican volunteer named Karen did. While Downing was at his hotel, Karen called Downing "several times" to tell him that poll challengers were stuck on a roof deck at the TCF Center, waiting to enter. A security guard wouldn't let them in. Downing, who had never been to the TCF Center before, drove over from his hotel room. He eventually found the group of about twenty people, who had gone to the roof, he later wrote in an affidavit, "because there was free parking."

Summoning random people to rush down to a location where they believed Democrats were perpetrating a criminal conspiracy, as it turns out, led to some problems. Many of those who arrived had no clue about the basics of how vote counting works or the procedures for challenging them. One GOP poll challenger who arrived went through "about 20 minutes of training" before being handed credentials. Lack of experience aside, there's a constant assumption of fraud throughout the hundreds of pages of affidavits later filed by the Trump campaign.

Calling these individuals "poll challengers" is a bit misleading. Some of the challengers did have experience and some remote sense of how the process worked. But many of them were just random people who saw a message on Facebook, thought the election was being stolen, and got a few minutes of training before being "certified"—it sounds more official than it is—as poll challengers by the Michigan Republican Party.

"I didn't quite process that, wait, these people are just regular, normal people off the street who walked in here, presented a badge with whatever party they were with and said 'I'm a challenger,'" recalled Tresa Baldas, a reporter with the *Detroit Free Press*.

"They had no idea how the system worked. They had no idea what they were there for," said Chris Thomas, the former elections director in Michigan who came out of retirement to assist with the 2020 election. "Many of them—not all of them, but many of them—they were on a mission. They clearly came in believing there was mass cheating going on in Detroit and they were on a mission to catch it."[34]

There wasn't a lot of nuance there. When Trump said it was important to watch for fraud in "certain areas" in 2016, or when he accidentally sparked a merchandising bonanza after Philly sports fans adopted his "bad things happen in Philadelphia" line during a 2020 presidential debate,[35] everybody understood what—who—he was talking about. He was talking about Black voters.

The notion that mass voter fraud is changing the outcome of elections "is a politically constructed myth" that capitalizes on "generally and widely held beliefs that are rooted in facts and real historical experience, notions such as corruption in party politics and government but also stereotypes and class—and racially biased preconceptions of corruption among groups long stigmatized by their marginal or minority status in US society," as political science professor Lorraine C. Minnite explained in her book *The Myth of Voter Fraud*. Minnite, even before Trump jump-started his political career by questioning Obama's birth certificate, wrote in the 2010 book that the voter fraud myth could be rejuvenated "with a wink and a nod—and maybe a little bullying." And rejuvenate it Trump did.

In Detroit, the racial dynamics were impossible to miss: a bunch of white Republicans had flooded into a majority-Black city on the premise that there was a massive criminal conspiracy. One challenger made sure to note in her affidavit that she was a "pregnant white woman." There were numerous complaints about Black Lives Matter masks. One woman complained she was intimidated by "poll workers wearing BLM face masks" as well as a "man of intimidating size with a BLM shirt on." When I checked her Facebook page, I found it littered with COVID-19 conspiracy theories, debunked right-wing memes, and a lengthy post about the rapture, the Antichrist, and microchips. There was also a painting of Trump, with Jesus Christ resting his hands on Trump's shoulders.[36]

One GOP watcher said that roughly 80 percent of the military ballots he saw processed went for Joe Biden, which he found suspicious. "I had always been told that military personnel tended to be more conservative, so this stuck out to me," Branden Gaicobazzi wrote.[37]

By early afternoon, the situation had gotten tense. Democrats, Republicans, and nonpartisan groups were each allowed 134 challengers inside the room under election rules, but the number had swelled past that, with hundreds of challengers "freely roaming" inside the hall without any screening process. Democrats and Republicans both exceeded their allotted numbers, so an official came outside the hall and told them that no additional partisan challengers would be admitted.[38]

"People started to lose their minds," said Baldas, the reporter.[39]

Republican challengers, or those who self-identified as challengers, weren't happy. Members of the crowd started pressing up against the windows, banging on the glass. One Twitter user compared it to a scene from *The Walking Dead*.

"Stop the count! Stop the count! Stop the count!" they chanted. Workers tried covering the windows to prevent the individuals from filming, but that only angered the crowd further.

"President Trump did not give up on Michigan!" one angry man, Ken Licari, yelled as he live streamed himself to his Facebook audience, standing feet in front of an officer protecting the counting room. "He worked his ASS off for us in Michigan! Get down here!"[40]

A crowd gathered outside the building too. "TREASON OR TRUMP" read one sign.

All of this had been unfolding while Downing was busy dealing with Karen's parking situation on the convention center roof. COVID-19 was raging, so a massive convention hall hadn't been the busiest place in recent months: the Detroit Auto Show was canceled, and the TCF Center had been chosen by FEMA as a field hospital.[41] The security guard said he had instructions to keep the rooftop door locked. Eventually, Downing spoke with the manager on Karen's behalf. There was another entrance they could use, in the basement parking garage. But by the time Downing got back up to the rooftop, Karen and co. were gone.

*The Guardian* had declared 2020 "The year of Karen."[42] The name Karen was a stand-in for a privileged, entitled, racist white woman. Even the *New York Times* wrote a piece on it.[43] Downing didn't read the *Times*, but he knew the Karen meme.

"That's unfortunate that that's her name," Downing later told me, noting that this Karen was a "lovely person" and that Michiganders were very nice. (Two years in, no one named Karen had yet been charged in connection with the Jan. 6 attack, but sleuths believed they had identified two Karens who could be charged. One was an elected official in California, the president of a community services district that provided sewer collection and streetlight services.[44])

The Karen name stuck inside the convention hall too. "Go back to the suburbs Karen," one woman from the suburbs, real name Jennifer Lindsey Cooper, said she was told. "The Democrat challengers would say things like 'Do you feel safe with this [woman] near you' and 'is this Karen bothering you?'"

Some of the Republican poll challengers engaged in what might be described as Karen-like behavior. Patty McMurray, the founder of the right-wing website 100 Percent FED Up, recorded a video complaining that poll workers "used this COVID B.S." to tell people to keep their distance. When she was told there were too many people in the room, McMurray disobeyed instructions not to enter and "kind of pushed my way through," she said.[45] Later, when McMurray decided she was ready to leave, she approached the door where the mob was gathered. The person guarding the door said they couldn't go through that way and asked them to go out another door. McMurray said she took off her mask out of spite so she could taunt the workers. Her friend flipped her mask around, revealing that it said "TRUMP" in big red letters. They walked through a crowd of "very pissed off workers," she said, as a "last hurrah."[46]

The entitlement that bled through the memos was astounding. James P. Frego, the founder of Michigan's largest bankruptcy law firm, used his affidavit to spell out precisely how he had obstructed a police officer trying to keep the mob out of the counting room.

"I put my foot in the doorway, which kept it from closing. The officer asked me to remove it," Frego wrote. "He insisted that I remove my foot, and I insisted I would do it as soon as I was given an exact COUNT of the number of challengers inside the room itself." Frego said that he "persisted" in asking for the numbers until a supervisor pointed at him, and other officers pulled him inside and placed him in handcuffs. Frego, after admitting that he obstructed a police officer with his foot, wrote that he was "consistently respectful" and had "never been arrested" in his life.

The "back the blue" crowd had turned on police. A Detroit Police officer emerged to try to calm down the crowd. "Hey, respect the police! You all respect the police, right?" he asked. "Respect the police, back up."

Licari, the Facebook user who live streamed himself yelling, wasn't ready to back down.

"This is crap! America, you need to stand up and quit being the silent majority!" Licari yelled. "If you think antifa got violent, you ain't seen nothing yet, my friend." Licari said he was going to get himself arrested. "I'm tired of being polite," he said. "I'm going to kick someone's butt, because this is ridiculous."[47]

By the time Downing arrived at the counting room, it was after 5:00 p.m., and the work was just about wrapped up. "It was pretty hectic," Downing later conceded. In his affidavit, Downing said he saw three officers: a white female and two Black males in "heavily armored gear." He told them he was an attorney and wanted to observe; they said their sergeant didn't want them to let anyone inside. Soon their work was done. Downing was upset.

"I felt very emotional that this country had gone so far downhill that we're almost like a banana republic," Downing later told me. "It's third world; it's horrendous. I was very emotional."

When we spoke more than two years later, the story had grown beyond what he'd written in his memo.

"My distinct memory is officers in flak jackets, and either machine guns or long arms," Downing said. "So there were at least three heavily vested and heavily armed guards at that window that everybody saw with the cardboard up on it." (Police do not carry machine guns, and if any officers inside the TCF Center were displaying long guns that evening, the pack of reporters there must have missed it.)

Downing doesn't trust the media and skips out on a lot of mainstream coverage. What he does see, he doesn't like. "I have AOL, and I use all my email from AOL," Downing said. "Every single day there was a negative story on Trump. . . . AOL was extremely biased."

He thought the police overreacted on Nov. 4 and said the people who assembled at the TCF Center just wanted to make sure things were running smoothly.

"Just like on Jan. 6," Downing said incorrectly, "there was not one person with a weapon."

All the pandemonium at the TCF Center obscured one surprising fact, something that really gutted all of Trump's fraud claims: Trump performed better in Detroit in 2020 than any other Republican candidate had in modern American history. That's what Attorney General William Barr told Trump weeks later, when the duo finally discussed the president's rage about his election loss and his refusal to accept the outcome.

"Did all the people complaining about it point out to you you actually did better in Detroit than you did last time?" Barr says he asked Trump on Nov. 23. "I mean, there's no indication of fraud in Detroit."[48]

There wasn't any counting going on at the TCF Center on Thursday, Nov. 5, or Friday, Nov. 6. Trump fans showed up anyway.[49] On Friday, there was a bomb threat.[50]

An antidemocratic mob, twisted up on online disinformation, had come together online and disrupted the election process.

They'd organized online, mostly on dozens of Facebook groups that could quickly reach hundreds of thousands of people.[51] It wouldn't be the last time.

———————————◆———————————

In the weeks before the chaos in Detroit, some of the top minds in the FBI put their heads together, trying to imagine worst-case scenarios. The final title of the Oct. 27, 2020, report was a bit academic and dry: "Alternative Analysis: Potential Scenarios for Reactions of Domestic Violent Extremists to a Disputed 2020 US Presidential Election." The FBI loves its acronyms: domestic violent extremists were referred to as DVEs. The heart of the exercise could be summed up with the help of a different acronym: *WTF happens next?*

For some insight, it would have helped to go back a couple of election cycles, when Donald Trump was a reality television star spreading another racist lie: that Barack Obama wasn't born in the United States.

In 2011, when Trump was mulling a run for the 2012 Republican nomination, he was best known as the host of *The Apprentice* and the de facto leader of the "birther" movement. That internet-fueled conspiracy theory certainly wasn't a disqualifier in a Republican primary: a 2011 survey found that most GOP primary voters thought Obama was foreign born, which would disqualify him from the Oval Office.[52] Even years later, at the end of Obama's second term, 72 percent of registered Republicans still had doubts about Obama's birthplace, which means they had doubts about the lawfulness of his presidency.[53]

Obama had released a copy of his birth certificate in the 2008 campaign. But the conspiracy wouldn't die: right-wing figures were now demanding a copy of his "long-form" birth certificate. For Trump, seeking a way to connect with the GOP base, it was a winning issue. He teased the issue in interviews in March 2011, delighting in how much attention he was grabbing, even as the media attempted, for the most part unsuccessfully, to cover the issue without spreading disinformation or lending legitimacy to his claims.[54]

By April the White House had had enough. Obama had a lawyer get a copy of the "long-form" birth certificate, in the hope of putting the questions to rest. America had too many real problems to solve for DC to be "distracted by a fake issue," wrote White House senior adviser Dan Pfeiffer in a blog post.[55] For some moderates, that was really the end of it. Rep. Paul Ryan was soon mailing out copies of the birth certificate to constituents who flooded his office with questions on the issue.[56]

Reaction from the birthers was mixed. World Net Daily editor Joseph Farah, who'd talked with Trump about the birth certificate, said the *real* issue was that Obama's dad wasn't an American citizen, pivoting to a new bogus claim a few weeks before his company was set to publish a book with the poorly timed title *Where's the Birth Certificate?*[57] Supporters of an army doctor who was locked up for refusing to deploy because he questioned Obama's legitimacy demanded "forensics testing to determine its authenticity."[58]

When I called one prominent birther after the release of the document, she suggested that maybe it wasn't legitimate because it said "African," when it should have said something else. "It sounds like it would be written today, in the age of political correctness, and not in 1961 when they wrote white or Asian or 'Negro,'" she said. Nonetheless, she knew who'd made the difference.

"I credit Donald Trump in pushing this issue," Orly Taitz told me.[59]

Trump was happy to take a victory lap. "I've accomplished something that nobody else has been able to accomplish," Trump said. "I am really honored frankly to have played such a big role in hopefully, *hopefully*, getting rid of this issue."[60] Then he suggested that the first Black president wasn't very smart and should release his school records too.

"How'd he get into Harvard, if you're not a good student?" Trump said. "I don't get why he doesn't release his records."[61]

Days later, Obama eviscerated Trump during the 2011 White House Correspondents' Association Dinner, comparing his birth certificate conspiracy theory to nutty beliefs about the moon landing and sarcastically praising Trump's "credentials and breadth of experience" before referencing Trump's rather unserious duties as a talk show host dealing with B-list celebrities who couldn't win at a cooking competition.

"There was a lot of blame to go around, but you, Mr. Trump, recognized that the real problem was a lack of leadership. And so ultimately, you didn't blame Lil Jon or Meatloaf. You fired Gary Busey. And these are the kind of decisions that would keep me up at night," Obama joked, as a crowd of important people in tuxedos and ballgowns laughed at Trump's expense, clapping with delight, and Trump sat there grimacing.

Former governor Chris Christie said Trump was "beside himself with fury" and "pissed off like I'd never seen him before" when they spoke after the dinner.[62] "I am not looking to laugh along with my enemies," Trump later told a reporter.[63]

What Obama didn't let on while dunking on Trump was that US Navy SEALs, at his command, were preparing for the operation that would kill Osama bin Laden.[64] Obama, who bore the politically unfortunate middle name Hussein, had even edited out a bin Laden joke from the routine, telling his team to change a joke featuring fake middle names of Republicans. Instead of "Tim bin Laden Pawlenty," Obama suggested, what about "Tim Hosni Pawlenty," as in Hosni Mubarak. "That's not as funny," speechwriter Jon Favreau recalled saying. "Trust me on this," Obama replied.[65]

A bit over eighteen months later, when Obama beat Mitt Romney, the Republican nominee conceded. Trump, on the other hand, couldn't bear the thought of Obama remaining in the White House for four more years. Romney was his candidate, the guy he'd endorsed at the Trump International Hotel and Tower in Las Vegas, and he wasn't about to take the loss lying down.[66]

"We can't let this happen. We should march on Washington and stop this travesty. Our nation is totally divided!" he wrote in one tweet on election night 2012.[67] "This election is a total sham and a travesty. We are not a democracy!" he wrote in another.[68]

"We should have a revolution in this country!" Trump tweeted.[69] Romney's loss meant it was time, Trump tweeted, to "fight like hell."[70] It would not be the last time he used those words after an election.

Trump never got around to launching a pro-Romney revolution or marching on the capital to "stop" the "total sham" of Obama's reelection. Instead, he went and won the 2016 Republican primary, vilifying Romney in the process.

The 2016 general election was supposed to be a different story. Polls and pundits gave Trump little chance of beating Hillary Clinton, but there was a lot of worry about what Trump would do after he lost. Trump was already working the refs, proclaiming weeks before Election Day that "large scale voter fraud" was happening.[71] The *New York Times* wrote, in the lede of an Oct. 16, 2016, story, that his claims were "setting off talk of rebellion among his supporters."[72]

At a debate, Trump refused to say that he'd concede if he lost. "I'll keep you in suspense. Okay?"

Clinton found his response "horrifying." At the final 2016 presidential debate in Las Vegas, she laid out a pattern of how Trump reacted when he lost, mentioning that Trump claimed he hadn't actually lost the Iowa caucus, which he claimed had been stolen by Ted Cruz.[73]

"Every time Donald thinks things are not going in his direction, he claims whatever it is, is rigged against him," Clinton said. "There was even a time when he didn't get an Emmy for his

TV program three years in a row and he started tweeting that the Emmys were rigged against him."

"Should have gotten it," Trump quipped.

Trump, of course, won the 2016 election, securing more than three hundred electoral college votes. Even so, his ego wouldn't allow him to concede that he'd lost the popular vote (which he lost to Hillary Clinton by nearly 2.9 million), insisting that he would've won that too had it not been for a massive, secret epidemic of voter fraud.

"I won the popular vote if you deduct the millions of people who voted illegally," he tweeted at the time. In the Oval Office, he'd go on to set up a doomed presidential commission on voter fraud intended to help soothe his feelings over his popular-vote defeat.[74] Yet the Justice Department, under the control of Trump appointees for four years, could barely manage to come up with evidence of dozens of potentially fraudulent votes, let alone hundreds, or thousands, or millions.[75] The biggest voter fraud scandal of the Trump era, in fact, involved a Republican congressional candidate in North Carolina, a man who had the backing of Trump himself.[76] Trump didn't seem too worried about the voter fraud in that case: Rep. Jim Jordan passed along a message from Trump, telling the candidate to "stand and fight."[77]

In 2016, the notion that Trump wouldn't concede an election loss was mostly about whether he'd shatter yet another political norm or how he'd be viewed in the history books—that is, *if* the history books noted him at all. Maybe he'd go on tweeting about how Hillary Clinton cheated, and maybe some of his followers would believe the first female president was an illegitimate president, just like her predecessor with the fake birth certificate. But surely even the hosts of *Fox & Friends* would grow tired of the shtick eventually. The country would move on.

That's not how it worked out. In 2020, the stakes were much higher. Trump was the commander in chief, sitting in the Oval

Office, with the power of the federal government at his disposal. Suddenly, the questions were consequential.

*Would he give up power? What if the military took his side? Does he really believe what he's spewing, or is he—like many Republican politicians—just spreading fear about voter fraud for political advantage? Does his real motivation even make a difference in the end?*

The first concrete suggestion from Trump that he might not concede if he lost in 2020 came in July, and it came paired with a refreshingly honest self-assessment of the president's psyche.

"I'm not a good loser, I don't like to lose," Trump said in an interview with Chris Wallace on *Fox News Sunday* three and a half months before the 2020 election. "I think mail-in voting is gonna rig the election, I really do."[78]

"Are you suggesting that you might not accept the results of the election?" Wallace asked.

"I have to see," Trump said. "I have to see, no, I'm not going to just say 'yes,' and I didn't last time either," he reminded the country.

Trump continued to spread fear about voter fraud and continued to refuse to commit to the peaceful transfer of power if he lost. "We're going to have to see what happens," he told reporters a few weeks before Election Day.[79]

———————◆———————

The FBI was in a precarious political position in 2020. It had been for five years, really, ever since the bureau opened "Midyear Exam," an investigation into the storage and transmission of classified information over Hillary Clinton's email servers, in July 2015.[80] The FBI had also launched an investigation, in July 2016, into whether individuals associated with Trump's campaign were coordinating with the Russian government to influence the 2016 election.[81]

The Clinton probe played out in public, with chants of "Lock her up" breaking out at Trump rallies. Trump, meanwhile, benefited tremendously from the FBI's efforts to keep the inquiry into his Russian associations quiet. Many Americans thought James Comey's actions in the 2016 election swung the election for Trump, and the evidence strongly suggests they're right.[82] Comey later said the notion that the bureau swung the election made him "mildly nauseous"[83] but conceded that operating "in an environment where Hillary Clinton was sure to be the next president" was a potential factor in his approach.[84]

Yet after Robert Mueller's investigation into Russian interference in the election and Trump's relentless attacks, the FBI had become a GOP punching bag.[85] As a law enforcement organization that is disproportionately white and male, the FBI is a generally conservative-leaning organization, full of former cops and military veterans. A lot had changed in five years, and by 2020 most Republicans thought the bureau was out to get Republicans.

If having an honest conversation about politics within the bureau had been difficult before, it was now nearly impossible. Several FBI employees had been scrutinized by the Justice Department's internal watchdog for their use of FBINet's instant message feature to have watercooler-type conversations with their colleagues. Some of them got a little heated and probably weren't the best things to type into an FBI system. One of the agents, who was talking with another FBI employee whom he was dating and later married, explained that they used the forum to "relieve stress, as a way to be jocular, as a way to exaggerate, as a way to blow off steam."[86] After the report, Wray repeatedly stressed the importance of avoiding "even the appearance" of political bias.

Trump had lashed out at Wray already, so Wray had to tread carefully during the 2020 campaign, while still reassuring Americans that elections were secure and that the FBI was on the lookout for any fraud.

"You should be confident that your vote counts," Wray said at an Oct. 21, 2020, press conference. "Early, unverified claims to the contrary should be viewed with a healthy dose of skepticism. We encourage everyone to seek election and voting information from reliable sources—namely, your state election officials. And to be thoughtful, careful, and discerning consumers of information online."[87]

The pivot that Wray made there was illustrative. Who was the loudest voice spreading those "unverified claims" that "should be viewed with a healthy dose of skepticism"? The bureau didn't say, directly. Instead of the bureau coming out and calling the president a liar—an inappropriate step, and one that certainly wouldn't have done anything to improve the FBI's reputation—Wray referred to the nebulous concept of internet disinformation. If state election officials were the "reliable sources" of information, who were the unreliable ones? Wray didn't specify. There were rules and consequences for referring to the commander in chief who was up for reelection, but there were no restrictions about referring, vaguely, to online bullshit.

That's the same approach the bureau had to take to their Oct. 27, 2020, analysis of potential reactions to a disputed election. They weren't about to go predicting future crimes the president *might* commit. Instead, they had to keep their analysis broad, ignoring the elephant in the room: the Republican presidential nominee and current resident of 1600 Pennsylvania Avenue.

Discussing election misinformation and extremism in 2020 without mentioning Donald Trump is a bit like discussing the early-1990s Chicago Bulls without mentioning Michael Jordan, or Super Bowl LII without the Philly Special. The bureau knew the role that President Trump played in inspiring extremists, including that 2016 anti-Muslim plot in Kansas. Trump-inspired plots unfolded during the Trump presidency too, including a cognitively impaired Florida strip club DJ who came to think of Trump as a "surrogate father"[88] and mailed bombs to the people Trump rhetorically attacked.[89]

Even just a few weeks before the Oct. 27, 2020, analysis, the bureau had announced that an FBI sting operation had netted thirteen men they say were involved in a plot to kill Michigan governor Gretchen Whitmer, a frequent target of Trump's ire.[90] "We want a revolutionary war. We want to get rid of this corrupt, tyrannical, fucking government," leader Adam Fox had told an undercover FBI employee in June. "The consensus is, as of right now, is taking the fucking Capitol by force, like with extreme heavy fucking prejudice toward our fucking government officials."[91]

Less than a month before the FBI analysis was finished, at the first presidential debate, the president told the Proud Boys to "stand back and stand by," a comment celebrated by the far-right group.[92] The same night, he ranted that the election was "rigged," called on his supporters "to go into the polls and watch very carefully," and made his race-baiting "bad things happen in Philadelphia" comment.[93]

Nonetheless, even internally, the FBI had to pretend Trump wasn't a factor. What if they did an honest assessment of the situation and a pro-Trumper within the government leaked it before the election? Worse yet, what would happen to the authors if Trump won again?

The FBI had distributed about a dozen "external intelligence products" to law enforcement partners throughout 2020 about domestic violent extremism, including many that related to the election.[94] This Oct. 27, 2020, analysis was different. It wasn't intended for external distribution, according to a summary I later reviewed. This was a "Red Cell" analysis aimed at anticipating potential outcomes.

After the intelligence failures at the turn of the century— missing intelligence about the Sept. 11 attacks and the incorrect assessment that Iraq possessed weapons of mass destruction—the intelligence community began using Red Cell teams to test their assumptions, to challenge conventional wisdom. The Directorate of Intelligence created a Devil's Advocate Program, while the Department of Homeland Security formed an Alternative Analysis Division and a Red Cell unit,[95] consulting with "futurists, philosophers, software programmers, a pop musician and a thriller writer."[96] The FBI doesn't publicize their Red Cell efforts, although an FBI Red Cell team was the subject of a short-lived, poorly reviewed *Criminal Minds* spin-off titled *Criminal Minds: Suspect Behavior*, starring Janeane Garofalo as Special Agent Beth Griffith.[97]

The FBI's Intelligence Council and the Red Cell team at headquarters teamed up with the Counterterrorism Division and the Boston Field Office to play out what could happen.

The Red Cell exercise was supposed to get the juices flowing, to spark creativity, to get FBI employees thinking outside the box. "This one did not," read the analysis I later reviewed. The FBI had "missed the forest for the trees."

The investigator who analyzed the document later wrote that the Red Cell report never considered "the rise of a mass movement that might support an aggrieved losing candidate."[98] None of the four scenarios that the FBI laid out considered that a broad right-wing movement could come together, that a crowd could be a threat. The FBI placed a major emphasis on "lone wolf" terrorism in the Red Cell report,[99] and the unpublished Oct. 27 product was "anchored by the lone offender bias," a later analysis found.

On top of the exercise's built-in flaws, the bureau was a bit late to the game. A bipartisan group called the Transition Integrity Project had run a set of "war games" to play out what could happen during the 2020 election.[100] They put out their report in August.

"The period from November 4th to December 14th sets the stage for a potential fight in the Congress on January 6th, 2021," that report states, positing a political fight in Congress, not a physical one.[101]

The media was also acting as a canary in a coal mine.

"There is a cohort of close observers of our presidential elections, scholars and lawyers and political strategists, who find themselves in the uneasy position of intelligence analysts in the months before 9/11," wrote Barton Gellman in an *Atlantic* feature published in late September.[102] "As November 3 approaches, their screens are blinking red, alight with warnings that the political system does not know how to absorb. They see the obvious signs that we all see, but they also know subtle things that most of us do not. Something dangerous has hove into view, and the nation is lurching into its path."

Three things were abundantly clear at the time: (1) Trump's rhetoric had inspired extremists before, (2) he had little regard for the law or political norms, and (3) he had a history of not accepting when he lost. But days before the election, the FBI had to operate in a different space, pretending that this was a normal situation and that there were two parties and candidates operating in a shared, common reality, who just happened to have different views on politics. Alarms were going off, but the bureau had on soundproof headphones, partially because of rules and norms, partially for self-preservation. Even if Trump lost, what if Republicans took the House or kept the Senate? Nobody wanted to be hauled before Congress or to become the target of online disinformation themselves. *The system will hold*, the thinking went. *Let's just make it to Inauguration Day.*

On election night, Trump acted in an entirely predictable manner. "Frankly, we did win this election," he declared after entering the East Room of the White House at 2:21 a.m., less than two hours after Fox News had called Arizona for Joe Biden.[103]

"Last night I was leading, often solidly, in many key States, in almost all instances Democrat run & controlled. Then, one by one, they started to magically disappear as surprise ballot dumps were counted," he tweeted at 10:04 a.m.[104]

Some Republican-appointed former top federal prosecutors spoke up, sounding the alarm over Trump's rhetoric and writing in an open letter saying that America's "very legitimacy as a nation of laws, not men" depended on the votes being counted without interference.

"Unfounded allegations of fraud and threats to initiate litigation aimed at stopping the vote count are clearly inappropriate and have the potential to undermine the rule of law as it applies to our electoral process," they wrote in the Nov. 5 letter. "Moreover, for any candidate, let alone the president of the United States, to claim victory, without justification, before each and every vote is counted is imprudent and irresponsible." They called on Trump to "avoid any further comments or other actions which can serve only to undermine our democracy."[105]

I spoke with David Iglesias, a former US attorney under President George W. Bush. He described himself as a former "true believer" in the myth of mass voter fraud. It was time for his fellow Republicans, he said, to "wake up and smell the coffee."[106]

"There is no voter fraud problem in this country," Iglesias said. "Time to move on."

Trump wasn't ready to move on. At a White House press conference on Thursday, Nov. 5, Trump did exactly what anyone paying attention expected him to do.

"If you count the legal votes, I easily win. If you count the illegal votes, they can try to steal the election from us," Trump said. "There's been a lot of shenanigans, and we can't stand for that in our country."[107]

Justice Department to step up. I fired off an email to the Justice Department's top spokeswoman. She had a law degree. *Clearly she couldn't possibly believe this bullshit*, I thought.

"It just really seems like a gut check moment here," I wrote. "Basically every U.S. Attorney in the country put out press releases about how closely they were watching for election fraud. We all knew this kind of attack was coming. The Justice Department really isn't going to say anything to reassure the American people about the integrity of the election?" I didn't get a response.

Meanwhile, across the country, millions of Americans believed Trump's ludicrous lies. Extremists were already organizing.

"It's time for fucking War if they steal this shit," wrote Joseph Biggs, a member of the Proud Boys, on Nov. 5.

"The media constantly accuses us of wanting to start a civil war," wrote Enrique Tarrio, the Proud Boys top official. "Careful what the fuck you ask for we don't want to start one . . . but we will sure as fuck finish one."[108]

"We aren't getting through this without a civil war. Too late for that. Prepare your mind, body, spirit," wrote Stewart Rhodes, the founder of the Oath Keepers.[109]

It wasn't just the organized groups; random right-wingers were popping off too.

"Omg. I fucking cant believe this but I can. Ugh he's gonna win," one woman named Anne wrote in a message to her Facebook friend, referring to Joe Biden's impending victory.

"I'm ready to stand with patriots to over throw this shit Live or dead," her friend Rodney Milstreed wrote. "We cannot walk away from this shit."[110]

Milstreed wanted a recount or revote for the entire 2020 presidential election. Do the whole thing over, he said. That was the only option. He talked about all the steroids he'd just purchased, how he felt he was in such good shape. He sent around photos of his guns.

"Watch and see what happens. These fucks are very crazy if they think we are gonna stand for this," he wrote. He wasn't alone.

"Democrats told us they were going to steal this election RIGHT TO OUR FACES . . . AND WE DID NOTHING," wrote Joe Pavlik, a retired Chicago firefighter who was associated with the Oath Keepers. He was ready to fight. "So get ready young guys because YOU WILL LIVE ON YOUR KNEES from this day forward THATS RIGHT because we will do nothing."[111]

But it wasn't just extremists and retirees with poor internet literacy who got sucked into Facebook algorithms who were ready to fight. Some of those "young guys" were ready too. Christian Secor, a political science major at the University of California, Los Angeles (UCLA), and a follower of the far-right podcaster Nick Fuentes and the America First movement, was taking a stand. "✓Standing Back ✓Standing By ✓Trusting the Plan," he tweeted. He suggested bringing back literacy tests.[112]

"Any Republican governor who doesn't begin drafting articles of secession if the election is stolen is a traitor to their constituents," Secor tweeted. He'd later storm the Capitol, sit in Mike Pence's seat inside the Senate chamber, and go to federal prison for three years.[113]

By Saturday, Nov. 7, while Trump was out golfing at his club in Sterling, Virginia, the results were in. The networks called it. Donald Trump would be a one-term president. Joe Biden was the president-elect.

In DC, the celebration kicked off quickly. In the city's Adams Morgan neighborhood, residents dining outdoors along Eighteenth Street NW cheered and danced as bikers rang their bells, drivers blew their horns, and one man blasted the song "FDT" ("Fuck Donald Trump"), by rapper YG (featuring Nipsey Hussle), from his vehicle. Parents loaded their kids up in strollers and bikes, flooding toward the White House, where someone used a large Bluetooth speaker to play a song that had topped the Billboard charts in 1969, when Richard Nixon was president, before Watergate brought him down: "Na na na na, na na na na, hey hey hey, goodbye."

Demonstrators were parading around with a giant balloon illustrating Donald Trump as a rat, holding two black balloons that read "LOSER." People cheered and snapped photos. High fives were exchanged. American flags were suddenly in vogue again. A man popped a bottle of champagne, spraying it over the crowd before chugging the remains and pretending to smash the bottle on the ground.

"You about to lose your job!" he yelled at the balloon, invoking a viral meme. The crowd cheered.[114]

---•◆•---

Mike Siravo wasn't cheering. Siravo was the thirty-two-year-old general manager of a Philadelphia landscaping company owned by his mother, Marie Siravo.

Mike Siravo had bought into lies about the election. On Facebook, he suggested that the Biden-Harris campaign sign had a hidden meaning that signaled support for Chinese socialism. The night before the press conference, in his Instagram story, he'd posted a meme that showed "Biden Harris" campaign signs in front of gravestones, suggesting that Democrats were casting ballots on behalf of dead people.

"America, where if you don't get your way the first time. You can lie, cheat and steal to get what you want," Mike Siravo wrote on Facebook that week.

"Family are definitely bigtime Trump supporters," someone who knew Mike and the Siravo family growing up told me.

Around 8:45 a.m., when his college buddy and director of sales was at a Bible study, someone in the Trump campaign had called to inquire about whether their property was available.[115] The Trump team's attempts to hold press conferences in downtown Philadelphia, near the convention hall where votes were being counted, had gone poorly: Biden supporters blasted "Party" by Beyoncé (featuring André 3000) as Pam Bondi yammered on.[116] They needed a new location.

Four Seasons Total Landscaping is located in northeast Philadelphia, right down the street from the city jail, the Curran-Fromhold Correctional Facility. There are parts of northeast Philadelphia, like sections of Juniata, that broke for Trump by a margin of thirty-three points.[117] It wasn't exactly known as a press conference venue, but it had a gate that could be secured to keep demonstrators out, and it was available on short notice.

So when the election was called, Giuliani and friends were at the landscaping place next to the sex shop, near the crematorium, with a group of GOPers that included a sex offender.[118]

Olympia Sonnier, an NBC News producer, informed Giuliani of the news. When Trump had tweeted out that there'd be a press conference at the Four Seasons, she'd mapped out directions to the Four Seasons Hotel in downtown. Then they checked again, and Trump had sent a new tweet. It was Four Seasons Total Landscaping. *Bizarre*, she thought. It got worse from there. They rushed over to the landscaping company, finding other reporters scampering to the scene. There were some macho security guards—it wasn't quite clear who they were—at the gates, and soon an overwhelmed woman with a clipboard started letting reporters and camera operators in. Protesters tried to get in too but were unsuccessful. There was a mad dash to get in position. It was a bit of a crazy morning, at the end of an insane campaign.

Now she was informing the former mayor of New York City—the guy who'd become a hero after Sept. 11 only to be reduced to, well, *this*—that it was a lost cause.

"It was called for Joe Biden," she said.

"They don't decide the election," Giuliani said. "Who was it called by?"

"All the networks," someone chimed in.

"Oh my goodness, all the networks!" Giuliani said mockingly, looking toward the sky and raising his arms as high as he could. "All the *netwoooorks*!"

The full story of how this press conference was gathered was never totally clear. One reporter, Olivia Nuzzi, spoke to thirty-seven sources in Trump world and still couldn't figure out precisely what happened.[119] But Siravo, his mom, and their business pivoted. They said they would've let any presidential candidate use their venue. They joined in on the memes, laughing at themselves and selling merch. Marie Siravo would appear in a Super Bowl commercial.

Not everyone was in on the joke at that point. "Folks. Be ready for war. Trump has refused to cede. Evidence shows fraud occurred and the Supreme Court cases will be successful. We blockchained and watermarked ballots in 16 states. Trump will prevail. Spread this message," wrote Robert Lemke, an air force veteran, on the day of the press conference. "FAITH my fellow Republicans. Do not give up. Keep an eye out for a variety of protests, and Stop The Steal Facebook groups for updates."[120]

Just over an hour after the networks made the call for Biden, another right-winger messaged Enrique Tarrio, the head of the Proud Boys, to check in.

"Hey brother, sad, sad news today," Shane Lamond texted Tarrio at 1:08 p.m., referring to Trump's election loss. "You all planning anything?"[121]

Lamond wasn't a Proud Boys supporter, per se; he was a member of the DC Metropolitan Police Department. Speaking with Tarrio was, nominally, a part of his job. But Lamond's messages seemed to stray beyond that. At the same time he was ostensibly gathering intelligence on what the Proud Boys were up to, Lamond was telling Tarrio what law enforcement officials were saying about him.

"Need to switch to encrypted," Lamond wrote at 2:10 p.m. "Alerts are being sent out to [law enforcement] that [social media] accounts belonging to your people are talking about mobilizing and 'taking back the country.' Getting people spun up. Just giving you a heads up." Lamond soon asked Tarrio to "keep this between you and me."

Lamond was working both sides. Not long after he gave Tarrio a heads-up about what law enforcement was saying about the

Proud Boys, he downplayed the Proud Boys' rhetoric in a discussion with a counterpart in the Department of Defense.

"Some chatter about PB mobilizing and calling for violent action. Spoke with my source—totally fake news and taken out of context," Lamond wrote at 3:20 p.m., adopting Trump's term for the media.[122]

While Giuliani was at Four Seasons, and while Tarrio was talking with a DC cop, one of Tarrio's group texts was blowing up. Some of the Right's most influential figures were members of the encrypted chat FOS—Friends of Stone, as in Roger—and were being pinged with a plan inspired by a "Serbian patriot" who was in contact with an American extremist.[123]

Round one was peaceful protests. Then came "complete civil disobedience." Then swarm the streets and confront opponents. Then march on the capital.

"There were no barricades strong enough to stop them, nor the police determined enough to stop them," the "Serbian patriot" had explained. "Police and Military aligned with the people after few hours of fist-fight—We stormed the Parliament."

Stewart Rhodes thought the "Serbian patriot" was right. The Oath Keepers founder had a plan.

"We need to march on DC right now, so President Trump knows he has our support as those around him whisper in his ear that it's no use and he should give up," he wrote on Nov. 7. "The final defense is us and our rifles."

I'd first met Rhodes at the Conservative Political Action Conference early in the Obama administration, when his organization was just taking off. "We're not a militia—we don't train, and we're not out in the woods or any of that stuff," Rhodes told me.[124]

During Trump's first election, he'd laid out a plan to watch out for voter fraud.[125] Rhodes told me he wanted members of the organization to go "incognito" and "blend in" at polling places.

"You won't even know they're there," Rhodes said at the time. "If someone is just going about their business, have a nice day."

Greg McWhirter, a member of the Oath Keepers' national board of directors, recorded a video that encouraged participants to be "friendly" and unarmed.

"The ideal would be to catch somebody—you know, a carload or a busload or a vanload—of people going from one polling place to another," Rhodes told me. "That is obviously a smoking gun video we'd like to have, but clearly us being out there is hopefully going to put a damper on those kinds of activities. So if nothing happens, then great, we have a boring day and just walk around and enjoy the outdoors."

In 2016, Rhodes wasn't yet the Trump diehard he'd become. He was driven more by his hatred of Hillary Clinton than by any allegiance to Trump, whose statements on guns concerned Rhodes. But by 2020, Rhodes was fully aboard the Trump train, talking with influential political advocates with strong ties to Trump. He kept networking. In the immediate aftermath of the election, Rhodes connected with Kellye SoRelle, a Texas family law attorney who was in Detroit as part of Trump's army of poll watchers.[126] SoRelle later told me that the Republican National Committee paid for her trip to Detroit.

SoRelle wasn't a Michigan resident, so she couldn't go into the area where they were counting ballots. But she was in communication with Andrew Giuliani, the son of Rudy Giuliani, and she just *knew* that something was wrong, that there was some sort of corruption afoot.[127]

Sitting in her car outside the TCF Center on election night, a bit before 3:00 a.m., she filmed a man removing a box from a white van. He placed it into a red wagon, the kind they make for kids. It soon became a viral video, tweeted out by Eric Trump.[128] She thought it might be a box of ballots. "SoRelle is raising alarms that the box may have been a ballot box that arrived long after all ballots were expected to have been received at the counting facility," said one conservative website. As it turns out, she was filming a member of a news crew, who was wheeling in media equipment.[129]

In Detroit, after her video came out, SoRelle was frightened. She got in touch with Rhodes. "She was doing a lot of legal activism," Rhodes later explained. He wanted to help her out, so he called up a former military man he knew, who'd worked security for him at prior events.

"I talked to her on the phone. She was scared. She was crying," said Michael Greene. "She was getting threats. People were showing up at her hotel and stuff like that." He headed out to help her.

Whether those threats were real or inflated was tough to nail down. SoRelle could be friendly, but she was also very conspiracy minded. Spend time on the phone with her, as I later did a lot, and you'd get a few good pieces of information, but you also might get a diatribe about how the Central Intelligence Agency (CIA) was hacking her brain.

This was the new Ballot Security Task Force of the digital era. It was random lawyers, like the guy from New Jersey, and fringe attorneys like SoRelle, with her Oath Keeper links. It was people like Mike Roman, who served as the Trump campaign's chief poll watcher in 2016 and 2020, which basically meant spreading fear about voters in majority-Black cities[130] and creating content that, as he wrote in one November 2020 email, "satisfies the narrative" that the Trump campaign was trying to spread.[131] It was folks like Linda A. Kerns, a family law attorney, who was the Trump campaign's lawyer in Philadelphia.[132]

During Obama's first election, she became a part of a minor incident involving members of the extremist New Black Panther Party, in which one of the men held a nightstick outside a polling place. The polling place was in a neighborhood in north Philadelphia, and the overwhelming majority of voters who cast ballots at that polling location were Black. No voter claimed they were intimidated, and the New Black Panther Party members said they were present because of the white poll watchers who had shown up to the polling place in search of fraud. Nevertheless, the minutes-long incident sparked a years-long controversy thanks to conservative media outlets like Fox News and folks like Roman, who published the video on his website.[133]

In prior election cycles, Kerns was one of the Republican attorneys who spent time in court trying to gum up the works, even trying to stop hospitalized voters from casting ballots.[134] One of those hospitalized voters Kerns opposed was future mayor Jim Kenney, then a city councilman, who was unexpectedly in the hospital for surgery.[135]

"I figured if I could slow this down, that's a win," Kerns later bragged about her efforts against sick voters. "And I did slow it down, and the other attorneys would be jumping up and down and talking about suppression of the vote and all this nonsense."[136]

———————— ◆ ————————

In the days after Trump's election loss, I called up Greg Brower, a former Republican elected official and FBI official, who wanted Trump to quit spreading misinformation and accept defeat.

"It's been happening for four years now, but it's still stunning to most of us, I think, that the president does absolutely nothing to try to tamp that down. In fact, he encourages it," Brower said. "I know it's very frustrating for law enforcement, but the president doesn't appear to be willing yet to face the facts, encourage his followers to remain calm and accept the fact that he has lost, and move on."

I also called up Bill Fulton, a former FBI informant who infiltrated a right-wing militia in Alaska in the early days of the Obama administration. Back then, his drinking buddies knew him as "Dropzone Bill," a military veteran who provided security, hunted down fugitives, and sold surplus military equipment. Working for Joe Miller during a 2010 senatorial campaign, he'd even handcuffed a journalist, which only helped bolster his background story, or "legend" as the FBI calls it. He'd help send a militia member away for decades.[137]

Fulton was in a different state now but was still working with the bureau as they kept tabs on the right-wing militia scene. He was worried about how things were shaping up.

"The latest round of conspiracy theories is just a continuation of the hell that the last four years has been if you work in a counter–domestic extremism capacity," Fulton told me. "This is nothing new, it's just a little more dangerous now.

"What we call it is walking them to the edge," Fulton said. "You have the president of the United States taking these people to the edge, and the second that something happens he's going to turn around and go, 'Well, I didn't tell them to do that.' It gives him that plausible deniability, and that's what's scary."

Republican leaders thought those fears were overblown. Not long before I spoke with Fulton, an anonymous senior Republican official assured the *Washington Post* that this was no big deal.

"What is the downside for humoring him for this little bit of time? No one seriously thinks the results will change," the official said. "He went golfing this weekend. It's not like he's plotting how to prevent Joe Biden from taking power on Jan. 20. He's tweeting about filing some lawsuits, those lawsuits will fail, then he'll tweet some more about how the election was stolen, and then he'll leave."[138]

Fulton knew better. "You can't keep people on the edge for that long," Fulton told me. "A lot of these guys think that they're like defending the United States. They think they're patriots. That is powerful. Patriots go to war for their country all the time. . . . When you start mixing that, you end up with really, really bad shit happening."[139]

Inside FOS, there was some bad shit happening. They were preparing for war.

"We are now where the Founders were in March, 1775," Rhodes wrote to the "Friends of Stone."

# PART II

# CHAPTER 4

# "A Call to Arms"

## Dec. 6, 2020

A month before the attack, the head of the Intelligence Division of the FBI's Washington Field Office (WFO) sent a mass email, issuing an important warning. Time was running out.

"We are T-11 days until the deadline for your submissions for the holiday virtual contests," Jennifer Moore, the Washington Field Office Intelligence Division chief, wrote on the night of Dec. 6. There were virtual contests: ugly holiday sweater, best holiday pet, best holiday light display, best gingerbread house, Elf on the Shelf. The photos and results would be posted for the office to view, unclassified. The stakes were high: winners would get gift cards.

"The pictures have started coming in and they are FABULOUS. Don't be left out," Moore wrote. "The emails truly brighten my day when I get them."[1]

Moore had been at the FBI since 1995, and she went to Quantico and became a special agent in 1998. She did stints in Dallas, Las Vegas, and Louisville and was named special agent in charge of the Intelligence Division at the WFO in 2019, overseeing about one hundred intelligence analysts and a total of about six hundred people in the Intelligence Division.

Then COVID-19 hit. A lot of white-collar and government workers were able to do their jobs from home, but that was a bit more complicated for the FBI's intelligence employees. They were mostly in the office, spreading out as much as they could.

The year 2020 was rough for everyone, and Moore tried to keep her FBI staffers cheery with her "Captain's Log" updates, which were sprinkled with jokes along with shout-outs and updates on the work of others in the Intelligence Division. Readers learned that the Fourth of July is Moore's favorite holiday, that she enjoys hot yoga, that she's originally from Florida, and that she likes corny jokes. "Do you know why ants never get sick? . . . because they have little antibodies!" she wrote in one 2020 email that mentioned how her son's college roommate came down with COVID-19.

Moore's email updates also offered a look into what the Intelligence Division was up to in 2020, including some news on the domestic extremism front. In May, there were updates about tactical intelligence reports on Patriot Front and a briefing on how racially motivated violent extremists were trying to circumvent law enforcement's monitoring. In June, at the direction of headquarters, they had to identify, rank, and map all federal monuments, statues, and buildings in the DC region, and twenty-nine employees worked on the project. In September, an intelligence analyst finalized a white paper on domestic terrorism that laid out how the "constant influx of visitors" to the DC region "poses a challenge in monitoring threats."[2] In October, when a Twitter user threatened to blow up the CIA, an intelligence analyst was able to comb through the user's account and "identify a nickname, DOB, phone carrier, school, likely city of residence, two additional social media accounts, and physical descriptors based on a review of the subject's Twitter posts, comments, and various content clues in photos the subject posted." And in November, there was a bureaucratically titled, election-related report on the *Impact of Potential Contested Results* on threats in the DC area. Later that month, an intelligence analyst presented a briefing on "the increasing use of the Parler social media platform and its potential impact on violent activity" to Moore and other top officials. "I tend to lean heavily on my kids to keep me up to date on the social media trends. I have Instagram down to a science, and my daughter and I have a 667 day streak on SnapChat (*Oh yeah, I am trendy*)," Moore wrote. "However, every now and then, a new social media platform comes along which might have a large impact that my kiddos are not familiar with."

Moore's Dec. 6 email started off with a *Finding Nemo* reference, some discussion of raking leaves, and a reminder for members of her unit to complete their Virtual Academy training before the holidays. "It makes for a true frowny face sort of day when I get a list that shows Intelligence Division folks are not compliant," she wrote.

"I'll close out with a reminder for #selfcare! Make sure you are taking time for yourself."

———————— ◆ ————————

Jason Dolan was #angry. The night Moore sent her email, he was in his garage in Florida yet again, drinking and scrolling.[3] He served nearly twenty years in the US Marine Corps, deploying five times and retiring as a staff sergeant, the civilian equivalent of middle management. Back in civilian life, he became a security guard but quit ahead of his hip replacement surgery, which didn't go the way he'd hoped. By the time the COVID-19 pandemic hit in 2020, Dolan was unemployed, spending a lot of time in his garage, downing a six-pack of beer or a half bottle of vodka every night, "trying to kind of kill the pain with booze."[4]

Dolan would lock into his phone, watching video after video about the 2020 election. He supported Donald Trump and thought the election had been stolen. Watching Fox News on election night, he found it "extremely odd" that the networks weren't calling states like Florida and Arizona, where Dolan believed Trump had won. (Dolan was correct about Florida, which the Associated Press called for Trump after midnight, but wrong about Arizona, which the AP called for Joe Biden at 2:51 a.m. Fox News had called Arizona for Biden much earlier, at 11:20 p.m.)[5,6]

"It just didn't seem right in my head," Dolan said. "I was pretty pissed.... It didn't seem possible that he was going to lose."

After Trump's loss, Dolan was looking for some way to vent. A friend told him about the Oath Keepers, and he contacted the Florida branch. They were "pretty unhappy" about the election results too.

The Oath Keepers asked him to download Signal, an encrypted messaging app. He went by the name Turmoil. He also joined in on the group's GoToMeeting gatherings every Friday, but the Signal chat is where he really had his chance to shine, to vent, to connect. Jason said he'd spoken with his neighbors—a fellow former marine lived just across the street, as did a sheriff's deputy—but thought they just didn't "give a shit" and weren't willing to step up and fight for Trump. In the Oath Keepers' encrypted chats, he found a community of people he thought were ready to stand up and do something.

"I didn't feel like I was alone. I don't have a lot of friends," Dolan explained. "It felt good to know that there are other people out there, I guess, that felt the same way I did."

On the night of Dec. 6, Dolan was in his usual spot in the garage. When a Google Street View vehicle drove past his home a few weeks later, the month of Joe Biden's inauguration, he'd have the American flag in his front yard flying upside down, signaling distress, and a black version of the Gadsden "Don't Tread on Me" flag affixed above his garage. But for now, he was sitting behind the garage door, very drunk and very online, preparing for war.

"Everyone has different circumstances," he typed. "I've been preparing my wife & daughter for the possibility of heading over-seas for the last couple months. We have that option but very few people do. I'm waiting on all legal remedies (from legitimate or illegitimate) courts to be solved. I have Christmas plans with the family but come the new year (for me anyway) I have to be men-tally prepared for however far I'm willing to go to stand for America, for the Constitution, for the President & for the survival of our ideals."[7]

There was more back-and-forth with members of the channel, including Kelly Meggs, a fifty-two-year-old who served both as the general manager of a car dealership outside of Tampa and the head of the Florida chapter of the Oath Keepers. Dolan told the other Florida Oath Keepers that he'd spent most of his time on the front lines, that he shouldn't be alive, that he was only on earth "by the grace of God." He didn't want to put his wife through another deployment after he'd left the military, even though he'd had some opportunities to work overseas with nongovernment organizations. Now, he said, he was asking her to let him fight another kind of war, on behalf of Donald Trump and, in his view, America.

*I'm asking her to put up with it again but this time there is no coming back, no pay, [no] awards, no homecomings, and if I'm lucky I get a prison sentence, tagged with treason, or a bullet from the very people I would protect. Yet I swore to defend this country against all enemies foreign & domestic. I think my biggest trouble is trying to convince myself to say good bye to my family, after all they had to endure, with the likelihood of never seeing them again ... again.*

Dolan apologized to his fellow Oath Keepers for ranting; he just couldn't help himself. "Things are getting closer to the point of no return," he wrote. "I can't help thinking about my decisions in the coming days. I may be a dumb grunt," he wrote, using a military term for a low-level infantry marine, "but I'm not stupid."

Later, Dolan would admit this was, in fact, "pretty stupid." He'd rethink those conspiracy theories about the election too. But at the time, he said, "I meant it literally." *Is this all just going to be talk?* Dolan wondered. *Or am I willing to back up my words with actions?*

———— ◆ ————

The FBI informant was #frustrated.

Kris Goldsmith is an army veteran. Under different circumstances, he and Jason Dolan might have been buddies. Goldsmith enlisted when he was eighteen, a couple of years after the Sept. 11 attack. By 2005 he was in Sadr City, Iraq, where he was assigned a role as an intelligence reporter. What that meant in Sadr City is that he went to mass graves—a dozen bodies in sewage and garbage—and took pictures of the corpses, some of which showed signs of torture. The smell stuck with him, "seeping into my bones, into the core of my being," he said. Symptoms of PTSD, posttraumatic stress disorder, showed up soon. When getting ready to deploy again, he felt as if he was having a heart attack and went to the hospital. Group therapy followed but didn't help much. He planned to end his own life. His roommate found his note and saved his life, and Goldsmith woke up in a psychiatric ward.[8]

Because of his suicide attempt, the military gave him a general discharge, not an honorable discharge, which would have allowed him to access educational benefits. He turned to booze, spent his pay "in six months at a bar," and ended up back in his childhood bedroom, feeling rudderless.

By December 2020, he was in a much better place. Years before, he'd leaned on his disability benefits from the VA, was diagnosed with PTSD, and eventually earned himself a spot at Columbia University. He wanted an honorable discharge from the military and began lobbying on Capitol Hill to change the system. The Vietnam Veterans of America (VVA), which had been working on the "bad papers" issue "since the '70s," took notice, and Goldsmith soon joined the organization as an employee. Poking around Facebook one day while managing the group's social media accounts, he noticed a bunch of fake Facebook pages that were set up to mimic the organization. One had twice as many followers as the group's real Facebook page. Veterans were being scammed and manipulated. Goldsmith now had a new role at VVA: chief investigator. From there he founded the High Ground Veterans Advocacy to help former military members advocate for themselves. Under public pressure, the Cybersecurity and Infrastructure Security Agency (CISA) partnered with VVA to launch the #Protect2020 campaign aimed at combatting election misinformation.[9]

CISA weren't the only feds Goldsmith was working with. He was now an FBI confidential human source, working out of the New York office. He first started talking to the bureau in 2020, a few months after he published the *Troll Report* with VVA and started talking to the press about how the FBI didn't seem to care about foreign entities targeting troops.

He first met with an FBI special agent in early 2020 to discuss how veterans were being targeted on social media and what the bureau could do about it. Afterward, Goldsmith reached out for a Gmail address so he could pass along his notes, screenshots, and spreadsheets. That was his first taste of FBI bureaucracy.

"I can mail you a return envelope, and a DVD to burn it to if you have that capability," an FBI special agent wrote him in an email. Goldsmith worked for an organization for people who served in war a half century ago. Average Vietnam veterans were in their midseventies and had been eligible for the AARP for decades. Even he hadn't used a DVD burner since the Bush administration, so he suggested a thumb drive instead. Months later, the FBI agent offered to return it or reimburse him for the cost. The thumb drive became a running joke.

When the VVA laid Goldsmith off during the pandemic in 2020, he told the FBI he'd be focusing his efforts on domestic actors like the Patriot Front, which he helped infiltrate and publish a report on ahead of the 2020 election.[10] Late in the year, the bureau set him up with a burner laptop and a phone but "always made a point of making clear they weren't directing or approving anything," Goldsmith said.

It could be a bit exasperating. Goldsmith would send them reports on all kinds of domestic actors, like the Proud Boys, but never got much feedback from the bureau. "Received" was the best he could hope for.

Goldsmith, like many extremism researchers and people with an internet connection, was worried about the spread of misinformation about the election and the potential for violence. Soon he decided to virtually infiltrate a militia that had been closely associated with the Stop the Steal movement.

Under a pseudonym, he checked in for an interview with the right-wing "III% Security Force" in Georgia, which had been attending Stop the Steal protests in that state, including a Dec. 5, 2020, event led by one of the most extreme Republican members of Congress, Rep. Andy Biggs.[11] Wearing a Trump hat and a red Nike sweatshirt at that event, III% Security Force leader Chris Hill—who went by "General Blood Agent" online—spread lies about the election and suggested it was time for more extreme measures.

"This is what happened in all of the contested states, at 2-o'clock in the morning, the ballot fairy comes around dropping off votes to help push things in Joe Biden's favor," Hill told reporter Ford Fischer. "We're literally witnessing our country being taken away from us, and we're just tired of playing softball with the cheats and with the communists, and it's time to take the gloves off and protect our country from a takeover."

Hill said that he hoped it wouldn't come to this point but that he would "refresh the tree of liberty with blood" and wouldn't "shirk at that responsibility," casually suggesting that he'd murder those he considered tyrants.

III% Security Force was founded in 2014 and in subsequent years received significant media attention that boosted their membership. In November, the group showed up at rallies outside the homes of Georgia's governor and secretary of state.

"Tar and feather tax collectors! Fuck around and find out!" one member yelled through a bullhorn outside of the governor's home. "Tyrants get the rope! Tyrants get the rope! Trump 2020!"

In late November, Goldsmith was incognito, using the call sign "Alex Hamilton 1776." On a Tuesday at around 8:30 p.m., he hopped into a "3SF" audio chat for his official interview before a group of militia members with names like Larcell and BiggKuntry1776. His military background and history in the state—Goldsmith had legitimately been stationed in Georgia—helped him maintain his loose cover as a thirty-five-year-old army veteran who was ready to move from New York to Georgia. He gave a fake name and said he didn't do social media.

Chris Hill ran most of the interview, but another member asked Goldsmith how he'd respond to antifascist protesters, whether he was willing to use deadly force. Goldsmith also assured them he'd keep things secret and not go talking about what the group was up to.

"I don't have anyone I'm blabbing to," the FBI confidential human source assured the militia members. "I've got my hobbies. I brew beer and I distill liquor, and if there's any feds listening, I don't do any of that."

A joke about the feds—he fit right in. Members quizzed him, asking him about his military history and what he'd do if shit

hit the fan. One participant complimented him on his "great bugout bag."

The vote was unanimous. Goldsmith was in.

"Welcome to the family," members said, one by one. "Guns up!"

———————— ♦ ————————

The passenger-side door of the white van slid open, and the man raised his weapon. Bullets zipped through the air, striking two security officers at the federal courthouse. One, David Patrick Underwood, was dead. A second security officer was seriously injured.[12] It was May 29, 2020, the same night the Secret Service rushed Donald Trump to the White House bunker after protesters faced off with law enforcement in Lafayette Square. Oakland, like many American cities, was facing civil unrest in the wake of George Floyd's death. Now, it seemed, someone backing the protests had assassinated an officer and tried to kill another.

"This antifa violent activity has to stop," Robert O'Brien, a White House national security adviser, said of Underwood's death in late May. On June 10, Rep. Jim Jordan blamed "the rioters in Oakland" for the officer's demise. The next day, the FBI arrested the man responsible for Underwood's killing: Steven Carrillo.[13]

Carrillo wasn't an antiracist activist, a criminal justice reform advocate, or a Black Lives Matter supporter. He was an active-duty member of the US Air Force, and he was a boogaloo "boi."

The boogaloo movement has been described by the *New York Times* as "America's Extremely Online Extremists." Its name is derived from a break-dancing movie that came out in 1984, before many of the adherents were born, called *Breakin' 2: Electric Boogaloo*, which became a meme applied to any sequel. In this instance, it would be a sequel to the Confederate rebellion of the 1800s, as in "Civil War 2: Electric Boogaloo."[14] The boogaloo movement doesn't cleanly fit into a political box, but it's progun, proviolence, anti–federal law enforcement, and generally right leaning.[15] Carrillo had met his accomplice, Robert Alvin Justus Jr., on Facebook, where the plot began to formulate. The plan was to use the riots as cover.

"Go to the riots and support our own cause. Show them the real targets. Use their anger to fuel our fire. Think outside the box. We have mobs of angry people to use to our advantage," Carrillo had written.[16]

That left-wing rioters weren't responsible for the officer's death became a politically inconvenient fact. Later, while accepting the GOP nomination for vice president, Mike Pence would tell a national audience that Underwood was "killed during the riots in Oakland," leaving out the messy fact that Underwood was killed by a far-right extremist in the military.

After his arrest, he was given access to a jail-issued tablet, a highly regulated device that allows inmates to perform a limited number of functions. Scrolling social media isn't one of them. But Carrillo was so hooked on social media that he spent his time "constantly" looking for the Facebook application "even though he knew it wasn't there and that no form of social media was accessible."

Brian Hughes, a radicalization and disinformation expert at American University who interviewed Carrillo, said the technical term for Carrillo's behavior in addiction terms is "transient compulsive foraging behavior," just as when crack cocaine addicts compulsively search their environment for potentially lost pieces of crack.

"My whole world was in my palm," Carrillo told Hughes, describing how his girlfriend would have to throw things at him to get his attention.[17]

Carrillo was a compulsive consumer of social media, especially when the pandemic hit. The videos he watched made him angry and frustrated, but he kept scrolling anyway, telling Hughes he'd oscillate between feelings of numbness and rage.

Social media companies used the same techniques as Las Vegas casinos to hook users and then served up "a stream of content and connections increasingly defined by hatred, fear, and loathing," Hughes wrote. It was a lethal combination.

"Save for the gun industry itself, social media is the only industry which predictably produces externalities in the form of mass murder," Hughes wrote. "And yet social media companies' annual expenditures do not reflect a serious commitment to ending this predictable externality."

———————◆———————

A few decades after his escapades in the New Jersey gubernatorial race, in 2016, Roger Stone started the hashtag #StopTheSteal, anticipating that Trump would lose to Hillary Clinton.[18] Trump won, of course, so there was no "steal" to stop. #StopTheSteal got another shot in 2020. In the days after the election, pro-Trump protesters chanted it outside the Maricopa County Elections Department. There were Stop the Steal events in Harrisburg, Pennsylvania, and Las Vegas. Ali Alexander, who used to be known as Ali Akbar, launched a website with information on Stop the Steal rallies.[19]

The real impact was on social media, where Stop the Steal content was virtual crack. With the help of a viral video showing the angry mob at the TCF Center in Detroit, a Stop the Steal Facebook group became one of the fastest growing in Facebook history, racking up 320,000 users in twenty-two hours, about 100 new members every ten seconds.[20]

Those were impressive numbers, so much so that some left-leaning Facebook users decided to have some fun. One member of a Facebook group dedicated to trolling explained to me how she and others set up a Stop the Steal Facebook group that gathered eighty thousand users before being shut down by Facebook.

"This was when the pandemic was still raging," she told me. "We all had a lot of free time."

Another member of that troll group hoped that it wouldn't be a pure troll and that perhaps with some moderate content they could make Trump supporters' "thinking shift more to center." Others just wanted to troll. But they never got a chance to test it out, because Facebook killed it so quickly. The troll group member said they weren't even that shocked by the numbers, given the interest that Trump supporters showed in the cause. "Everybody was trying to join these groups," they said.

They weren't the only ones with trolling hopes: another Stop the Steal group racked up sixty thousand members before rebranding as "Gay Communists for Socialism." Members who joined the group because they thought the election was stolen were confused. "What in the hell?? I would never invite anyone to anything such named," one Facebook user said. "How the hell did I start seeing posts from this sick group?" another Facebook user asked. "I am soooooo confused right now," wrote one commenter.[21]

Democrats didn't need to trick Trump supporters to get some laughs. Rudy Giuliani's infamous Four Seasons Total Landscaping trip was followed by a presser at the Republican National Committee, where hair dye (or was it mascara?) dripped down his sweaty face.[22] That was only slightly less humiliating than when Giuliani, a former federal prosecutor, appeared in court in what a judge described as an attempt to disenfranchise "every single voter in the commonwealth" of Pennsylvania[23] and didn't seem to understand a first-year-of-law-school question about the standard of scrutiny that the judge should apply in the case.[24] A judge effectively called the Trump campaign's arguments a hot pile of garbage and expressed shock at their attempt to invalidate the votes of millions of Pennsylvania residents.[25]

At the RNC press conference, Jenna Ellis—who a few years earlier had been fired as a prosecutor working traffic court cases[26]—declared herself a member of an "elite strike force" of lawyers assembled by the Trump campaign.[27] Ellis repeatedly claimed that Trump "won in a landslide," a claim about as absurd as Giuliani's unwitting role in the *Borat* movie sequel.

Even Orly Taitz, a dentist-lawyer who was one of the most prominent voices in the birther movement until Trump came along, wasn't so sure about that.

"I don't know, I wouldn't go that far," Taitz told me of Ellis's "landslide" claim.[28]

William Barr knew the "bullshit" that Trump lawyers, whom he thought of as buffoons, were spewing was having a major impact. Millions of people, including the president, couldn't or wouldn't see what a joke the Trump campaign's effort was.[29] Yet the Justice Department remained silent.

"I've got to keep asking," I'd written to the Justice Department spokeswoman in late November. "The Justice Department had federal monitors in jurisdictions across the country. Every U.S. Attorney said they were on the hunt for voter fraud. Has the Justice Department detected any evidence at all of mass voter fraud schemes to steal hundreds of thousands of votes across multiple states? Any plans to reassure the American public here?" No response.

In 2005, on the debut episode of his new show,[30] political satirist Stephen Colbert launched into a segment[31] on what would become American Dialect Society's word of the year: truthiness.[32] (In 2010, when Colbert cohosted the "Rally to Restore Sanity and/or Fear" in DC, he also wildly overestimated the size of the crowd, foreshadowing the first controversy of the Trump administration, when Trump did the same with his inauguration.[33])

Fifteen years later, the year that Merriam-Webster finally added "truthiness" to its dictionary,[34] the concept had taken over the 2020 election aftermath. There are no facts to support the notion that Democrats engaged in a massive, multistate criminal conspiracy to steal the 2020 election from Donald Trump. But there were millions who just *knew* and *felt* it was true, in their guts.

During the campaign, Barr *felt* as if mass voter fraud *could* happen and gave interview after interview that gave the look of legitimacy to Trump's entirely predictable claims that he'd won the election, *actually*, and that Biden had stolen it from him.[35]

But the claims of fraud Barr saw "were completely bogus and silly and usually based on complete misinformation," he said. In his view, there was a much more logical explanation. Months before the election, in April, Barr visited Trump in the White House to offer some advice "as a friend" who wanted Trump and his administration to succeed. Barr told Trump that suburban voters were giving up on his candidacy because Trump was—as Barr phrased it—an "asshole." Barr told Trump he was headed for defeat and needed to temper his tone to win back those "middle-class suburban voters."[36] Here, Barr's predictions came true.

"It just didn't look to me like the results of the election were the result of fraud. It looked to me that the difference in the vote were in the suburbs, where everyone expected the president to be hurt, and that's what happened, in my mind," Barr said about what he realized after the election. "I didn't see in cities like Philadelphia some, you know, unexpected upsurge in votes." He thought Trump would keep claiming fraud anyway.

"I actually told my staff very soon after the election . . . I didn't think the president would ever admit that he lost the election, and he would blame it on fraud, and then he would blame the actions and evidence on the Department of Justice," Barr said.

Now, nearly a month after Trump lost, Barr decided it had gone too far. It might finally be time to admit that the emperor (or MAGA king) had no clothes. It was Trump's rant on Maria Bartiromo's Fox Business show that was Barr's final breaking point.

In the 1990s, when open sexism in the media was a bit more common, Bartiromo was nicknamed the Money Honey and was a well-known presence on cable television.[37] Former Fox News kingpin Roger Ailes had hired Bartiromo in her midtwenties and soon had her delivering updates from the floor of the New York Stock Exchange for CNBC, which he ran before departing for Fox. "I got into screaming matches with guys who could have been my father or grandfather," she later said[58] of her experiences on the "testosterone filled NYSE trading floor."[39]

Bartiromo was effective on the medium, becoming so popular that she inspired a song by punk rocker Joey Ramone of the Ramones. "I watch you on the TV every single day," he sang. "Those eyes make everything okay."[40] She remained loyal to Ailes until his death,[41] publicly declaring he was "nothing but a professional"[42] after the first sexual harassment allegations emerged, setting off a tidal wave of accusations that led to his departure from the network.

By 2020, Bartiromo's star wasn't shining so brightly anymore. Ailes had been out of the company for four years,[43] dead for three.[44] Fox had renewed her reported $6 million contract in 2019, but now there were questions of whether she was worth the cost and how low she was willing to go for ratings. "What happened to Maria Bartiromo?" read one headline.[45] Some of the talk came from within Fox itself: a senior vice president at Fox News said Bartiromo had "gop conspiracy theorists in her ear" and was saying "crazy wrong shit," while Fox Business News president Lauren Petterson said Bartiromo should "get off social [media] all together."[46]

Bartiromo believed it was "easier to get good ratings when you are giving your audience something they want to hear." She agreed with her producer, who said after the election that Bartiromo's audience "doesn't want to hear about a peaceful transition."

On the day of the Four Seasons Total Landscaping presser, Bartiromo received and read an email from someone who claimed they were a "ghost" who could "time-travel in a semi-conscious state" and received messages from "the wind." She later called it "kooky" and "nonsense."

The "source" claimed that the Dominion voting company was the "one common thread" in the "voting irregularities in a number of states," that the late justice Antonin Scalia was killed during a "human hunting expedition," and that Bartiromo's mentor Roger Ailes—who'd been dead for three and a half years at this point—was alive and well and secretly conspiring to "portray Mr. Trump as badly as possible."

The email was forwarded to Bartiromo by conspiracy theorist Sidney Powell. The next day, Nov. 8, Powell was on Bartiromo's show, claiming that Dominion was flipping votes from Trump to Biden. The following week, Nov. 15, Powell returned to the program alongside Rudy Giuliani.

By the end of the month, it was Trump's turn. Bartiromo had landed Trump's first postelection interview, if you can call it that.[47] Bartiromo essentially acted as Trump's moral support as the president rambled on for forty-five minutes. Barr wasn't a fan.

"I was annoyed," Barr said. "This got under my skin, but I also felt it was time for me to say something."[48]

So finally, on Dec. 1, 2020, Barr admitted to a reporter what any serious person had known for weeks:[49] that there was no evidence of a massive criminal conspiracy across several states and that Joe Biden had won the election.[50] Inside his office, over lunch—Italian chicken soup with arborio rice, steak with a red wine and shallot reduction, a twice-baked potato, a mélange of vegetables, and flourless chocolate cake—he laid out his conclusion to Mike Balsamo, the Justice Department beat reporter for the Associated Press.[51]

The news hit the wire and soon was the leading story in the country.[52]

*WASHINGTON (AP)—Disputing President Donald Trump's persistent, baseless claims, Attorney General William Barr declared Tuesday the U.S. Justice Department has uncovered no evidence of widespread voter fraud that could change the outcome of the 2020 election.*[53]

Six blocks up Pennsylvania Avenue, White House aide Cassidy Hutchinson heard a commotion down the hallway, near Trump's dining room. The valet was headed her way, looking for her boss, White House chief of staff Mark Meadows. Hutchinson soon walked down to the dining room to find the valet changing the tablecloth in Trump's dining room, the spot where he'd watch television just outside the Oval Office. He motioned for her to come inside and pointed toward the fireplace mantel and the TV.

"There was ketchup dripping down the wall," Hutchinson recalled, "and there's a shattered porcelain plate on the floor."

The valet told Hutchinson that Trump was "extremely angry" at Barr's AP interview and "had thrown his lunch against the wall," Hutchinson recalled. She grabbed a towel and started wiping the wall down.

"I would stay clear of him for right now," Hutchinson recalled the valet saying. "He's really, really ticked off about this."[54]

Barr had a meeting prescheduled with Meadows already and figured he might end up getting fired. He told his executive assistant at the Justice Department that they might have to pack up his things.

Barr arrived at the White House and found Trump back in the dining room off of the Oval Office. The TV was on One America News Network (OAN), which was showing a Michigan state senate committee focused on election issues. If there were any remnants of sugary tomato goo on the wall, Barr didn't clock them. He stood at the table.

At the Michigan hearing, Mellissa Carone, a temp agency staffer contracted by Dominion Voting Systems "for one day to clean glass on machines and complete other menial tasks,"[55] had recently finished up her testimony. Carone had worked at the TCF Center in Detroit, and now she wanted to speak to the manager.

"They told me that I would be parking in a parking lot and I would be shuttled in through a shuttle. I called my mother and I told my mother about this, and my mom said 'No, absolutely not, you're not doing that.' I also had a concern about that, because I do have two small children at home, and I needed access to my vehicle," Carone said. She quickly segued from complaining about transportation to calling everybody at the TCF Center criminals. "What I witnessed at the TCF Center was complete fraud, the whole 27 hours I was there."

Carone's claims were bizarre. She claimed that a truck that she was told was bringing in food was actually filled with ballots—maybe; she didn't see. She claimed that ballots were being counted over and over again, which would easily show up if it were true.

"*Everything* that happened at that TCF Center was fraud. *Every single thing.* Every avenue was taken to commit it," she declared. "This is fraud."

Carone was perhaps not the most reliable witness. She was freshly off probation as part of a plea deal. Under the name Mellissa Wright, she was originally charged with first-degree obscenity and using a computer to commit a crime after being accused of sending pornographic videos to her fiancé's ex-wife and framing the woman by claiming she'd stolen them. As part of a plea agreement, Carone had her charge reduced to disorderly conduct and received twelve months of probation. (Carone said it was her fiancé, Matthew Stackpoole, who sent the explicit videos from her phone and that she confessed to a crime she didn't commit because she didn't want to spend any more time in court.)[56]

Chief Judge Timothy M. Kenny of the Third Judicial Circuit Court of Michigan found that Carone's allegations about the TCF Center "simply are not credible," much like her claims that her partner's ex tried to hack their sex tapes. It would later emerge that in 2016 and 2020 Carone never even voted for Trump, or any candidate, for that matter.[57]

At another hearing in Michigan the next day, Carone got even more heated when arguing with lawmakers. She was seated next to Giuliani, who repeatedly tried to calm her down as she talked down to lawmakers who had at least rudimentary understanding of basic civics. With her hair pulled into a bun and a scarf around her neck, she was prepared for battle with the reality-based community.

Rep. Steven Johnson, a Republican, tried to pull Carone back to planet Earth to get a sense of what she was actually claiming.

"We're not seeing the poll book off by 30,000 votes, that's not the case," Johnson said.

"What'd you guys do, take it and do something crazy to it?" Carone said, biting her lips and raising her eyebrows.[58]

Pretending as though Carone was a credible witness wasn't Giuliani's most embarrassing moment of the day. While seated next to Ellis, he also appeared to very loudly flatulate, drawing a visible reaction from his seatmate. I tweeted the moment, leaving out any commentary, and it racked up millions of views and became a late-night television sensation.[59]

"I can tell you that making a cameo on @jimmykimmel for questioning Rudy Giuliani during an evidence-free hearing and startling him so much that he farted definitely wasn't on my 2020 bingo card 💨," wrote state representative Darrin Camilleri, who'd been grilling Giuliani at the time of the incident.[60]

It was hard to fathom that anyone could possibly take Carone seriously. But Mark Meadows had. In November, he texted Barr a copy of the Carone affidavit, the one in which she declared she'd witnessed "nothing but fraudulent actions" at the TCF Center.[61]

"Still not convinced this isn't a SNL character," I'd tweeted when she testified the first day. On Saturday, she was. Trump tweeted in Carone's support as soon as the *SNL* episode ended.

"Melissa is great!" Trump tweeted at 1:02 a.m. on Dec. 6.[62] He apparently thought Carone was credible too.

———◆———

Back in the dining room on Dec. 1, Trump was fuming. He'd initially refused to look at Barr when he entered. Then he shoved a copy of the AP article in his face.

"Did you say this?"

"Yes, I did say it."

"Why?"

"Well, because it's true," Barr replied.

Barr told Trump that he needed a crackerjack legal team and that instead he'd wheeled out "a clown car" of lawyers.

"I told him that the stuff that his people were shoveling out to the public was bullshit, I mean, that the claims of fraud were bullshit. And, you know, he was indignant about that. And I reiterated that they wasted a whole month on these claims on the Dominion voting machines and they were idiotic claims," Barr later explained.

Eventually Trump tired of the voter fraud complaints and brought up his other gripes. Barr offered up his resignation.

"He pounded the table very hard," Barr said. "Everyone sort of jumped. And he said, 'Accepted.' And then he repeated, 'Accepted.'" Barr went to leave, but Trump had two lawyers follow him out. They relayed that Trump wanted him to stay on. Barr said they'd talk about it the next day. They'd come up with a plan on Dec. 2.

Down in Florida, right-wing extremist Jeremy Liggett was making plans too.[63]

"Now, I think it would be hysterical if you got morale patches that said 'plan B' or 'B squad' because I think it's one of the top 3 funniest things I've personally ever heard from politicians as they try to dance around the M word lmao," an associate wrote on Dec. 2, talking about lawmakers avoiding talking about militias.

"Hahahahaha," Liggett said. "I am going to name DC operation plan B."[64]

———————— ◆ ————————

The federal government had spent more than two decades going after one kind of terrorism, so it should come as no surprise that it has struggled to pivot to handling the growing threat of domestic terrorism. Private organizations, unmoored to regulations about monitoring groups engaged only in protected free speech, have stepped into the void. For the Middle East Media Research Institute (MEMRI), an organization cofounded by a retired colonel in the Israel Defense Force Intelligence Corps, looking at American hate groups wasn't plan A. But the counterterrorism space had shifted since its formation in 1998, with terrorists associated with "domestic" ideologies now killing more people than extremists associated with "foreign" ones.[65]

So a few years back, MEMRI launched the Domestic and Transnational Terrorism Threat Monitor (DTTTM) to scrutinize "the online activity of neo-Nazi and white supremacist groups and individuals who are circulating hateful and antisemitic content, with a focus on incitement, calls to action, and violent threats against Jewish organizations and institutions, blacks, people of color, LGBTQ, political and other prominent figures, government officials and agencies, and the public at large, in the U.S. and around the world." Later, in August 2022, MEMRI's DTTTM team would tip off the FBI Social Media Exploitation (SMEX) team at the National Threat Operations Section (NTOS) about a Gab user who threatened FBI agents after the raid of Donald Trump's estate in Mar-a-Lago. "If You Work For The FBI Then You Deserve To Die," the user wrote.[66]

MEMRI, like many outside organizations, was sending up warning signs ahead of the Capitol attack, starting right after Trump lost the 2020 election.

"Neo-Nazis, White Supremacists on Facebook, Telegram, 4Chan, Instagram, Gab, React to U.S. Elections, Anticipate Coming 'Civil War,'" read one report from Nov. 6.[67]

"On Parler, White Supremacists, Neo-Nazis, QAnon Adherents Call For Violence Against Politicians," another warned on Nov. 20.

For an outside organization like MEMRI, it's relatively easier to pivot, to readjust, to expand a mandate. It was a different story within the FBI. The whole federal government can't just turn on a dime; there are laws and budgets and bureaucracies in place. Investigations into domestic terrorism were more difficult, and less appealing as well.

"Look, international terrorism is looked at more sexy. I mean, I don't want to sound ridiculous, but it's just true," said Michael Sherwin, the former acting US attorney in DC. "You know, you're assigned to an international terrorism squad; it looks more prestigious or sexy. You know, I'm going after, you know, ISIS or al-Qaida AP, than going after, you know, the Michigan militia in Plymouth, Michigan."[68]

Add in the politics. Even if agents don't personally lean right, as many FBI agents do, stepping into the domestic terrorism space is setting yourself up for blowback from the legislative branch, and maybe pushback from your family and friends too. Radical Islamists don't control a voting bloc in Congress. Radical right-wingers do.

At the beginning of President Barack Obama's first term in office, a Department of Homeland Security official named Daryl Johnson helped with a report that stated what now seems like basic common sense. The 2009 thesis has been borne out by the last fourteen years of American history.

"Rightwing extremists have capitalized on the election of the first African American president, and are focusing their efforts to recruit new members, mobilize existing supporters, and broaden their scope and appeal through propaganda," it stated. The conditions were similar to those in the 1990s, but the next few years were looking grim, the 2009 report stated.

> *Unlike the earlier period, the advent of the Internet and other information age technologies since the 1990s has given domestic extremists greater access to information related to bomb-making, weapons training, and tactics, as well as targeting of individuals, organizations, and facilities, potentially making extremist individuals and groups more dangerous and the consequences of their violence more severe. New technologies also permit domestic extremists to send and receive encrypted communications and to network with other extremists throughout the country and abroad, making it much more difficult for law enforcement to deter, prevent, or preempt a violent extremist attack.*[69]

In hindsight, the analysis seems exceedingly obvious. *No shit,* you might say. Johnson thought it was a solid report based on extensive evidence. So did his supervisors.

"We wrote up the report, it went through 23 revisions, got vetted at all levels, coordinated it with the FBI, went and briefed Janet Napolitano a couple of days before it was disseminated. When I was headed home that day, my branch chief, my immediate supervisor, talked about what a great briefing it was, and she was hearing a lot of positive feedback," Johnson said. "Then three days later, all hell broke loose."[70]

Once the report was distributed to local law enforcement, it was quickly sent to a conservative radio host who wrote for the birther-conspiracy website World Net Daily,[71] as well as to a libertarian blogger.[72] In Congress, the backlash wasn't just from the usual suspects, like then representative Michele Bachmann (R-Minn.). It was from the top Republican in the House too.

"When you look at this report on right-wing extremism, it includes . . . about two-thirds of Americans, who, you know, who might go to church, who may have served in the military, who may be involved in community activities," House Speaker John Boehner said. "I and my colleagues are trying to understand who wrote this report, why wasn't it edited or—I just don't understand how our government can look at the American people and say, 'You're all potential terrorist threats.'"[73]

Twelve years before a right-wing mob stormed the Capitol chanting "Hang Mike Pence," the then congressman and Republican Conference chairman cosponsored a resolution demanding to know how the Department of Homeland Security (DHS) report had been put together.[74] Tea Party activists started wearing "Domestic Terrorist" T-shirts to rallies. The Obama administration caved, with DHS repudiating Johnson's analysis and then dissolving his team when he left the next year.[75]

"I would say it's never too late to reconsider your stance on how serious this threat is," Johnson told me years later, after a number of right-wing terror attacks but years before extremists would storm the Capitol. "We've seen the threat grow and grow year after year . . . and incident after incident has pretty much validated the analysis in that report. What is it going to take for them to step up and actually recognize the threat for what it is?"[76]

For bureaucrats, actually recognizing the threat is risky. Even acknowledging basic reality is difficult, with the temptation to add in a political counterweight, an "on the other hand," a false equivalence, a caveat. They'd seen what happened to the people Trump thought crossed him, to those who said the wrong thing in a report, to those who discussed their personal political views with their future spouse over an FBI chat platform they didn't realize was permanently stored. The caution, the fear, permeated throughout the bureau's pre–Jan. 6 discussions and even stuck around long after Trump left office.

"I will definitely say all rhetoric, again, increased as we got closer to January 6th," Moore, the head of the intelligence office in Washington, later told investigators.

"On both sides," she added, echoing Trump's famous remark after the Charlottesville Unite the Right rally broke out into lethal violence.[77]

Social media was not, in fact, flooded with Joe Biden supporters gung ho on storming the Capitol and executing politicians to stop them from locking in the election for their preferred candidate. That would have made very little sense. The threat was the thousands of Trump supporters who believed that the country was being stolen, that Joe Biden was an illegitimate president, and that the lies they'd been fed about the 2020 election were true.

In late 2020, the man who rose to political prominence on the back of a racist lie about the first Black president was in the Oval Office, spreading another lie that posed an existential threat to the country's national security. But those in charge of ensuring the nation's security couldn't speak plainly about it, lest they be targeted too.

To be fair, the FBI had been through years of this crap by this point, almost inoculated against violent rhetoric against politicians, particularly those Trump attacked. There were a massive number of tips. How do you sort out what's real? What's protected speech? What's "keyboard bravado," as some at the FBI called it?

"It was the same actors that say crazy shit all the time," one person involved in the FBI's domestic terrorism work told me. "Years of these idiots saying the same shit." The FBI also had a mandate, from the president's attorney general and the president himself, to go after those he ideologically opposed.

"The politicians and the appointees are the ones who set the priorities. So those priorities were going into, you know, antifa," they said. "When the attorney general says, 'This is the FBI's priority,' we don't have a choice."

It was the volume that stood out. In November, the head of the FBI's National Threat Operations Center, which operates out of West Virginia, said NTOC had been dealing with an "unprecedented volume" of calls in 2020, first with calls about riots after the death of George Floyd and then with tips about violent threats surrounding the 2020 election.[78]

Once again, the bureau took the so-called lone wolf approach, treating each individual reported threat as the work of a single isolated person. In the bureau, at least, there wasn't a big-picture look at what it meant, for national security, that millions of Americans thought the presidency was about to be stolen because they bought the bogus lies propagated by the commander in chief.

The approach the FBI took—churning through a virtual geyser of social media posts individually, closing out those that didn't meet a very specific set of criteria—was like putting out a forest fire one tree at a time. It might work on specific trees, but it did almost nothing to prevent the real problem.

"The bureau is looking to charge cases," one former prosecutor familiar with the FBI investigation said. "When you read something and you ask yourself *What's the charge, what's the federal violation I'm going after?* there is none. But if you look at it as *This is a threat, a potential datapoint that would serve a bigger intel bulletin*, that's a different analysis."

Take the case of Alan Hostetter, a former police officer turned yogi later charged in connection with the Capitol attack. Not long after the election, Hostetter was driving alone in his truck in the middle of nowhere when he started recording. He was on his way to the Million MAGA March in DC in November, and he had some thoughts on his mind.

"Some people at the highest levels need to be made an example of with an execution or two or three. Because when you commit treason against this country and you disenfranchise the voters of this country and you take away their ability to make decisions for themselves, you strip them of their Constitutional rights. That's not hyperbole when we call it tyranny, that's fuck-ing tyranny. And tyrants and traitors need to be executed as an example."[79] Hostetter called for "an execution or two or three," a few tyrants and traitors "executed as an example so nobody pulls this shit again."[80]

Bad? Yes. Chargeable? No. Actionable? Probably not. Hostetter wasn't saying *he* was going to execute any tyrants or traitors. Nor was he saying *who* those tyrants or traitors were (nor was he directly threatening those tyrants or traitors, via interstate com-munication). He also wasn't specifying *when* those tyrants and traitors would be executed or inviting his audience to join with him in an articulated criminal conspiracy to execute specific tyrants or traitors at a set date and time. He was just saying, as a general matter, that President Trump's enemies should be mur-dered in public, at some yet-to-be-determined point.

Another tip that the bureau didn't take any action on came into the FBI in November from a man named Abdullah Rasheed, previously known as Sergei Neklovech, who had used at least five other aliases over twenty years. He was a former marine and reg-istered sex offender who joined the right-wing Oath Keepers after he moved to West Virginia. On Nov. 9, he got an invite to a GoToMeeting with other members of the Oath Keepers.

"I was expecting to hear, yeah, Biden bad; Trump good, you know," he later said. The tone was different. "It sounded like we were going to war against the United States government, and I wasn't comfortable, so I started to record it."

He sent in a tip to the FBI, but nothing happened. It was months after the Capitol attack when he finally heard from the feds.

"The FBI gets thousands of tips a day into the National Threat Operations Center in West Virginia," an FBI special agent later explained. This particular tip "was not sent out to the field for further action."[81]

*No further action* was a running theme in tips the federal gov-ernment received in the lead-up to Jan. 6. The Department of

Homeland Security's Office of Intelligence Analysis (I&A)—the same unit that produced that paper on right-wing extremism that caused a firestorm in 2009—was charged with providing intelligence to state and local officials across the country.

Statutorily, I&A is only allowed to look at publicly available—open-source—information, an ever-expanding universe in the social media era. DHS policy says that information in intelligence products should contain true threats or incitement to violence, not hyperbole; enhance I&A's understanding of "known threat actors"; or illustrate a risk of violence "during a heightened threat environment." There's an extensive internal review process within I&A for writing an open-source intelligence report: first a peer must review the report, then a senior collector, then a supervisor. If there's a disagreement, they can bring in the Intelligence Law Division (ILD).

Right before the election, the Trump administration imposed yet another layer of bureaucracy. Anything related to the election had to be cleared by the I&A's Intelligence Oversight officer as well before it was published on the Homeland Security Information Network, which made it available to state and local law enforcement.

On top of all those layers of reviews, there was a general hesitancy to rock the boat. Over the summer I&A had a major screwup: it published reports featuring the names of journalists who "engaged in ordinary journalism" and posted leaked information.[82] "What If J. Edgar Hoover Had Been a Moron?" one of the journalists, Benjamin Wittes, wrote of the effort.[83]

I&A already didn't have a great reputation within DHS, where many thought of them as a clipping service, or within the broader intelligence community. One experienced national security journalist wrote that the office had "for years been the butt of jokes among larger, more established agencies like the CIA and the FBI, who liken it to a team of junior-varsity athletes."[84] Adding insult to injury, a top DHS intelligence official later testified that they had never heard that critique printed in the *Washington Post*, nor had they ever even encountered anyone or any information that gave the *perception* that I&A wasn't respected.[85]

They'd also had a lot of turnover in 2019, when the Open Source Collections Operations branch moved to a twenty-four-hour schedule, with inconvenient shift changes at 5:00 a.m., 1:00 p.m., and 9:00 p.m. Those inconvenient start times weren't a recipe for success at a government agency like DHS, which has "had low employee morale and low employee engagement—an employee's sense of purpose and commitment—since it began operations in 2003."[86]

Inside DHS, Open Source Collections did not believe that storming the Capitol was possible and thought that the Trump supporters were just using hyperbole. There was a "very high" threshold for reporting threats ahead of Jan. 6, and one collector said they were too nervous to report anything. After I&A's intelligence efforts in Portland came under scrutiny, collectors worried they would be punished either by I&A leadership or by Congress. One explained there was a "chilling effect" on their reporting. In personal chats, however, collectors thought there was a threat and expressed fear for their own personal safety, encouraging one another to stay safe on Jan. 6.

"I found a map of all the exits and entrances to the capitol building. I feel like people are actually going to try and hurt politicians," one collector messaged on Jan. 2. "Jan 6th is gonna be crazy."[87]

"Like there's these people talking about hanging Democrats from ropes like wtf," one wrote.

"I mean people are talking about storming Congress, bringing guns, willing to die for the cause, hanging politicians with ropes," wrote one collector on Jan. 3.

An internal report later found that DHS "rapidly hired inexperienced open source collectors in the months leading up to January 6, 2021." On Jan. 6, sixteen out of the twenty-one collectors had less than a year of experience, and many hadn't had any intelligence or federal government experience when they were hired.[88]

In early December, inside the bureau, most of the worry was about potential threats on Inauguration Day, Jan. 20. A Dec. 8, 2020, report from the FBI highlighted the potential threat of the Million Militia March on Jan. 20,[89] and multiple FBI confidential human sources told me the bureau leaned on them for information on that potential event.

But if the government had missed signs in the early weeks of December, they were right about this much: the threat of Jan. 6 did not begin in earnest until after Dec. 11, when the Supreme Court rejected[90] what was thought at the time to be Trump's last-ditch effort to overturn the election results; and it grew even more serious after Dec. 14, when electors met in statehouses across the country to formally cast their votes in the electoral college.[91] At that moment, Trump and his constituents reached a new level of desperation.

# CHAPTER 5

# "Will Be Wild"

A week after his Supreme Court loss, Trump welcomed guests to the White House for what was later described as the "craziest meeting of the Trump presidency."[1] Sidney Powell was there. So was the disgraced former national security adviser Michael Flynn, by that time a publicly avowed adherent to the QAnon conspiracy theory. There was talk of smart thermostats being hooked up to the internet, communicating IP addresses all over the world, international interference Dominion voting machines, crazy shit.

"What they were proposing, I thought was nuts," said Eric Herschmann, an attorney who represented Trump in his first impeachment trial in 2019 before joining the Trump White House as a senior adviser. "The theory was also completely nuts, right? It was a combination of the Italians and the Germans. I mean, it had different things that had been floating around as to who was involved. I remember Hugo Chavez and Venezuela, and she had an affidavit from somebody that says they wrote a software in, and something with the Philippines. Just all over the radar."[2]

Giuliani eventually arrived, and Herschmann let slip that Powell had been shit-talking him.

"Sidney, why don't you tell Rudy what the fuck you just said, that he's a dumb ass, and he doesn't know what the hell he's doing," Herschmann said. "Tell him!"

Trump wasn't happy with his approach. But Herschmann accomplished what he'd wanted, driving a wedge between Powell and Giuliani.

Everyone went home after a long night. At 1:42 a.m., Trump tapped out a tweet.

"Statistically impossible to have lost the 2020 Election," he wrote. "Big protest in D.C. on January 6th. Be there, will be wild!"[3]

At the National Capital Region Threat Intelligence Center (NTIC), a fusion center funded by the Department of Homeland Security, Donnell Harvin was under the impression that Jan. 6 was going to be "a nothing burger." They'd done this before, in November and December, when Trump fans showed up in DC to protest the election results. There were violent clashes between opposing groups in December, when the Proud Boys marched through the streets, but nothing law enforcement couldn't handle.

He assigned an analyst to it anyway, but his most junior one, the one he'd hired earlier in the year, who'd been a bartender.

"I use the basketball analogy: It's garbage time now because it's the end of the game, quite frankly, from a political standpoint," Harvin said. "And what's the importance of January 6th?"[4]

Harvin soon learned that Trump's tweets changed the analysis "exponentially." The analyst was soon briefing him on the significance of Jan. 6, which hadn't been a big day historically. A lot of people online thought it was the final day to stand up, to stop Joe Biden from taking office.

*Well, that's interesting,* Harvin thought. *Little conspiracy there that you can stop the election.*

Trump's promotion of Jan. 6 set it apart from the events that the intelligence world described as MAGA 1 and MAGA 2.

"November and December weren't promoted by the president, although he did, you know, I'll use the term 'flirt' with the crowd, if you will. There was a flyover with Marine One in one of them, and there was a drive through in the motorcade in another," Harvin said. Harvin said Trump's Dec. 19 "wild" tweet was an inflection point.

"So if you can distill that down: This is a flawed election, a stolen election, and you should be out there that day," Harvin said. "And so that was the beginning of the very, very heavy drumbeat that we were tracking for about a week with increased numbers like you wouldn't believe as far as level of interest, retweets. I mean, in the hundreds of thousands."

The tone of Trump's supporters completely changed at that point, Harvin said. They completely switched on police, he said. "There were plenty of tweets and online posts from groups basically telling law enforcement: We're coming, and if you're not with us, you're against us." Lots of posts about bringing firearms into DC too.

"You can say just about anything you want to online so long as you're not being credible or specific about a threat, right? So an individual can say: I'm going to Washington, DC, and I'm going to kick some butt and who's with me and, you know, so forth and so on. And that's protected free speech," Harvin said.

"So there wasn't a lot of credible, specific threats, but there was a vast amount of noncredible, unspecific threats, and that should have prompted something, right?" Harvin said. The Capitol should have been better protected, he said, for "the worst-case scenario."

"Daddy Says Be in DC on Jan. 6th" read a popular thread on TheDonald.win forum, which had migrated over to its own website after getting too hot for Reddit.[5] Trump supporters wrote that they'd said their goodbyes to their families, prepared to die to keep Trump in office.[6]

"I want it to be peaceful but if our 'leaders' do the wrong thing and we have to storm the Capitol, I am going to do it," read one post.

"I look forward to standing with you on the front lines. If it came down to it I wouldn't think twice about dying just so my family and friends can live free," read another.

After Jan. 6, a forum user named Jose Padilla would use The-Donald to brag about his actions. He'd later testify that his violent rhetoric was all for "internet cool points."[7]

Hours after the Trump tweet, on a Saturday, yet another FBI confidential human source emailed their bureau contact, saying that the Right considered Trump's tweet "a call to arms in DC on Jan 6." The FBI was mostly focused at the time on what could happen on Inauguration Day on Jan. 20, but the FBI source noted that Jan. 6 might end up being just as big of a threat.[8]

I later reviewed other emails that had been sent to the FBI by the source. "We must all join/link forces and be ready to leave our lives behind. We must pool resources and fight like there's no tomorrow! The Constitution still lives and so we must preserve it. Blood is the price of freedom," one Dec. 21 message read.

In Kansas, an FBI confidential human source embedded within the Proud Boys said that the rally in DC on Jan. 6 should be a concern. They shared screenshots of Trump's tweet.

In California, Trump supporters who'd joined a Telegram group called Patriots 45 MAGA Gang and attended rallies in support of Trump in Beverly Hills and Huntington Beach started making plans.

Daniel "DJ" Rodriquez was *deep* into conspiracy theories. The California resident had been watching InfoWars since he was in his midtwenties. But it was Trump, the guy who felt "almost like an old friend," the guy he knew from *The Apprentice*, who got Rodriguez really excited about politics. One day in 2016 he went to a Trump protest at a park in Beverly Hills and stood there arguing with anti-Trump protesters.

"I argued with two hundred, three hundred people there for eight and a half hours, nine hours without water or food, and I just annihilated every single one of them," Rodriguez bragged. "I thought that was, like, one of my greatest days of my life."

He was thirty-five when Trump was elected, and he tried to join the army, even wearing a Trump shirt to the recruiting office. He had a criminal record, though, and was rejected. Globalists, elitists, and unelected officials had taken over the country, he thought. Coronavirus was "meant more to kill the economy than to kill people," he thought, to hurt Trump's reputation. Joe Biden didn't have anyone coming to his rallies; he didn't even hold any rallies, so he clearly didn't win the election, Rodriguez believed.[9]

"You don't have to be that bright to see that it's—it was— something wasn't right and it was rigged," Rodriquez said. When he saw Trump's Dec. 19 tweet, he was ecstatic. This was his moment.

"Trump called us to DC," he said. He thought the commander in chief was calling for his help, and he could see Trump speak in person.

*I need to be there*, Rodriguez thought. He believed he was going to save America. In the Patriots 45 MAGA Gang chat, he let it rip.

"We gotta go handle this shit in DC so the crooked politicians don't have an army of thugs threatening violence to back their malevolent cabal ways," he wrote on Dec. 23.

"Congress can hang. I'll do it. Please let us get these people dear God," he wrote on Dec. 29.[10]

In Florida, Bruno Cua, an eighteen-year-old high school senior, saw Trump's tweet as a signal too and took to Parler to vent.[11]

ATTENTION PRESIDENT #TRUMP IS CALLING ALL #PATRIOTS TO RALLY IN #WASHINGTONDC JANUARY 6TH!! This is extremely important because this is the day congress will finalize the election! If they don't do what's right MILLIONS of us will BREAK DOWN THEIR DOORS! #LIVEFREEORDIE #FREEDOM.

He followed up on Parler two days later. "Bring your guns. We're gonna need them. It ain't illegal if there's an army of us. BRING YOUR GUNS. ALL OF THEM." He messaged a friend saying they could "storm the freaking senate/house" and that's why they needed to bring weapons. "Holding signs is useless," he wrote in an Instagram message. "We have to forcefully take our freedom back on Jan. 6th." The day after Christmas, he posted about wanting the "blood of politicians" to be spilled.

The former head of the Justice Department's National Security Division, now in the private sector, emailed the deputy assistant director in the Counterterrorism Division at FBI headquarters, warning him on Dec. 23 that someone in an Oath Keepers chat was saying that "fucking bullets" were the only way to achieve their goals.

Just ahead of the holidays, the Washington Field Office got hit with another whopper at the end of a rough year. Saul "Seahawk" Tocker, who started his career as a security officer at FBI headquarters and rose to be a supervisory investigative specialist with the WFO Special Surveillance Group, died on Dec. 20. The cause was the gallbladder cancer he'd been diagnosed with after working twelve-hour shifts sifting through debris at the Pentagon after Sept. 11.[12]

On top of everything else, the bureau was also dealing with the fallout and ramifications of the Russian hack of SolarWinds, which had clients in the Centers for Disease Control and Prevention, the State Department, the Justice Department, and even some components of the Pentagon.[13] One US official said that the hackers "got into everything" and that it was shaping up to be "the worst hacking case in the history of America."[14]

The government was under siege, penetrated electronically by foreign adversaries, while attempting to survive an insider threat emanating from the White House. The physical attack was looming, but the government's top officials were distracted, their security tools compromised. Perhaps Christmas would provide at least a couple of days to reset, recharge.

"Hope everyone has a great holiday and some rest before 1/6," read one email in the FBI system on Dec. 23.

On Christmas Eve, nineteen-year-old Jackson Reffitt was in his room, looking up how to send tips to the FBI. His dad had been making him nervous. He was tied up in a militia and was talking a lot of big talk about what he had planned for his trip to DC on Jan. 6.[15] Still, tipping off the FBI about his own father left him with a bit of a queasy feeling.

"When are you going to realize the Democrats are using race to keep minorities suppressed? We The People are willing to die to protect their safety. We did it in the Civil War and now we are doing it again. It's the government that is going to be destroyed in this fight," Guy Reffitt wrote in a text to his family. "Hold my beer and I'll show you."

"I didn't know what I was doing. I just felt gross. I don't think I can explain it. I just felt uncomfortable," Jackson Reffitt said of his decision to send in the tip. He decided to "shut off" his day by watching TV and scrolling on his phone. He didn't want to dwell on the matter.

Something big was coming, Guy told Jackson. Congress had made a "fatal" mistake. "Time to remove them. That's why I'm going to DC. Promise, I'm not alone," he wrote.

"VOTER FRAUD IS NOT A CONSPIRACY THEORY, IT IS A FACT!!!" Trump tweeted that night.[16]

"MERRY CHRISTMAS!" he'd tweet hours later.[17]

———— ◆ ————

It was Christmas in Nashville, and Crystal Deck was opening presents at her brother's house. "Merry Christmas!" she texted her best friend, Anthony Q. Warner, who'd just been diagnosed with a fatal form of cancer.[18]

It had been an awful year for businesses, and the Nashville entertainment district along Second Avenue was expected to be quiet at 6:00 a.m. on Christmas morning. But gunshots had rung out, first around 4:30 a.m. and then again around 5:30 a.m.[19] At 6:00 a.m., an RV began blasting the 1964 song "Downtown" by Petula Clark. The uplifting lyric "Downtown everything's waiting for you" suddenly seemed like a threat.

*Boom.* At 6:30 a.m., a massive fireball shot into the sky, destroying trees, annihilating windows, taking off the face of a building, twisting up and blowing out the sign of the nearby Hooters restaurant. The bomb went off in front of the AT&T hub, and cell phone and internet service were affected for two days across multiple states.

The FBI did what it does best: surged resources, flooded the zone. There were 130 FBI employees involved in the investigation out of the FBI Memphis Field Office, aided by 147 others from twenty other field offices, along with folks from headquarters. The Justice Department's Counterterrorism Section sent down staffers from DC, who manned the FBI command post in Nashville in the days after the attack.[20] The investigation involved 3,000 pounds of evidence, over 2,500 tips, and more than 250 interviews.

As the FBI investigated the Nashville attack, Trump had other issues on his mind. "The 'Justice' Department and the FBI have done nothing about the 2020 Presidential Election Voter Fraud, the biggest SCAM in our nation's history, despite overwhelming evidence. They should be ashamed. History will remember. Never give up. See everyone in D.C. on January 6th," Trump tweeted on Dec. 26.[21]

Anthony Quinn Warner, the Nashville bombing investigation found, acted alone, intending to end his own life and make an impact but hoping to avoid harming humans. He was the only person killed. "Downtown" was accompanied by a recording warning people to get out of the area, that a vehicle was going to explode. The FBI determined that Warner's actions were "driven in part by a totality of life stressors—including paranoia, long-held individualized beliefs adopted from several eccentric conspiracy theories, and the loss of stabilizing anchors and deteriorating interpersonal relationships."

Warner was a sixty-three-year-old conspiracy theorist with a deep mistrust of government and was obsessed with shape-shifting alien lizards. "9/11 is what did it for me," he told his friend Deck, who said the navy veteran believed that Sept. 11 was an inside job.

After his death, Warner's fellow conspiracy theorists turned the conspiracy theories on him, writing that the bomb went off at AT&T because the company was going to do a forensic audit of the Dominion Voting Systems.[22]

Donoghue, the acting no. 2 at DOJ, was back in New York for Christmas and was at his home in Long Island when he got the call on his government cell phone on Dec. 27 about another conspiracy theorist who had reached a breaking point and was lashing out. Rosen was calling, and Trump was on the other line. "We've been on for about thirty minutes. He's talking about some of this election stuff," Rosen said. They joined a conference call.

The country was up in arms over corruption, Trump bellowed. The president tossed around some nonsense he'd been told, like there were 205,000 more votes than voters in Pennsylvania. The president thought it was clear fraud. "We're like a Third World country," Trump said. "He had this stream of allegations, which were clearly being fed to him by a number of people, that he would keep referring our way," Donoghue said. The president referred to Ruby Freeman, a Black poll worker in Georgia receiving death threats because of misinformation, as a "huckster" and "election scammer."[23]

"You guys may not be following the internet the way I do," Trump said.

"Just say that the election was corrupt and leave the rest to me and the Republican congressman," Trump told them. He wanted them to hold a press conference; Donoghue and Rosen shut that down. Trump kept at it. Dead people are voting, he said. "Indians are getting paid to vote," Trump said, which Donoghue interpreted as a reference to Native American reservations.

Most of the stories Trump was telling Justice Department officials were just things that "people are telling him and/or he's seeing on TV or whatever," Rosen said. "There's not new information there," Rosen said. "News flash: President Trump doesn't accept the election result."

Donoghue told Trump flat out that the information he was getting was false and not supported by the evidence. The facts didn't matter.

"We have an obligation to tell people that this was an illegal, corrupt election," Trump said. He then brought up the idea of replacing DOJ leadership, meaning them. "People tell me Jeff Clark is great," Trump said.[24]

There was that name again. Trump had brought up Clark a few days earlier, on a Christmas Eve call with Rosen. Clark worked at Rosen's old law firm, and they had a professional relationship. *That's odd*, Rosen thought. *How does the president know Jeff Clark?*[25]

Jan. 6 was approaching, the online chatter was increasing, and the top leaders at the Justice Department—in between fending off pressure from the White House to help overturn the results of the election—wanted to keep tabs on preparations. After Christmas, Donoghue tried to set up a meeting with Mike Sherwin, who was the acting top federal prosecutor in DC.

"We were concerned about the potential for violence just because we had seen that people were very upset about the election," Donoghue said. "We were well aware of the political tension in the country. And we knew that if you have tens of thousands of very upset people showing up in Washington, D.C., that there was potential for violence."[26]

As Jan. 6 approached, public information suggested that anti-Trump groups were going to stay away, which came as a relief to Justice Department officials. "There was some relief to hear that it would essentially be a one-sided protest, because it lessens the potential for violence," said Donoghue. He called the five law enforcement agencies within the Justice Department and had them send tactical teams to DC: FBI SWAT and Hostage Rescue, ATF special operators, the US Marshals' Special Operations Group, and a Bureau of Prisons Special Operations Response Team. The Drug Enforcement Administration's Special Operations Team would be in reserve, on a twenty-four-hour lag.[27]

"I think that we all understood the danger that was posed," Donoghue said. "Everyone was confident that everyone else would be ready, willing, and able to provide assistance if that was needed."

Publicly, the bureau wasn't saying much. In late December, the FBI received a media inquiry on Jan. 6, asking whether the bureau "has information about planned violence" and whether the FBI would be issuing any bulletins on the topic. The bureau declined to comment.[28]

Internally, the FBI was keeping tabs on things. An intelligence analyst, in an email on Tuesday, Dec. 29, noted that officials on a coordination call that morning were anticipating "large numbers at the 6th January event, especially as POTUS has discussed and promoted it," though they noted that it "does not appear at this time that he or the family was invited to these events or will attend."[29]

Anticipating that Trump would want to speak before an adoring crowd he'd summoned to DC didn't require any advanced predictive modeling, and plenty of people outside of government could speak plainly about the danger lurking ahead. On Dec. 29, the same day as the FBI intelligence analyst's email, a former aide to Vice President Mike Pence went on MSNBC and said she was worried about violence on Jan. 6.[30] "The president himself is encouraging it," said Olivia Troye. "This is what he does. He tweets, he incites it, he gets followers and supporters to behave in this manner." Trump, she said, "continues to stoke the flames of these far-right people who show up and think that he's calling out to them."

Inside the FBI, yet another bureaucratic snafu was about to rip away an essential tool. As 2020 became 2021—in the critical days just before the attack—the bureau would lose access to the service it used to monitor social media threats.

A year earlier, the folks in the FBI's procurement section had sent out a request for proposals, offering up a contract for a company that could provide a tool that would help them "obtain early alerts on ongoing national security and public safety-related events through lawfully collected/acquired social media data." The $14 million contract was awarded on Dec. 30, 2020, to CMA Technology in Vienna, Virginia, for subscription licenses to ZeroFOX, a social media monitoring tool.[31] The system they were used to, DataMinr, was on the way out.[32]

"We have an urgent need for the DataMinr replacement to be on and active starting on January 4th in support of some potential issues in the D.C. area," an FBI analyst wrote in a Dec. 31 email. "The sudden discontinuation is most untimely as much of our crisis response funnels through DataMinr."[33]

The New Year's Eve email got attention quickly. Steven D'Antuono, the assistant director in charge of the Washington Field Office, wrote in a follow-up email that there were "serious concerns" about the switch and said the analytical cadre did not receive accounts and there was a "large concern regarding familiarity with the new system." D'Antuono said it was "extremely concerning" that they didn't have access.

"We need the sign ons this weekend to effectively do our jobs," he wrote at 4:00 p.m. on Dec. 31. "I'll cross my fingers and toes that the company that we are paying for a service starting January 1st can get us sign ons for their service by this weekend. Doesn't make sense to me." Another official said they would try to prioritize the Washington Field Office, but that accounts would be available to everyone at the FBI in just a few more days. By Wednesday, Jan. 6.[34]

## New Year's Day, 2021

*The BIG Protest Rally in Washington, D.C., will take place at 11.00 A.M. on January 6th. Locational details to follow. StopTheSteal!*—@realDonaldTrump, 2:53 p.m.[35]

*Massive amounts of evidence will be presented on the 6th. We won, BIG!*—@realDonaldTrump, 3:10 p.m.[36]

*January 6th. See you in D.C.*—@realDonaldTrump, 6:38 p.m.[37]

*Just a reminder, there's strength in numbers. They can't arrest all of us. Don't be afraid."*—Bruno Cua on Parler

Donoghue was back in Washington, at his DC apartment, when he heard from Rosen. Mark Meadows had sent over a batty conspiracy YouTube video, flagging it for the nation's top law enforcement official.

"Pure insanity," Donoghue wrote to Rosen.[38]

"It was about a 20-minute YouTube video by this individual who identified himself as Brad Johnson. And it had this very conclusory explanation of how multiple intelligence agencies—U.S., British, and others—conspired to use Italian military satellites to change vote tabulations in the U.S. Presidential election," Donoghue said. "And it just seemed to me to be completely off the wall. It was not evidence. It was an internet conspiracy theory that was presented in a very conclusory manner, saying this, in fact, is what happened, without citing to any evidence that supported it."[39]

Meadows forwarded a bunch of other stuff that afternoon too. Donoghue said it seemed like it was just passing things along, maybe getting them off his desk. "I think he didn't have a firm opinion one way or the other about these matters," Donoghue said. "They were landing on his desk, and he was sending them off."

Inside a Proud Boys chat, the right-wing extremists were rethinking backing the blue.

"Our disposition towards the police needs to be reevaluated," wrote one member, who went by "Aaron of the Bloody East."

"Not sure what to do about it though because it would be an escalation that we would never be able to back away from," said Johnny Blackbeard.

Aaron of the Bloody East saw it as an inevitability.

"We can/should start adopting black bloc style tactics. Not necessarily all black like Antifa, but we should make an effort to hide our identities when in public," he wrote. "Masks, standardized attire, etc. Same thing black bloc does to make individual identification as difficult as possible."

"I wanna fuck shit up," said Joe Biggs. "#fucktheblue," Noble-Beard chimed in. "Agree, they chose their fucking side so let's get this done," Johnny Blackbeard wrote.

## Saturday, Jan. 2

*An attempt to steal a landslide win. Can't let it happen!—*
@realDonaldTrump, 6:15 p.m.[40]

Steven D'Antuono, the head of the FBI's Washington Field Office, spoke with Deputy Director Bromwich "and told him what we had for intel and our plans regarding the 6th." Meanwhile, within the bureau, aging technology was once again getting in the way of the bureau's efforts to stop some domestic extremists from traveling to DC.

"Our ridiculous phones won't let me open the contact card. Can you cut and paste into an email?" they wrote in an email about an RFI—a request for information—in connection with one of the extremists the FBI was worried about traveling to DC on Jan. 6. "FBI . . . making the simple hard for over 100 years lol."

In Kentucky, Michael Starks was planning a trip to DC with his coworkers. "This time we are going to shut it down," another coworker heard Starks say. He'd previously reposted Trump's Dec. 30 Facebook post about Jan. 6, informing his friends that he'd be there that day.[41] Now he was telling them he was willing to put his life on the line.

"The votes were stolen. You will have time to fight for what you believe in," Sparks wrote on Facebook on Saturday. "As for me I believe in the constitution so I'll die fir it [sic]. Trump is my president."[42] Four days later, Sparks would be the very first rioter to jump though a broken window, leading a pack of rioters into the Capitol.

A North Carolina man was preparing to drive from that state with his grandmother and another family member to DC for the president's speech. "If they want to raid Congress, sign me up, I'll be brave heart in that bitch!" Matthew Mark Wood texted an associate.[43] He was the tenth person through the window on Jan. 6.

In the evening, Oath Keeper Kelly Meggs posted a map of DC in a Signal chat for Oath Keeper leaders. They had a backup plan in case the bridges were shut down.

"1 if by land[,] North side of Lincoln Memorial[,] 2 if by sea[,] Corner of west basin and Ohio is a water transport landing !!" Kelly Meggs continued: "QRF rally points[.] Water if the bridges get closed."[44]

Enrique Tarrio fired up an encrypted chat for Proud Boys who had been accepted into what he dubbed the "Ministry of Self Defense." A few days earlier, they had held a video chat making sure that recruits knew they had to follow the commands of

Proud Boy leadership. At least sixty-five members were allowed into the chat.

"Open for business," Tarrio wrote.[45]

## Sunday, Jan. 3

*Florida is taking Roger Stone for 2 Days when he moves full kits. Insurrection act should be [what] he is presenting to America. . . . The natives are very restless. Tell your friend this isn't a Rally!!*—Kelly Meggs, Oath Keeper, 5:15 a.m.[46]

*I spoke to Secretary of State Brad Raffensperger yesterday about Fulton County and voter fraud in Georgia. He was unwilling, or unable, to answer questions such as the "ballots under table" scam, ballot destruction, out of state "voters", dead voters, and more. He has no clue!*—@realDonaldTrump, 8:57 a.m.[47]

*"I will be there. Historic day!"*—@realDonaldTrump, 10:27 a.m.[48]

Donoghue had come into the office on the first Sunday of the new year for an important meeting with the Defense Department and other law enforcement partners about their preparations for Jan. 6. The meeting was supposed to get under way at 1:00 p.m. on Jan. 3, and there was plenty to discuss. Donoghue arrived at the Robert F. Kennedy Building at 950 Pennsylvania Avenue NW around 12:30 p.m., maybe 12:45, and headed up to Jeffrey Rosen's office.[49]

After Trump mentioned Clark in their Christmas Eve phone call, Rosen put a pin in it. The day after Christmas, he rang up Clark and got to chatting, seeing why a DOJ environmental lawyer was suddenly on the president's radar.

"Jeff, anything going on that you think I should know about?" he asked. Clark wasn't exactly forthcoming, but after some to-and-fro, Clark told Rosen that he'd been at a meeting at the Oval Office, in violation of the Justice Department's policy on contact with the White House.[50]

Clark was apologetic and indicated that it wouldn't happen again. But a couple of days later he dropped an email that was just . . . well, bonkers. Clark thought that Dominion voting machines had accessed the internet through smart thermostats, with a net connection trail leading back to China.

Soon Rosen was asking the same type of question about the man he'd known for twenty years as Maria Bartiromo's Fox colleagues were asking about her. "What's going on with Jeff Clark?"[51]

"He said he got that off the internet," Rosen recalled. "So this was peculiar."[52]

When people think of Jeffrey Clark now, if they know the name at all, they might think of him as the guy who attempted a coup or the guy who'd later answer the door to law enforcement officers with a search warrant when he wasn't wearing any pants. At the time, nobody gave Clark much thought. Nobody ever really did. But he had everyone's attention now.

The day before the prep call with the Department of Defense (DOD), on Saturday, Jan. 2, Rosen called up Donoghue and asked him to get down to headquarters to meet Clark. They headed up to the sixth-floor SCIF, a sensitive compartmented information facility that is built to certain specifications to allow for the discussion of classified information. They left their cell phones outside and sat down, just the three of them.

Clark, thinking it was just going to be a one-on-one conversation, was upset that Donoghue was there, but Rosen said it was important for his no. 2 to be there.

"Well, the president has asked me to be acting attorney general," Clark informed them. "I told him I would give him a response on Monday. I'm going to think about what I'm going to do and I'll let you know."[53]

It was a heated meeting, and the Justice Department was now in full crisis mode. Donoghue and Rosen looped in another Justice Department official, talked for a while, and then headed home that Saturday.

Sunday was another quiet day at Justice Department headquarters. Over at the FBI, an email had landed in the inbox of top bureau officials at 10:01 a.m., marked in red letters: "This is for FBI internal use only."

The email started off with a standard warning, the type of language inserted out of an abundance of caution:

> *FBI WFO does not have any information to suggest these events will involve anything other than first amendment protected activity and is being distributed for situational awareness. Their inclusion here is not intended to associate the protected activity with criminality or a threat to national security, or to infer that such protected activity itself violates federal law. However, based on known intelligence and/or specific, historical observations, it is possible the protected activity could invite a violent reaction towards the subject individual or others in retaliation or with the goal of stopping the protected activity from occurring in the first instance. In the event no violent reaction occurs, FBI policy and federal law dictates that no further record be made of the protected activity.*

Finally, the content: "A review of open source, law enforcement, and liaison information suggests a sizable number of people plan to attend a series of loosely affiliated rallies in Washington, DC on 6 January 2021. Counterprotesters are aware of many of these plans and continue to organize events in opposition. To date, FBI Washington Field Office (WFO) is tracking four predicated domestic terrorism suspects traveling to the [area] for unknown purposes as well as a collection of unsubstantiated reports of threats to the city, protest participants, and/or US Government officials." The report listed the staffing plans of DC Metropolitan Police, the Secret Service, Supreme Court Police, US Park Police, and the Metro Transit Authority Police, but nothing from the Capitol Police.[54]

The report also relayed sixteen "Guardian" reports that the WFO was tracking, including one based off a tip that came into the FBI's National Threat Operations Center about Enrique Tarrio and the Proud Boys. "These men are coming for violence. They will cause mass unrest, destruction, and potentially kill many people in the streets of DC on January 6th," the complainant said. "They will roam the streets incognito attacking anyone they deem as antifa."

The Washington Field Office closed out the Guardian reports. "A review of the Tarrio posts submitted did not reveal any call for violence." It lacked specificity.

Donoghue was ready for the Jan. 6 prep call. He walked into Rosen's office to find the head of the Justice Department exasperated.[55]

"I spoke to Jeff Clark," Rosen said. "Clark says that he decided to take the president up on his offer, that he's going to be acting AG, but he wants to have one last face-to-face conversation with me before he tells the president."

The head of the Justice Department thought he was about to get canned, and his no. 2 thought he'd be right behind him. The future of the Justice Department was up in the air. But it was time for Clark and Rosen to dial into the conference call, as top officials in the government coordinated to prepare for the flood of conspiracy theorists the president had summoned to the nation's capital. Trump, back at the White House, tweeted through it.

*The Swing States did not even come close to following the dictates of their State Legislatures. These States "election laws" were made up by local judges & politicians, not by their Legislatures, & are therefore, before even getting to irregularities & fraud, UNCONSTITUTIONAL!*—@realDonaldTrump, 1:24 p.m.[56]

*Sorry, but the number of votes in the Swing States that we are talking about is VERY LARGE and totally OUTCOME DETERMINATIVE! Only the Democrats and some RINO'S would dare dispute this—even though they know it is true!*—@realDonaldTrump, 1:45 p.m.[57]

Meanwhile, in an encrypted Proud Boys chat, members thought Trump was giving them the go-ahead.

"1776 flag flying over the White House last night," Gabriel PB typed in a Telegram chat. "Gonna be war soon . . ." he wrote, trailing off.

Trump gave the Proud Boys plenty of signals, but this one was imaginary. Dan Scavino, Trump's former golf caddie who became the White House director of social media, had posted an image on Facebook on Jan. 1 that showed him taking a photo of the White House.[58] It was actually a recycled image that he first posted on Instagram back in 2019, but a right-wing website claimed that the blurry, washed-out flag flying atop the White House in the background of Scavino's photo was "the 1776 flag" with thirteen white stars arranged in a circle, representing the rebelling colonies. "It is the signal of REVOLUTION from our President! The time is here. The storm is upon us," the right-wing site read. "Rise-up, America. Rise up against the fraudulent election."[59]

The Proud Boys believed this really was a sign. Another chimed in that it was "time to stack those bodies in front of Capitol Hill." They started discussing specifics.

"So are the normies and 'other' attendees going to push thru police lines and storm the capitol buildings? A few million vs. A few hundred coptifa should be enough. I saw a few normie groups rush police lines on the 12th."

Someone joked that they should cue up the song "Bodies" by Drowning Pool: "Let the bodies hit the floor, let the bodies hit the floor, let the bodies hit the floor."[60]

Back at the Justice Department, the interagency call on Jan. 6 preparations was wrapping up.

"So what do you want to do with Jeff Clark?" Donoghue asked Rosen. The acting attorney general said he'd meet with his would-be replacement one more time.

Donoghue went back to his office and worked on some things. Rosen went and met with Clark, who told Rosen that Trump was going to replace him that day.[61]

"Well, here's the thing, Jeff Clark, my subordinates don't get to fire me. So I'm not accepting what you're telling me, that you're going to replace me," Rosen later recalled telling Clark. Rosen said he planned to contact the president directly.[62] "I don't get to be fired by someone who works for me," Rosen said.[63] Then he went to talk to Donoghue.

"He asked me to stay on as his DAG [deputy attorney general] once he becomes the attorney general," Rosen told him. "Can you believe the nerve of this guy?"

"Well, I guess we're done," Donoghue said. "Are we going to find out in a tweet?"[64] He started taking things off the wall of his office.

"I had plaques hung up. I figured we were going to be out of there within hours or a day or something like that," he said. "I took out boxes and began packing up things, because I was going to resign as soon as Jeff Rosen was removed from the seat."

Rosen came back and said he'd talked to Pat Cipollone, the White House counsel. Cipollone thought that this wasn't a done deal, that there was still time to stop the Justice Department from entering a free-fall crisis.

"It's time to broaden the circle, because we don't know how this is going to end," Rosen said. They got some of the Justice

Department's leaders on a conference call. Everyone was on the same page.

"It was unanimous; everyone was going to resign if Jeff Rosen was removed from the seat," Donoghue said. Only the head of national security would stay on; they felt that role, at least, needed some stability.[65]

The plan was to meet with the president at 6:15 p.m. Rosen wanted Donoghue there.

"I was not dressed for the White House. I was in jeans, boots, covered in mud because I had been walking on the Mall earlier in the day. I had an Army t-shirt on. I hadn't shaven in four days," Donoghue recalled. There was no time to change.[66]

Donoghue and Rosen headed over to the White House, and Donoghue took a seat outside in his boots and army T-shirt. The news was on, and the story about the president's call to Georgia's secretary of state, Brad Raffensperger, was all anyone could talk about. The *Washington Post* had obtained a recording of the phone call, which took place Saturday afternoon, and it was all over cable news.[67]

"I just want to find 11,780 votes, which is one more than we have because we won the state," Trump said. "So what are we going to do here, folks? I only need 11,000 votes. Fellas, I need 11,000 votes. Give me a break. You know, we have that in spades already."[68]

Meanwhile, down in Florida, Roger Stone was attending a rally urging Sens. Rick Scott and Marco Rubio to "vote with the president" and stop the certification on Wednesday.[69]

Daniel Stone, a Proud Boy who went by the name Milkshake, was providing security for Roger Stone, who wasn't the twenty-something he'd been in the early Reagan years. Stone was now pushing seventy, though still in decent shape and happy to show off the tattoo of Richard Nixon he has on his back. Still, Milkshake helped him up a ladder to make sure he didn't fall as the crowd cheered.

"All right. This is amazing," Stone said over the bullhorn.

"The jackals in the fake news media tell us over and over again that there is no evidence of either election fraud or cyber manipulation in the 2020 election," Stone said.

"Bullshit!" someone yelled out in the crowd.

"The evidence is growing, overwhelming, and compelling!" Stone said. "By any measure whatsoever, Donald J. Trump won a majority of the legal votes cast."[70]

"Four more years! Four more years!" the crowd cheered.

Stone said Donald Trump was the greatest president since Abraham Lincoln. "Rick Scott has a fundamental choice," Stone said. "He will either stand up for the constitution . . ."

"Or give him the rope!" Milkshake yelled.[71]

Back in Washington, a Fox News alert hit an FBI official's inbox.

"Proud Boys flock to Washington 'incognito' for Jan. 6 protests," read the Jan. 3 email, repeating a Fox News headline. "The far-right, male-only group, the Proud Boys, will descend onto Washington, D.C., in protest the same day Congress plans to certify Electoral College votes, officially confirming President-elect Joe Biden's win."[72]

A top DOJ official, Associate Deputy Attorney General Patrick Hovakimian, prepared a draft email to send to Justice Department staffers.[73] It would be one for the history books, the biggest Justice Department scandal since the Saturday Night Massacre, when Attorney General Elliot Richardson refused to fire Archibald Cox, the Watergate special prosecutor. "Mr. President, we may not see this in exactly the same terms, but I would like at least to be understood as acting in the light of what I believe is the national interest," Richardson told Nixon.[74]

History looked kindly on Richardson. The official Justice Department portrait of the Republican lawyer adorned the walls of the attorney general's office even during Democratic administrations. When Trump's first attorney general, Jeff Sessions, took office, Richardson's portrait was shuffled off to a hidden nook in the Justice Department's library.[75]

Figuring out who were the heroes and villains of an attempted coup—or *autogolpe*, a self-coup—wasn't rocket science, exactly. Trump appointees wouldn't even be giving up that much, with only a few weeks left on the job until the new administration was supposed to take office, if Trump's efforts failed. Working for Trump was already a blemish in some legal circles. What kind of law firm would hire a coconspirator in a plot to overthrow the American democratic system? The morally correct decision here was also the financially advantageous one.

Hovakimian, ten years out of Stanford Law School, spent Sunday afternoon composing an email that would define his legal career, be quoted in stories across the globe, lead off his obituary, and maybe even earn him a Wikipedia page.

*Dear all—*

*Apologies for the impersonal nature of this e-mail.*

*This evening, after Acting Attorney General Jeff Rosen over the course of the last week repeatedly refused the President's direct instructions to utilize the Department of Justice's law enforcement powers for improper ends, the President removed Jeff from the Department.*

*PADAG [Principal Associate Deputy Attorney General] Rich Donoghue and I resign from the Department, effective immediately.*

*Jeff loves the Department of Justice, as we all do. Preserving and defending the institutional integrity of the Department remains Jeff's paramount concern. The decision of whether and when to resign and whether the ends of justice are best served by resigning is a highly individual question, informed by personal and family circumstances. Jeff asked me to pass on to each of you that whatever your own decision, he knows you will adhere always to the highest standards of justice and act always—and only—in the interests of the United States.*

*It has been a high honor to serve with each of you.*

*Best,*

*Pat*
*Patrick Hovakimian*
*Associate Deputy Attorney General*
*United States Department of Justice*

Claire Murray, a Yale Law School alumnus who had served as acting associate attorney general, had been at the White House as an associate White House counsel, and already had her own Wikipedia page, was game too.

"Team Rosen. Justice is our client," she texted Hovakimian. "If the DAG gets fired for not publicly espousing a falsehood, I walk."

There was no need. Rosen texted with the news at 9:03 p.m.

"I have only limited visibility into this," Hovakimian wrote at 9:07 p.m., "but it sounds like Rosen and the cause of justice won."

Claire Murray texted Steve Engel, breathing a sigh of relief. "Great outcome. Congrats, and thanks for everything you did to push things in the right direction."

"Thanks!" Engel replied. "Glad I was able to help. We need to get points for the things that are not done."[76]

With the leadership of the Justice Department thoroughly distracted, the plotting for Jan. 3 continued in public and in private.

*We are on a mission to save America. Lone wolf attacks are the way to go. Stay anonymous. Stay alive. Guns up patriots!!"*—Parler user, Jan. 3[77]

"Biden won't be inaugurated. We will ensure that on the 6th," wrote a retired air force lieutenant colonel on Facebook, where he'd gotten sucked into conspiracy theories about the 2020 election. His posts had grown increasingly unhinged since Biden was declared the winner on Nov. 7, when he wrote that a "revolution every now and then is a good thing." The "traitors" who stole the election needed to be executed, and "that includes the leaders of the media and social media aiding and abetting the coup plotters." By December 2020, he was adding #civilwar2021 to his posts. In a message on Christmas Eve, he laid out a plan.

"Seize all democratic politicians and Biden key staff and select Republicans (Thune and McConnell). Begin interrogations using measures we used on [al Qaeda] to gain evidence on the coup," he wrote.

"Do not kill LEO"—law enforcement officers—"unless necessary," he wrote. "Gas would assist in this if we can get it." They should also attempt "to capture Democrats with knowledge of coup" and shoot and destroy "enemy communication nodes and key personnel.

"Seize national media assets and key personnel. Zuck, Jack, CNN lead and talking heads, seize WAPO and NYT editors. Eliminate them. Media silence except for White House communications," he wrote.[78]

In a couple of days, Larry Brock would be on the floor of the US Senate, carrying zip ties. Two years later, ahead of his sentencing, a retired air force lieutenant colonel who served alongside Brock and had Brock as a groomsman in his wedding compared Brock to the "Tank Man" who stood in front of tanks at Tiananmen Square in 1989, after the Chinese government's crackdown on protests.

"Larry is a man of principle, in much the same way our founding fathers were men of principle, and as such he is at times shunned by society (as were the founding fathers)," he wrote.[79]

## Jan. 4

*The "Surrender Caucus" within the Republican Party will go down in infamy as weak and ineffective "guardians" of our Nation, who were willing to accept the certification of fraudulent presidential numbers!—@realDonaldTrump, 10:45 a.m.*[80]

On Monday, FBI and Justice Department officials got a report from the SITE Intelligence Group that highlighted threats of attacks on Democratic and Republican politicians, as well as calls to occupy federal buildings and invade the Capitol.[81]

They also received a report from Mary McCord, the former Justice Department national security official, who had been sending in worrying posts that her team was seeing. "Surround the Swamp," a graphic stated, mapping out Capitol access tunnels and suggesting that rioters form a blockade of the entire Capitol complex.[82]

Over at the Capitol Police, an official warned during a meeting that Trump supporters "see this as the last opportunity to overturn the election." The head of the Capitol Police wasn't there as the official warned that Trump supporters "have nothing left to lose" and that the target was not counterprotesters. "The target is Congress," she said.

At the National Capital Region Threat Intelligence Consortium (NTIC), Donell Harvin held a conference call to discuss real-time open-source analysis of threats related to Jan. 6 and the inauguration. They'd picked up chatter about extremists planning to occupy the Capitol to influence lawmakers to change election results: to come with guns, to be prepared to battle, for election deniers to exercise Second Amendment rights. There was mention of a West Virginia militia, the Oath Keepers, the Three Percenters, Atomwaffen, Stormfront, Occupy the Capitol, MAGA Drag.[83]

Inside the Proud Boys chat, the conversation picked up again on Jan. 4 after Sen. Pat Toomey said he was voting to certify the vote in Pennsylvania.

"What would they do [if] 1 million patriots stormed and took the capital building. Shoot into the crowd? I think not," wrote a user named BrotherHunter Jake Phillips.

"They would do nothing because they can do nothing," replied Johnny Blackbeard.

There was plenty of talk about storming the Capitol out in public too. *Washingtonian* magazine even published a list of "5 Reasons to Avoid the January 6 Pro-Trump Marches in DC." Included on the list: No. 1: "The Proud Boys will be back." No. 2: "The events reflect an apocalyptic mindset among Trump fans." No. 4: "Many would-be protesters hope to defy DC's gun laws."[84]

"The noise seemed to be growing as we got closer to the event, and it reached a point of we just didn't know what was going to happen," said Bowdich, a top FBI official.[85]

It was "almost common sense" that there was a risk of violence on Jan. 6, another official said. "There's reports that a lot of people are coming to a rally who are unhappy about the election," Rosen later said. It was "kind of obvious" that law enforcement needed to be prepared.

Moore, the Washington Field Office intelligence chief, later insisted the information wasn't specific enough. "There was some rhetoric out there that we should, you know, *storm the Capitol*, but it wasn't like, *Let's go storm the Capitol, we are going to storm the Capitol.*"[86] It lacked specificity.

With the attack on the Capitol forty-eight hours away, some officials within the Washington Field Office were asking their bosses for permission to loop others in on the intelligence they had in their possession. They were getting a lot of questions, both from within the FBI and from other law enforcement partners.

"Let's wait till after the SAC meeting tomorrow. We may need to sanitize the non-pertinent info before releasing to partners if that's the direction the SACs decide we will take," an assistant special agent in charge in the Washington Field Office wrote Jan. 3, referring to the special agents in charge.

The FBI had certainly gone overboard in responding to social media posts before. In California, during the 2020 civil unrest, someone within the FBI created a situational information report (or "SITREP") built on a single tweet from the Long Beach Anarchist Collective that read "see a blue lives matter flag, destroy a blue lives matter flag challenge." The "SITREP" solemnly reported that "as of 29 May 2020, 26 users had liked the Tweet and four had shared the Tweet on their own accounts." The FBI report was then distributed to thousands of law enforcement officials.[87]

In June 2020, the FBI Boston Field Office—citing a blog post by a pro-Trump biker who calls himself "The Wolfman"—distributed a memo stating there were "rumors of stacks of bricks and stones that have been placed strategically throughout protests" and that a social media post "indicated that 'we think this is all part of the big plan.'"[88] The notion that *someone* was staging piles of bricks was an easily debunkable claim, and numerous fact-check organizations pointed out the ridiculousness of the claims and hunted down information on the construction projects the bricks were being used for.[89] "No evidence suggests a coordinated effort existed on behalf of government entities or billionaires to place bricks on protest routes to incite violence during the demonstrations," wrote Snopes. "Also, evidence showed that in many cases 'suspicious' bricks depicted on social media were on streets for ongoing construction projects, not 'planted' for protesters."[90] Nevertheless, the rumors about pallets of bricks were spread by the Trump White House and by the FBI.[91]

One-off, vague tweets and silly social media rumors were a far cry from what was available online in the lead-up to the Capitol attack, when the volume of threats and rhetoric centered on a specific time and place—Jan. 6, in Washington—was off the charts. The FBI, however, didn't aggregate that information into any kind of comprehensive report, and officials were concerned about distributing intelligence too widely for fear of leaks.

Moore, the intelligence chief in DC, later explained that the FBI SITREPs about Jan. 6 were kept "inside on what we call a red system, a classified system" that was available to dozens of law enforcement agencies on the FBI's Joint Terrorism Task Force in DC.

"So all our partners are getting it, all of our federal partners are getting it, our state and our local partners are getting it, Department of Justice is getting it," she said. "But we're trying to prevent—we're trying to at least control some of the dissemination" to prevent leaks to the media and compromise sources.

A senior FBI official said the bureau felt constrained from creating an intelligence bulletin due to First Amendment concerns.[92] But the elephant in the room seemed hard to ignore: an FBI intelligence bulletin warning that Trump's rally was a national security threat, if leaked, would draw intense ire from both the sitting president and his allies on Capitol Hill. *Let's just make it to Inauguration Day*, the thinking went.

By Monday, Jan. 4, the final call had been made, and the circle had tightened even further.

"WFO will not be releasing anything outside of WFO [executive management] regarding January 6th events so if you get inquiries please convey this," an email sent at 12:22 p.m. said. The decision, the email said, had been made by the assistant director in charge, D'Antuono, and the special agents in charge.[93] (Within eight months of the Capitol attack, four of the SACs who were in charge on Jan. 6 were replaced, while a fifth shifted to a new position.[94] D'Antuono retired in late 2022.)

Meanwhile, on Jan. 4, the distracted leaders of the Justice Department had another conference call with the Defense Department. Rosen, who had no Justice Department experience until he joined DOJ from the Transportation Department barely a year and a half earlier, was now in charge of overseeing the federal government response to Jan. 6 while dealing with an internal crisis as Trump tried to stay in power.

## Jan. 5

*See you in D.C.*—@realDonaldTrump, 10:27 a.m.[95]
*The Vice President has the power to reject fraudulently chosen electors.*—@realDonaldTrump, 11:06 a.m.[96]

An FBI official was set to meet with the bureau's deputy director ahead of Jan. 6. So they hit up the Park Police for info.

"I have to meet with the Deputy Director later today and was wondering if your intel folks have developed any information you could share about the upcoming activities," the official wrote. They didn't receive a reply that day, but they got one the next morning.

"On leave this week so trying to disconnect a little," Deputy Chief Steven Booker of the US Park Police Intel Unit wrote in an email at 11:24 a.m. "This week's events will mirror what has happened the past two months. The same groups seeking permits and the same counter groups (with affinity groups) will attend with largely the same desires and actions."[97]

James Mault, a military veteran and iron worker from New York,[98] was finalizing his plans. He was about ten months into his sobriety, visiting a Veterans Outreach Center in Rochester for an Alcoholics Anonymous meeting.[99] A few years earlier he'd been convicted of driving while impaired, took a seven-week driver awareness course, and realized he'd been overusing alcohol since he returned from his deployment in 2014. The former high school wrestler was now married and had a young kid at home, with another on the way.

At about 12:30 p.m., Mault hit up the group text. Bring long sleeves, he wrote, maybe gloves, a baton, pepper spray, a helmet, eye protection, and clothes they weren't worried about losing. Don't forget the "asskicking boots."[100]

At 1:05 p.m., Acting Attorney General Rosen got a text from Ken Paxton, the Texas attorney general who had sued to disenfranchise millions of voters in Pennsylvania, Georgia, Michigan, and Wisconsin. He was going to speak at the Stop the Steal rally in less than twenty-four hours and also wanted to meet with the acting attorney general afterward. "Hi Jeff. This is Ken Paxton. I am flying to DC tonight for the rally. Do you have time to meet after that or on Thursday?" he wrote.[101]

Tens of thousands of people began arriving in DC, and there was more violent rhetoric being promoted online. "If all of these people are kind of saying the same thing, when do you start taking it seriously?" one official familiar with the FBI investigation asked.

One source told me there was "battle fatigue" inside the bureau, with officials "walking on eggshells" because they didn't know what the president could do. "You just don't know what fuel he is going to put on this fire," they said. "No one was looking at it in aggregate," they said. "No one was tasked to look at the big picture."

At the Washington Field Office, one FBI supervisory intelligence analyst wrote an email saying they would be taking off at 2:00 p.m. and out on Jan. 6.[102] Down in the FBI's Norfolk Division, an official prepared a short memo. "Potential for Violence in Washington, D.C. Area in Connection with Planned 'StopTheSteal' Protest on 6 January 2021," it read.

"An online thread discussed specific calls for violence to include stating 'Be ready to fight. Congress needs to hear glass breaking, doors being kicked in, and blood from their BLM and Pantifa slave soldiers being spilled. Get violent . . . stop calling this a march, or rally, or a protest. Go there ready for war. We get our President or we die. NOTHING else will achieve this goal,'" the report relayed.[103]

DHS had picked up on some info. One collector found information about a person talking about arriving in the DC area and looking for a location for armed individuals to park their cars and prepared an open-source intelligence report. A colleague conducted a peer review and said that the information didn't rise to the level of a report, but the collector brought the issue to the Office of General Counsel's Intelligence Law Division, which signed off on it at 12:16 a.m. on Jan. 6.

Yet another bureaucratic hurdle stood in the way. Supervisors within Open Source Collection Operations, under a policy put in place for election-related intelligence, had to get permission from I&A's Intelligence Oversight officer. They didn't put that request in for fifteen hours, on the evening of Jan. 6, after the building was breached. The intelligence product—now rendered utterly useless—was finally published on Jan. 8.[104]

As Proud Boys chairman Enrique Tarrio's Uber crossed into DC after arriving at Ronald Reagan National Airport across the river in Arlington, sirens went off behind him. He was on the phone with *USA Today* extremism reporter Will Carless at the time. "They're for me," Tarrio said, referring to the police lights. He asked the driver to pull over. Tarrio didn't let on at the time, but he knew that a warrant had been signed for his arrest. Lamond, his contact at MPD, had tipped him off. The pair had met up at the Dubliner Irish Pub for drinks in December 2020, and Lamond set an autodelete timer on their encrypted chat. He had sensitive intel to pass along to the subject of a pending criminal investigation.

"Just a heads up, [the Criminal Investigation Division] had me ID you from a photo you posted on Parler kneeling down next to the banner so they may be submitting an arrest warrant to US Attorney's Office," Lamond wrote on Christmas Day, alerting Tarrio to his potential forthcoming arrest. He passed along confirmation of the warrant as Tarrio hopped on his plane to DC, and Tarrio relayed the news. "Here's something to write about,"

Tarrio told Carless as law enforcement prepared to take him into custody. Tarrio hung up.[105]

It seemed obvious: Jan. 6 was going to be messy. "Washington is bracing for unrest on Wednesday, when a motley crew of far-right extremist groups, including Tarrio's Proud Boys, are set to descend on the capital as Congress is scheduled to certify the Electoral College votes from November's presidential election and declare President-elect Joe Biden the winner," read a story I cowrote about Tarrio's arrest that relayed what the Proud Boys president had publicly proclaimed on social media. "Tarrio wrote recently on Parler that Proud Boys would attend the protests 'in record numbers.' He also claimed they would not be wearing their signature black and gold uniforms, but instead would go 'incognito' and 'spread across downtown DC in smaller teams.'"[106]

The president, meanwhile, was soaking it all in. He saw the crowds flooding into the city, and he liked the numbers.

*Washington is being inundated with people who don't want to see an election victory stolen by emboldened Radical Left Democrats. Our Country has had enough, they won't take it anymore! We hear you (and love you) from the Oval Office. MAKE AMERICA GREAT AGAIN!*—@realDonald Trump, 5:05 p.m.[107]

*I hope the Democrats, and even more importantly, the weak and ineffective RINO section of the Republican Party, are looking at the thousands of people pouring into D.C. They won't stand for a landslide election victory to be stolen.* @senatemajldr @JohnCornyn @SenJohnThune—@realDonaldTrump, 5:12 p.m.[108]

*Antifa is a Terrorist Organization, stay out of Washington. Law enforcement is watching you very closely!* @DeptofDefense @TheJusticeDept @DHSgov @DHS_Wolf @SecBernhardt @SecretService @FBI—@realDonald Trump 5:25 p.m.[109]

*I will be speaking at the SAVE AMERICA RALLY
tomorrow on the Ellipse at 11AM Eastern. Arrive early—
doors open at 7AM Eastern. BIG CROWDS!—@real
DonaldTrump, 5:43 p.m.*[110]

Trump supporters gathered that night at Black Lives Matter
(BLM) Plaza. Among the crowd was Ricky Shiffer. About a year
and a half later, after the FBI searched former president Donald
Trump's Mar-a-Lago estate, Shiffer would attack an FBI field
office in Cincinnati with a nail gun and would be shot and killed
after he raised his gun at police.[111]

At BLM Plaza on the eve of the insurrection, Trump supporters tried to rush a police line and get into the plaza as police did
their best to keep the groups separate.[112] Fights broke out. "Nazis!"
a Trump supporter yelled at police. "You sold your soul!"

———— • ————

The folks at the Hotel Harrington had seen the signs. A few days
after Christmas, the 106-year-old hotel and Harry's Bar—one of
the few places left near FBI headquarters that still served up
moderately priced beer—said it would shut down on Jan. 4, 5,
and 6 for the "safety of visitors and employees."[113] The hotel and
restaurant became a major hangout for the Proud Boys during
their excursions in the city in November and December. I happened to be biking past Harry's during the Million MAGA
March on Nov. 14, on my way to the DC fish market for one last
order of steamed blue crabs before the season came to an end for
the winter. When I stopped to take a photo of the gathering at
Harry's, a man carrying a flagpole and wearing a Proud Boys
hat, American flag gaiter, and "KYLE RITTENHOUSE DID
NOTHING WRONG" T-shirt yelled "Fuck antifa!" at me. I
looked down, realizing I may have leaned into the black athlei-
surewear look a bit hard during the pandemic.

Ahead of Jan. 6, Black Lives Matter DC wanted other hotels in the region to follow Hotel Harrington's lead. The organization asked J. W. Marriot and Holiday Inns in the area to close due to the threats being made against the group. "Your silence is violence," said Nee Nee Taylor, who urged BLM supporters to stay home and not counterprotest on Jan. 6.

"We will be totally outnumbered," she said. "Their president said, 'stand back and stand by' and this is the standby."[114]

Trump supporters at The Donald forum didn't appreciate that Harry's backed down. "HOTEL HARRINGTON GOT CUCKED BY A BUNCH OF LOSER LIBERALS," one wrote.[115]

Black Lives Matter DC was not the only left-wing group that saw trouble ahead on Jan. 6. Antifascists even hung posters about the arrival of Jeremy Liggett's B-Squad. "DANGER ARMED FASCISTS IN DC Jan. 6," the posters read. Liggett spotted one of the posters when he went to speak at an event at Freedom Plaza on Jan. 5, where a speaker introduced him as "antifa's worst nightmare" and Liggett led the crowd in a chant of "fuck antifa." Liggett posed for a photo next to one of the signs, mugging for the camera while throwing up three fingers, symbolizing the Three Percenters movement, and wearing his B-Squad patch.[116]

The tips were flooding into the bureau too, including a specific warning about what the Proud Boys were up to. "They think that they will have a large enough group to march into DC armed and will outnumber the police so they can't be stopped," the tipster wrote. "They believe that since the election was 'stolen' that it's their constitutional right to overtake the government and during this coup no US laws apply. . . . Their plan is to literally kill people. Please please take this tip seriously and investigate further." Asked on the web form to describe what kind of attack was being reported, the tipster wrote: "Attempted coup/terrorist attack on Jan 6th."[117]

Beyond the incoming intelligence, the threat just seemed sort of obvious, even to top people in the government who were simply paying attention to what was happening around them. "You didn't need an intelligence report to know that thousands of angry people were going to be showing up at the Capitol that day who were upset about the election and who wanted to disrupt the congressional proceedings," Donoghue said.[118]

Former FBI director James Comey said the attack on the Capitol was "painful" because of the lack of preparation, saying he was angry there wasn't better security ahead of the obvious threat. "It wasn't a failure of imagination, it was just a failure to see a threat that was in bright daylight coming at you," Comey said less than a week after the Capitol attack. Comey said it would be "really important" for there to be some kind of commission to understand what happened.[119]

Moore, the FBI Washington Field Office intelligence chief, said that they'd put personnel on Jan. 6 intelligence, but that they were being pulled in a lot of directions at the time.

"There's still the rest of the world going on as we're preparing for January 5th and 6th," Moore later said. "And while it's all hands on deck, and most certainly we have the bulk of our resources there, we just have onesies and twosies out there that are still handling your bank robberies or your transnational organized crime type things as well as public corruption."[120]

Michael Sherwin, the former US attorney in DC, thought there was more they should have done, in retrospect, and that the FBI wasn't up to speed on social media:

*I think there is a type of institutional arrogance that we don't—I don't need to rely upon this guy's Twitter feed for intel. We have our sources. We know what's going to go on. This is unvetted, unverified, where I think the Bureau has to plus up and better discern how to use social media and how to join up with partners like that, because sometimes that's incredibly good intelligence, even though it's not coming from a Bureau undercover or a Bureau source. And I think it's just a function of how social media has just grown. And that's a gap that probably needs to be addressed by the Bureau.*[121]

Tim Heaphy—a former US attorney for the Western District of Virginia who had previously conducted an independent review of the deadly events in Charlottesville in 2017[122] and found that law enforcement "could not have been reasonably surprised" by the events there—found the same thing had happened ahead of Jan. 6.[123]

"When you make mistakes, ideally, you'll learn from them," Heaphy, the top investigator on the Jan. 6 committee, later said. "And this was a mistake."[124]

Over the course of the Trump presidency, those in the domestic terrorism space had seemingly built up an immunity to the alarm bells. The constant drumbeat of threats could be exhausting, and an environment that once seemed unimaginable had become commonplace. The heat had been slowly rising for years, and now law enforcement were the metaphorical frog in the boiling water. "Were we forwarding out intel like we always do? Yeah. But were we screaming it from the rooftops? No. None of us were," an FBI informant later told me.

———— ◆ ————

Rodriguez, the MAGA fanatic from California, had made it with his friend Ed Badalian. Driving in a passenger van they had picked up at the Enterprise at LAX, they met up with other MAGA supporters in Louisville, Kentucky, in the parking lot of a Kohl's, and joined a caravan of cars headed toward DC. They used a radio cell phone app to talk with others in the caravan and stopped at an overpass so they could display a "Fuck Biden" flag to passersby.

Badalian had been getting people ready for months, ever since Election Day, when he suggested that "patriots" swarm CNN, surround it, and only let people out, not in. He said the country needed a revote, not a recount, and that voters should only be able to cast their ballots in person, with photo ID. After the election was called, he said Biden was "definitely guilty" of treason and that the punishment for treason is death.

It was time for patriots to start training, Badalian wrote to his group of Trump supporters in an encrypted chat. They needed to organize so they could effectuate arrests. "Surgical force not brute force," he said. "The Constitution works for us and it's the ultimate law, but no one is willing to enforce it. We the people have to." He organized a "Patriot Paintball" session to help them train and urged others to stay focused on the mission. "Warped Paintball is where its at . . . another drunken late night wont lead to a day of solid training for the coming unrest. Weve been slipping," he wrote on Nov. 29. After Trump's "will be wild" tweet, it was on.

"Our duly elected leader has called his marching orders," Badalian wrote. "Its time to stop giving a fuck about optics. Were not trying to win a GODDAMN ELECTION ANYMORE. Its not about a goddamn popularity contest. If you know the Constitution is whats right, then its time to fight for it."

Rodriguez agreed. "The time for a quick strike is coming. Patriots will have a chance to make history and save the greatest nation in world history," he wrote. "Don't give up the ship!"[125]

They came prepared with gear—weed too—and had been preparing for this moment. They arrived at their Airbnb on Washington Boulevard in Arlington, Virginia.

"We don't want to fight antifa lol we want to arrest traitors," Badalian wrote in a chat that day.

"There will be blood," Rodriguez wrote the night before Jan. 6. "Welcome to the revolution."

# CHAPTER 6

# "The Storm"

The email from the FBI's Strategic Information and Operations Center (SIOC) hit inboxes at 5:55 a.m. Subject: "CIRG'S DAILY SITREP," meaning the Critical Incident Response Group's daily situational report. NC3—the National Crisis Coordination Center at the FBI's J. Edgar Hoover Building on Pennsylvania Avenue—would be in a "supportive role" to aid "efforts Preventing Violence and Criminal Activity (PVCA) today" in the NCR (National Capital Region), where the weather would be "mostly sunny and breezy, with a high/low of 46°/33°."

The FBI Public Affairs News Briefing hit inboxes thirty-four minutes later, compiled by Bulletin Intelligence, a firm that started in 1990 as a midday newsletter for Capitol Hill insiders and now marketed itself as the "gold standard in providing intelligence to senior executives and government officials." (They keep their client list confidential.[1]) "Officials, Law Enforcement Prepare for Possible Violence as Trump Supporters Flock to DC for Protests," read one summary of a news story. "Far-Right Forums Urge Violence During Protests," read another summary of a *Washington Post* story.[2] "Legislators Advised to Use Tunnels amid Protests," read another summary, highlighting reporting from the Associated Press.

What it didn't include were Donald Trump's tweets from a few hours earlier.

> *Get smart Republicans. FIGHT!*—@realDonaldTrump, 12:43 a.m.[3]
>
> *If Vice President @Mike_Pence comes through for us, we will win the Presidency. Many States want to decertify the mistake they made in certifying incorrect & even fraudulent numbers in a process NOT approved by their State Legislatures (which it must be). Mike can send it back!*—@realDonaldTrump, 1:00 a.m.[4]

An FBI email for "internal executive management only"—limited distribution, to avoid any leaks—pointed to a huge spike in tips to the National Threat Operations Center. As of 2019, the NTOC in West Virginia was handling roughly 3,100 phone calls and electronic tips per day.[5] The volume at the NTOC had been triple that the day before Jan. 6. "A total of 10,250 complaints were processed yesterday by NTOC with 5,189 being related to social media posts which were related to political matters," an 8:31 a.m. email said. There was a threat against law enforcement from an individual who posted on YouTube that they planned to bring a gun on Jan. 6; online threats made by a TikTok user against Kamala Harris and Nancy Pelosi; and, from an individual unlawfully possessing firearms, "concerning statements" about revolution because Trump wasn't reelected.[6] Another email stated that a congressional staffer at the Hyatt Regency Hotel had heard a man wearing fatigues talk about a plan to storm the FBI at 2:00 p.m.

Mike Sherwin, DC's acting top federal prosecutor, gave Attorney General Rosen and Deputy Attorney General Donoghue his forecast via email. There were eight arrests Tuesday night: three for illegal possession of firearms, two for assault of police officers, and the rest "simple assault/battery charges (one a possible hate-crime)." He expected a "spike in activity" that day and into the night, but things were "all good at the moment," he wrote. "All estimates are still trending under 10,000," Sherwin wrote, referring to the expected crowd. "I'll continue to push you updates throughout the day."

Sherwin really thought they were in a solid place. There'd been a "litany of conference calls" between law enforcement entities to prepare for the event, and the FBI and federal prosecutors took some action to "mitigate, to the best we could, some elements that may too obviously cause trouble in the district."[7]

One of those mitigations took place two days earlier, when Tarrio was arrested in the Uber. The *USA Today* reporter he was on the phone with at the time tweeted about the arrest, and the news soon landed in the Proud Boys' encrypted chats. "Fuck these pigs," wrote a Proud Boy known as Aaron of the Bloody East. "Time to use black bloc tactics." Others said not to worry: Trump had their backs and would pardon him and maybe get him out of jail before Wednesday. "Either way, everyone should be pissed off and ready to fight," Aaron said.

The arrest wasn't news to all the Proud Boys, of course. Tarrio knew, from his contact in MPD. After being picked up Monday, Tarrio made an appearance in DC Superior Court on Tuesday and was released after a judge ordered him, at the government's request, to leave the city. Assistant US Attorney Paul Courtney said Tarrio needed to "stay away from the District of Columbia, in its entirety" to "ensure the safety of the residents of the District of Columbia."[8]

After Tarrio was released, he briefly went to the Phoenix Park Hotel, which is located above the Dubliner pub where he'd met up with the MPD officer. He was accompanied by Nick Quested, a filmmaker who was embedded with the Proud Boys. They then went to the parking garage of the nearby Hall of States Building, home of C-SPAN and the DC bureaus of Fox News and NBC News, and Tarrio briefly chatted with Stewart Rhodes, the head of the Oath Keepers. Rhodes allowed for his photo to be taken by Amy Harris, a photographer who had shifted to covering the Proud Boys and had what Quested called a "very strong connection" to Tarrio. But Tarrio brushed away Quested for a private conversation with Rhodes and others, including SoRelle and a man who'd been arrested on gun charges outside the Philadelphia vote-counting location in November.

Later, as Rhodes worked his phone, Quested joined in a conversation with SoRelle and a masked man accompanying the Oath Keepers' leader. Quested, who's filmed music videos for some of rap's greatest artists and then made films about soldiers in Afghanistan and Mexican drug cartels, tried to make small talk.

"People always joke, buy gold in times of struggle. It's like gold doesn't get you through checkpoints. Whiskey and cigarettes gets you through checkpoints," Quested joked.

"Speaking of," said the masked man, pulling a pack from his breast pocket. Then he said what Quested later described as "the most seditious thing" he'd heard that day.

"It's inevitable what's going to happen. We just gotta do it as a team, together. Strong. Hard. Fast," he said.

Tarrio headed out of the city, navigating to a "shabby art deco hotel that was quite economical" with Quested. They arrived about 8:00 p.m., and Tarrio headed to a drugstore to get a new phone. Quested hung around for a while. There was a dog running around the hallway, a "super cute" and sweet pit bull, who apparently got into leftover pizza and defecated in the hallway. "So that's the sort of hotel we were in," Quested said.[9]

The next morning, Jan. 6, Quested went out to film the Proud Boys, joining them near the Washington Monument and walking east toward the Capitol. There were hundreds of them. "Be advised, the Proud Boys group is now crossing Third Street onto US Capitol property, numbers still approximately two hundred," the Capitol Police radio cracked at 11:07 a.m. A law enforcement report at 11:15 indicated that there were "approximately 300 Proud Boys dressed in black, some with orange hats."

Quested is used to bonding with even unlikable subjects as part of his job.[10]

"Look, I am a funny guy. I tell jokes," Quested later told me over beers at Kelly's Irish Times, the bar next door to the Dubliner, where Tarrio had met the MPD official. "I'm always joking and probing and finding commonality. It's like my way of being extroverted. I always got upset at my friends who could go and talk to girls in bars, because I couldn't do it. But I can go talk to militias!"[11]

But on Jan. 6, he said, the tone was different from what it had been in his previous encounters with the Proud Boys.

"I'm just going to make light of the situation, like this isn't serious. *We are all friends here. Don't worry about the camera, it's all good. I'm just one of you guys,*" he said. "But it felt much darker."

Eventually they made their way to the east side of the Capitol, where the Proud Boys got into formation and took a group photo.

"Let's take the fucking Capitol!" Milkshake yelled. "Let's not fucking yell that, alright?" another Proud Boy said. "It was Milkshake, man, you know," said Proud Boy Ethan Nordean. "Idiot."

"Don't yell it," another Proud Boy chimed in, "do it."[12]

———◆———

Sen. Todd Young, a former marine who wore a military-style jacket over his suit, was outside a Senate office building, addressing his constituents. He'd spent more than twenty minutes talking with some of his constituents from Indiana. First a man prayed over Young, asking God to bring him and his fellow senators "fortitude and strength" along with "wisdom and sound judgment." Young told them he knew that he'd been on their prayer lists and that he'd felt strong and felt conviction as a constitutional conservative.

"I took an oath, first as a United States Navy sailor, and then as a United States Marine, and then as a member of the US House of Representatives to our constitution," Young said, pulling at his "One Nation Under God" facemask. "I took an oath as a United States senator. It's covenant."

Bob Croddy, who told Young he was a regional chairman with the Trump campaign, warned Young that hundreds of thousands of grassroots organizers were "watching everything you do." A woman was near tears. *Please*, she cried out, *we need you.*

"I've looked at all the facts; I've looked at the law," Young said. "You upset with me, Bob? You don't want me to follow politics; you want me to follow the constitution."

"We want you to fight, Todd," Croddy said.

Young told them he was going to respect the separation of powers, the Twelfth Amendment to the Constitution, and the Electoral Count Act of 1887. "That's what you all asked me to do. Dammit, that's what I'm gonna do."

Not realizing what he was saying—that he was going to certify the electoral college vote and that Biden would be president—they started cheering. He was going to have to dumb it down a bit.

"I believe in the separation of powers; do you believe in the separation of powers?" Young asked. Yes, they replied. "You know what that means?" he said, delving into a brief lesson straight out of Schoolhouse Rock. "That means we have an executive branch, a president of the United States. That means we have a legislative branch. That means we have a federal court system, which, I have to say, is populated with all kinds of conservative judges in the wake of the Trump presidency." Young told the crowd he didn't think there was "a vast judicial conspiracy by sixty courts," referring to the cases lost by the Trump campaign.

"I wanted Trump to win," Young said.

"He did win!" said the crowd. "He won!"

Young delivered the news in the simplest terms he could. "I will not be joining the Cruz effort; I think it's a hijacking of the

Constitution," Young said. Soon he gave the floor—a sidewalk, in this case—to his colleague from Indiana, Sen. Mike Braun. When Sen. Ted Cruz rose to object to the counting of Arizona's electoral college votes at 1:12 p.m., many of his colleagues stood up and clapped, while Young sat, stoic, with his hands in his lap.

Sen. Mike Braun didn't have as much on the line as Young did: he wouldn't be up for reelection until 2024, whereas Young's race was 2022 (spoiler alert: he won). Yet Braun took a different approach with a group of frustrated Hoosiers. He told them what they wanted to hear. A few weeks earlier, before Trump's "will be wild" tweet, Braun had issued a statement saying he was "disappointed" by the results of the electoral college vote on Dec. 14 but that the day "marks a watershed moment where we must put aside politics and respect the constitutional process that determines the winner of our Presidential election."[13] It wasn't long before his calculus changed. On Jan. 2, he joined in with Sens. Ted Cruz (R-Texas), Ron Johnson (R-Wis.), James Lankford (R-Okla.), Steve Daines (R-Mont.), John Kennedy (R-La.), and Marsha Blackburn (R-Tenn.)—along with four senators-elect— on a statement announcing their intention to "reject the electors from disputed states as not 'regularly given' and 'lawfully certified' (the statutory requisite), unless and until" there was "an emergency 10-day audit of the election returns in the disputed states."[14]

Braun told the crowd that he would be objecting to make a point but added that this wasn't going to "solve what happened on Nov. 3, and that's not going to occur, but that doesn't mean we remain silent about it, and we should be doing everything to make sure there's election integrity." Braun told them they should "not let it lay" after Jan. 6.

Though he'd announced days earlier that he would be objecting, that was apparently news to many in the crowd, who reacted as though they were hearing it for the first time. They were also surprised to learn it was pretty much a done deal.

"The electors are going to be certified," Braun said. To his right, a woman named Mary Beth, who was wearing a Trump beanie, closed her eyes, flinching as if hearing about the death of a loved one. "No," another woman sighed. "You'll have a big problem if you certify the election today," one man chimed in.

"Even people that have been around politics a long time have been texting me and emailing me about what can be done, and I knew today was going to be a big disappointment," Braun said.

He tossed some red meat to his constituents, telling them that Trump had been treated unfairly from the start, from the Mueller investigation to the impeachment.

"It was relentless. And then the sanctimony of some to say that just because you're speaking out for election integrity that it's seditious," Braun said. "That really tees me off."

After his speech, Braun shook hands with a young woman, telling her something about the importance of her generation getting involved in politics.[15]

"Speaking and listening to Hoosier @realDonaldTrump supporters who came to DC from Indiana about why I will object today and support an emergency audit into irregularities in the 2020 election," he tweeted at 12:20 p.m., being sure to tag the president.[16]

Within a few hours, that young woman Braun spoke with was at the front of the police barricades as a mob swarmed the building, had climbed on top of a BearCat SWAT vehicle while holding a "STOP THE STEAL" sign, and then joined the pandemonium at the rotunda doors, where the mob dragged officers away.

Two other Indiana men in the photos Braun posted, Mark and David, were farther up those stairs, leading the mob into the rotunda around 2:27 p.m. A South Carolina man named George Amos Tenney III, who'd later be sentenced to three years in federal prison,[17] had busted open the doors from the inside after rioters broke the glass. Mark went in first, holding a flag and watching as a marine veteran named Kaleb Dillard grabbed an officer by the neck and threw them to the ground.[18] Others went over to help, but Mark mostly just watched.

As the cop was on the ground, David surged in. As soon as other rioters used violence against officers, David was encouraging them to give up, escorting them away from the door. "It's okay, officer," he said to one. "You're doing the right thing." He waved the mob inside and went to convince another officer being crushed by the mob to give up the doorway that Lincoln's body had once passed through. As of this writing, neither of the men has been arrested.

A couple of hours earlier, as the Proud Boys grabbed lunch at some food trucks, videographer Quested sneaked off to call his lawyer. The Proud Boys, he thought, "were looking for a scuffle," and he thought there could be some arrests. If he got locked up along with them, he wanted to make sure that someone came and found him, that he wasn't stuck at the bottom of a pile of four hundred or five hundred people. At the time, though, he was surprised that the police staffing levels were "paltry," considering the crowd that would soon be headed from the president's speech.

"I thought it would be a different chaos," he said. "I thought we were about to see a potential medieval battle on the Mall. Like, you know, a freeform melee."

———◆———

Sherwin knew a thing or two about criminal organizations, and taking out the head of the Proud Boys was a good way to disable the organization, or at least temporarily stun it. Before he came to DC, Sherwin had spent twelve years as a federal prosecutor in South Florida, where he worked drug cartel cases that came out of South and Central America as well as foreign corruption cases and health-care fraud. He'd joined DOJ's National Security Division in 2018 and advised Barr and then Rosen on national security matters.

Heading the US Attorney's Office for the District of Columbia—
the largest federal prosecutor's office in the nation—was a different beast. Given DC's unique statute and lack of statehood, the US
Attorney's Office in DC handles both state and local crimes. That
means DC voters don't get to decide law enforcement priorities—
the entire country does, by way of the electoral college. Often, that
means the top prosecutor in DC isn't someone residents would have
chosen for the job.

In the 1980s, for example, the office was headed by Joe diGenova, who went to war with then mayor Marion Barry over
corruption in his administration and drug use. The federal government spent years trying to find a charge they could get to
stick to Barry, setting up a fake consulting firm and going
through his tax records and credit card bills, and the US Attorney's Office freely leaked to the media about Barry.[19] Finally, not
long after diGenova resigned, the FBI ran a sting on Barry, using
an old girlfriend to lure him into using crack cocaine. "We talked
about how the easiest way to get Barry was with a woman," one
senior law enforcement official said. Barry thought he was there
for sex, lounging on the bed and fondling his ex-girlfriend's
breast and leg. "Can we make love before you leave, before you
leave town?" What he didn't know was that FBI special agents
were babysitting the woman's three children as they waited for
their mother to seduce Barry. The FBI agents swarmed in.[20]
"Bitch set me up," Barry said during his arrest, a quote that
helped make him the butt of late-night jokes.[21] Years later, New
York mayor Rudy Giuliani—a longtime friend of diGenova—
would help poke fun at Barry during a *Saturday Night Live*
sketch on a 1997 episode Giuliani hosted.[22]

After a few years at a law firm, diGenova and his wife, former
Reagan Justice Department official and fellow Republican congressional staffer Victoria Toensing, struck out on their own in
1995 and soon became regulars on the television circuit. In the
weeks after the news broke that Bill Clinton had had an affair

with Monica Lewinsky, you could scarcely turn on a TV without seeing them. At a time when defending Donald Trump on Fox News was the best way to get yourself hired by the president, that's exactly what they did. Their conflicts of interest in Robert Mueller's special counsel investigation kept them off the Trump team formally, but they'd get their chance soon enough. The duo was hired by a Ukrainian oligarch and then teamed up with Rudy Giuliani to dig up dirt on Joe Biden ahead of the 2020 election.[23] That landed them in the middle of the scandal that led to Trump's first impeachment, after he threatened to withhold aid from Ukraine, which ended in a Senate acquittal in February 2020.

By November 2020, after Trump lost the election, diGenova was too crazy for Fox News—which hadn't had him on since he said George Soros controlled the State Department and the FBI—but not too nuts for Trump's "elite strike force" of lawyers who couldn't or wouldn't accept the fact that Trump lost.

"Rudy Giuliani, Joseph diGenova, Victoria Toensing, Sidney Powell, and Jenna Ellis, a truly great team, added to our other wonderful lawyers and representatives!" Trump tweeted.[24] DiGenova joined the whole crew over at the RNC, where black liquid ran down Giuliani's face and Sidney Powell went on an unhinged rant.

"What we are really dealing with here and uncovering more by the day is the massive influence of communist money through Venezuela, Cuba and likely China and the interference with our elections here in the United States," Powell declared. "President Trump won by a landslide." Later, after she was sued, her lawyers said that "no reasonable person" would think that what Powell said were "truly statements of fact."

Within a few weeks of joining the team, former US attorney diGenova was appearing on a conservative podcast when he suggested the violent murder of a government official for stating the truth about the election. Trump had fired Christopher Krebs, his administration's senior-most cybersecurity official, for daring to state that the 2020 election was secure.[25] DiGenova said firing Krebs wasn't enough.

"He should be drawn and quartered," diGenova said, sounding as if he were in a Proud Boy or Oath Keeper chat, or perhaps on TheDonald.win. "Taken out at dawn and shot."

Soon, diGenova was forced to resign from the elite Gridiron Club, where he'd been a "ringer" for their annual dinner because of his singing chops. "I guess I was canceled," diGenova said.[26]

Now Sherwin was in diGenova's old role, trying to assess how much of a danger people making comments like his predecessor were. He wasn't the outright partisan diGenova was, but his arrival was met with a lot of suspicion at the US Attorney's Office in DC.

Barr named Sherwin as acting US attorney for the District of Columbia on May 19, 2020.[27] The move came after a rough fifteen-week run by Timothy Shea, a top Barr aide who—on his first day as interim US attorney—told federal prosecutors he wanted a more lenient sentencing recommendation for (here's that name again) Roger Stone.[28] Months earlier, a jury had convicted Stone on seven charges after he lied to congressional investigators looking into Russian interference in the 2016 campaign.[29] Justice Department whistleblowers soon testified before the House, saying that Barr had improperly politicized the DOJ.[30] Stone was ultimately sentenced to more than three years in federal prison, but he dodged serving his time after Trump granted him clemency on July 10, 2020.[31]

Upon Sherwin's appointment, a former Barr aide who'd served short stints as acting attorney general in the 1990s said Barr's actions amounted to a "political coup," with the president improperly intervening to get desired outcomes in cases against his friends and allies.

"This represents a politicization of the U.S. attorney's office of the District of Columbia that is remarkable and unique and unprecedented," Stuart M. Gerson said.

Career prosecutors held on to some hope given Sherwin's background as a career prosecutor himself, and the office regained some stability, so much as that was possible in 2020, in the year of COVID-19 and Trump's reelection campaign.

———•◆•———

The morning of Jan. 6, Sherwin hit the streets. He joined up with the Metropolitan Police Department and headed to some of the hot spots, the inflection points. They went to Black Lives Matter Plaza, an area on Sixteenth Street just north of the White House where Mayor Muriel Bowser had had DC employees work overnight in June to paint "Black Lives Matter" in thirty-five-foot-tall yellow letters that stretched two blocks after federal law enforcement agencies took over the area in the wake of George Floyd's death. "There was a dispute this week about whose street it is, and Mayor Bowser wanted to make it abundantly clear," her chief of staff said.[32]

Sherwin was surprised to find BLM Plaza basically deserted. "I was like, wow, this is eerily quiet. This is odd. Where is the— where's the protester-counterprotester violence?" he said. "That still sticks out in my mind, how BLM Plaza was empty."

Law enforcement had been gathering crowd estimates based on hotel reservations and flights, but their numbers were starting to look off as people flooded in on buses. Still, Sherwin said he didn't see a threat from the crowd, which he called carnival-like. "It was kind of surreal. There was like people selling T-shirts and popcorn. Like, literally, I think I saw a cotton candy vendor. It was ridiculous. And a lot of hooting and hollering, but no— there was no violence per se. There was no like—there were no assaults. I didn't see anyone throwing stuff at cops."

The police officers Sherwin accompanied were getting high fives from the crowd, in fact. The crowd was boisterous, but they were friendly to the cops, at least.

There were soon signs of trouble. Near the site of the president's speech, which was only about halfway over, men were in handcuffs. They'd been holding what looked like weapons. The bike cops rolled in. The crowd grew angry, asking cops how many Black Lives Matter activists they'd arrested, saying that arresting "patriots" looked bad.

"Oath breakers! Oath breakers!" yelled a man in a helmet as police lined up and created a barrier. "Those are symbols, not weapons!" one said. The crowd grew. Later, a man said they'd only brought their "uppers," meaning the guns weren't capable of shooting. Another said they were pellet guns.

Sherwin soon arrived on the scene and was talking with cops.

"I was like, these guys are f'ing idiots because they're walking around with an upper, which is not a rifle. An upper is just a barrel and a bolt. It's not a lower receiver. If you just have an upper on a sling, it looks like a rifle . . . you look menacing, but it's not," Sherwin said.

He told the cops to confiscate the uppers but said they couldn't arrest the men because it didn't meet the definition of a firearm. "They're going to get shot or they're going to create chaos, because it looked like they were walking around with automatic weapons, and it was like an AR-15 upper," Sherwin said. "So I was like, take it from these guys, just confiscate them. And that was the first thing that got my attention, that some of these fools were walking around."

A woman named Samantha who had traveled with a group of fellow Trump supporters on a bus screamed at the police, upset her fellow bus riders had been detained. She had a pink Trump hat on her head, a Trump flag tied to her back over a black coat with a puffy hood, and pearls on her ears. She was pissed.

"I'm a woman, and I still got more balls!" Samantha said in a vaguely Philly accent. "I took an oath and I take that oath seriously. It has no end, no expiration date."

Samantha was a military veteran and the recent widow of a plumber, and days earlier she had brought her three children to the Palm Beach County Convention Center, not far from where her mother lived, to try to catch a glimpse of Vice President Mike Pence. The VP was in town to speak to the young conservative group Turning Points USA, the group that employed a "veteran right-wing meme maker" who called Donald Trump Jr. "a meme general in the meme wars" who was "commanding the D-Day invasion."[33]

"As our election contest continues, I'll make you a promise," Pence said in that Dec. 22 speech. "We're going to keep fighting until every legal vote is counted. We're going to keep fighting until every illegal vote is thrown out."[34]

Word for word, that was the same thing Pence had said more than a month earlier, at a rally in Georgia. "We're going to keep fighting until every legal vote is counted," he promised those voters, before the Supreme Court rejected the Trump campaign's attempt to disenfranchise millions and throw the country into chaos. "We're going to keep fighting until every illegal vote is thrown out."[35]

By Pence's own account, he told Jared Kushner on the day of the Four Seasons Total Landscaping press conference that he "wasn't convinced" that voter fraud had cost the Trump-Pence ticket the election. But he appeased Trump and the base anyway, feeding a narrative that there was still hope they could overturn an election they had lost.

If you parse the scripted line Pence repeated in November and December, it was easy to see what he was doing. He wasn't saying directly that the election results were fake or that Trump had actually won; he was just saying they would keep fighting. On Dec. 19, when Pence first heard about the idea of a rally on Jan. 6 from Trump, he said he thought it "might be useful as a way to call even more attention to the proceeding on the floor of the

House and Senate." Pence told a senator that the Trump campaign hadn't "gotten our day in court" but that they would "get
our day in Congress." He fed into the idea that something could
still change or that they could at least injure the incoming
administration.[36]

By Jan. 6 Trump supporters were still holding on to the hope
that Trump—and Pence—had given them. Now, Trump told
millions, it was all up to Pence to save the country.

> *States want to correct their votes, which they now know were*
> *based on irregularities and fraud, plus corrupt process never*
> *received legislative approval. All Mike Pence has to do is*
> *send them back to the States, AND WE WIN. Do it Mike,*
> *this is a time for extreme courage!*—@realDonaldTrump,
> 8:17 a.m.[37]
>
> *THE REPUBLICAN PARTY AND, MORE IMPOR-*
> *TANTLY, OUR COUNTRY, NEEDS THE PRESI-*
> *DENCY MORE THAN EVER BEFORE—THE*
> *POWER OF THE VETO. STAY STRONG!*—@real
> DonaldTrump, 8:22 a.m.[38]
>
> *The States want to redo their votes. They found out they*
> *voted on a FRAUD. Legislatures never approved. Let them*
> *do it. BE STRONG!*—@realDonaldTrump, 9:15 a.m.[39]

In the final days, Trump kept up the pressure on Pence, telling
the vice president that he was trending on Twitter, and at one
point he told him he shouldn't participate in the certification process at all. "If you want to be popular," Trump said, "don't do it."
Trump told Pence that he was "too honest" and that "hundreds
of thousands" would hate his guts, that people "are gonna think
you're stupid."

In a final phone call that morning, Trump put the pressure
on again.

"You'll go down as a wimp," Trump said. "If you do that, I made a big mistake five years ago!"[40]

———————•◆•———————

Back in 1992, Rudy Giuliani had helped rile up[41] a mob of angry, drunk cops against New York City's first Black mayor over his call for a civilian complaint board and a commission to investigate NYPD misconduct.[42] Thousands of demonstrators blocked traffic, assaulted bystanders, destroyed property, and tried to storm City Hall, and officers inside had to slam the doors shut and put a bar across them. Many in the mob were armed, and they used racial slurs with ease. Even an internal strategy memo for Giuliani's 1993 campaign said he should "acknowledge and criticize the underlying racial nature of the protest."[43] He never did.

Now, instead of a truck bed outside City Hall, Giuliani was on a platform near the White House, using the same tactics.

"If we're wrong, we will be made fools of," Giuliani told the crowd at the Ellipse on Jan. 6. "But if we're right, a lot of them will go to jail." The crowd roared, and Giuliani banged his hand on the "SAVE AMERICA" lectern. "Let's have trial by combat!" Giuliani said. "I'm willing to stake my reputation, the president is willing to stake his reputation, on the fact that we're going to find criminality there."

After his warm-up acts, it was Trump's turn. His speech at the Ellipse was a master class in political demagoguery. "You're stronger, you're smarter, you've got more going than anybody," he told the crowd. "You're the real people, you're the people that built this nation."

That made Dustin Thompson feel "good." The then thirty-six-year-old had been laid off from his job at the beginning of the pandemic and then "went down the rabbit hole on the internet," he later said. His wife had packed him some food for the trip and was at home "just enjoying the house being quiet" as Thompson and a friend went to DC. Soon, he'd storm into the Capitol, steal a bottle of liquor and a coat rack, and then try to convince jurors he was following what he felt were presidential orders. "I can't let other people tell me what to do," he'd later say. "Even if they're the president."[44]

Onstage, Trump was once again deploying a logical fallacy known as *argumentum ad populum*, or an appeal to the people. It's also known as the "everybody knows" fallacy. When you don't have a factual basis to make a claim, you just point out that everybody believes it.

"Democrats attempted the most brazen and outrageous election theft and there's never been anything like this," he said. "Everybody knows it."

In the same way that he'd endorsed police brutality in a speech before cops during his presidency—suggesting that they should hit arrestees' heads on police car doors by telling them "don't be too nice"[45]—Trump implied that the mob wouldn't be so friendly to his political enemies, saying "we're probably not going to be cheering so much for some of them."[46]

But he got more explicit too, telling the rioters that everything was on the line.

"We fight like hell," Trump said. "And if you don't fight like hell, you're not going to have a country anymore."

———◆———

As Trump spoke, Samantha was busy yelling at cops, and Mike Pence was moments away from sending a tweet that would result in a mob storming the building searching for him, seeking to lynch him.

"My oath to support and defend the Constitution constrains me from claiming unilateral authority to determine which electoral votes should be counted and which should not," Pence wrote in a letter posted at 1:02 p.m.

Samantha wouldn't learn the news about Pence yet, but she was quite upset, yelling at cops that it might be time for Trump supporters to switch tactics.

"We're not terrorizing anybody. We don't break laws, we abide by them!" Samantha yelled. "But maybe it's time that has to change. Maybe it's time that we have to start breaking some laws to be heard."

Soon, Samantha did break some laws. She headed over to the Capitol with an angry mob, many of whom articulated their plans quite clearly.

Wearing a camo bucket hat, Ryan Nichols recorded himself on his phone as he and the mob walked east toward the Capitol.

"I'm hearing reports that Pence caved. I'm telling you, if Pence caved, we're gonna drag motherfuckers through the streets," Nichols said, angrily pointing his finger at the screen. "You fucking politicians are gonna get drug through the streets, because we're not gonna have our fucking shit stolen.[47]

"This is the second fucking revolution, and we're fucking done. I'm telling you right now, Ryan Nichols said it, if you voted for fucking treason, we're going to drag your fucking ass through the street," Nichols said. "Let the patriots find out that you fucking treasoned this country, we're going to drag your fucking ass through the street."[48]

Up ahead, at the Capitol, the Proud Boys had prepared to rush to the barricades. Leading the pack from Peace Circle up to the violence was Ryan Samsel, who had a history of beating women to the point they lost consciousness, leaving them with chipped and missing teeth. Once, he'd allegedly smashed a hot pizza into a woman's face before beating her, pouring a beer over her head, and throwing her into a canal, where he held her head underwater. "I was afraid he was going to kill me," his victim said.[49]

On Jan. 6, Samsel walked with his girlfriend, Raechel Genco, toward the Capitol hand in hand, with Samsel holding an enormous flag that depicted Trump as Rambo.[50] He spoke with Joe Biggs, the Proud Boys leader, and then approached the police line. As a small number of greatly outnumbered officers stood behind barricades, Samsel began rocking the fence. Soon the whole mob joined in and began brawling with officers on the front lines.

Near the middle of the front of the Capitol lawn, Bryna Makowka ascended a stone wall built in the 1800s under a plan led by Central Park landscape architect Fredrick Law Olmsted, and grabbed a bullhorn. The blond California woman was set to turn forty the next day, on Jan. 7, and she was fresh off a breakup with her ex-boyfriend Ed Badalian, who was in his midtwenties. The pair met at a Beverly Hills pro-Trump rally, and Makowka heard that Badalian had a crush on her. *Oh, he's really cute*, Makowka thought. They had dinner as part of a group; she had a bungalow in the Hollywood Hills where MAGA folks would meet up, and things just clicked.

At some point in December, after their breakup, Badalian kicked Makowka out of the Patriots 45 MAGA Gang chat they belonged to, where LA-area Trump supporters would plan, commiserate, and share fake news about COVID-19 and pedophiles.

After Makowka left the group, and after Trump promoted the Jan. 6 events on Dec. 19, the MAGA Gang started making their plans for DC. They collected weapons, including a Taser, pepper spray, a baseball bat, gas masks, and walkie-talkies. They rented a vehicle to drive across the country. In the MAGA Gang chat, other members were reacting to the Pence news. "The establishment has won," someone wrote. A couple of members of the group, Badalian and Rodriguez, were oddly silent.

———◆———

"Priority!" the police radio cracked at 12:53 p.m. "We just had protesters at Peace Circle breach the line. We need backup!"

"Patriots! Patriots! Patriots! Go!" Makowka yelled, standing in front of an "Area Closed" sign and urging rioters to jump the wall, rip down the green snow fencing, and run toward the Capitol. "What are you fucking waiting for?"

Rioters flooded over the retaining fence, joining up with others as Samsel broke through the police line. Makowka didn't join them.

"The minute that I did that, I turned around and was facing the Capitol, facing the lawn, and I swear to you, it was like I got figuratively slapped across my face," Makowka later said. "I got this slap in the face, like, *Oh shit, we shouldn't go up into the Capitol. What are we doing?*"

Someone in the crowd was confused, wondering why Makowka had urged the mob forward but hadn't gone herself.

"I felt this guilt all of the sudden, like *Oh, fuck, what did I just do?*" she said. She ended up staying farther back on the lawn as thousands of others flooded toward the inauguration stage.

Samantha, the mom in the puffy black coat, did rush forward, and she was facing down officers near the inauguration platform at the front of the line and yelling about them betraying their country. Joe Biggs, a member of the Proud Boys, was at the front of the line too, watching and cheering.

"This is the craziest fucking day ever in American history," Biggs said. "The people are taking the fucking Capitol back. This is insane. Man, this is awesome, man. Like to be here right now, I have a fucking boner."

Rioters broke down one line and then another. Samantha put on a gas mask as she stormed the inauguration platform with the mob. She was among the first to break through a police line as her fellow Trump supporters assaulted cops.

"I hope you all fucking die!" one fellow supporter yelled at cops as she and others broke the line. "You better watch your head, fucking bitch! . . . You better run, cops!"

"God's gonna judge everyone of y'all; everyone of you is going to go to eternal damnation," yelled a man who recorded himself chasing down officers in a video he posted to YouTube. "Do you rape little babies too, you sick bastard? . . . I rebuke you in the name of Jesus! I rebuke you Satan!"

On her way up to the top of the inauguration platform, Samantha filmed a TikTok video celebrating her work. A bearded man with long hair popped up in the background. He seemed to be everywhere that day, leading the charge toward the stage and then inside the West Terrace tunnel. The sleuths—the online community that would spring up and identify hundreds of Capitol rioters—would call him #SwedishScarf.

———◆◆———

Sherwin was walking east back to the US Attorney's Office when he saw that the crowds were getting rowdy. "I start seeing people climb up on the scaffolding that was set up for the inauguration. Then I knew, it was like, this isn't going to—nothing good's going to come from here. Because all those dudes need to be arrested for trespass, because you know you can't climb up the scaffolding," he said. "At that point I wasn't thinking of, like, breach of the Capitol. I was like, these guys are going to kill each other because the thing's going to fall and kill a hundred people."

Over at Justice Department headquarters, the acting no. 2 was trying to get a sense of what the hell was happening. It wasn't supposed to go like this. There'd even been an "outbriefing" session for political appointees set for 10:00 a.m. to help Trump appointees plan their impending departure. Acting Deputy Attorney General Donoghue had learned from a television in Acting Attorney General Rosen's office that the Capitol had been breached. Rosen was trying to get a hold of the FBI, and Donoghue decided he'd just walk over himself. He and his security detail crossed Pennsylvania Avenue and walked into the Hoover Building, and nobody there knew much.

"They didn't have a lot of information," Donoghue said. "They had the screens showing people marching through the rotunda as well, but they didn't have a lot of information as to exactly what was going on at the Capitol."

Bowdich, the deputy at the FBI, had headed over to the FBI's Washington Field Office, so Donoghue decided to head over too. WFO had the televisions on too, and people were working the phones, shouting information.

At one point there was a discussion of doing a "virtual command post," Bowdich said, given that the inauguration was coming up and "COVID is ripping through the country." The new assistant director in charge of the Washington Field Office, Steven M. D'Antuono, who had been on the job for just a few months,[51] was worried everyone would get sick if they were "coalescing in one room, which is a valid concern," said Bowdich. "Although I understood the reason why he was thinking that way, I told him no, I want a full command post inside WFO," Bowdich said.[52]

Donoghue arrived at WFO and found Bowdich in the back, in a conference room.

"Hey, I'm going to go over to the Capitol," Bowdich told Donoghue.

"I'll go with you," Donoghue replied, thinking they'd be "better off" if they could avoid information lags and see "what was really going on." He stayed until the Senate gaveled back into session that night.[53]

Bowdich started getting emails and texts, including from Sen. Mark Warner (D-VA). "This is a mess," Warner told him. They had the vast majority of the Senate in one room.

FBI special agent Michael Palian, who came to the bureau after graduating with a PhD in bioorganic chemistry, was assigned to the health-care fraud squad on Jan. 6. He was working from home that day when he got an email asking him to come in and assist with the Capitol. He was dispatched to a Senate office building where the senators were being held and stayed with them for about three hours. It was chaotic, and he saw senators crying.

Hours later, about seventy FBI special agents eventually escorted the lawmakers back to the Senate through the underground tunnel, finding the area outside the chamber in disarray. "It looked like a bomb had gone off in there. There was pepper spray, teargas everywhere, on the floor and on the walls. Walls—windows broken, doors broken, lots of debris in the hallways," Palian said. "If you took your mask off, you started to inhale the pepper spray that had been sprayed all through the day. So we all had to keep our COVID masks on to make sure that we didn't start coughing and hacking ourselves."[54]

For now, the senators were still inside their holding location in a Senate office building, and pure chaos was unfolding on the west side of the Capitol. "West Terrace entrance," the radio crackled at 2:42 p.m. "They've broken a window."

"We need everybody. We need everyone," a commander yelled. "Let's go. They're not getting into this fucking building."

Officers headed into the gauntlet, the West Terrace tunnel, where rioters had been viciously battling cops. Unbeknownst to this group of MPD officers, the Capitol had already been breached, and a rioter would soon be shot and killed trying to jump through a broken window and into the House Speaker's Lobby. But right now, they thought this was the last line of defense.

"We are not losing the US Capitol today!"

"Never been defeated. Not gonna happen today!"

"They want some, give it to 'em! Gotta hold this door."

Commander Robert Glover was the head of the Metropolitan Police Department's Special Operations Division, which comprises many different sections and handles everything from Click It or Ticket programs to SWAT responses, waterway patrols, and civil disturbance responses.

It had been a long few months for Glover. The night that Trump held up a Bible during a photo op after federal law enforcement cleared the streets near the White House, Glover was overseeing the MPD response. A lawsuit alleged that it was Glover who'd ordered cops to kettle protesters on a residential street, trapping them for hours.[55] Some residents of the street let demonstrators into their homes to help them avoid arrest. I spoke with one woman as she huddled up in the home of an acquaintance she hadn't even realized lived on the street; she described getting pinned in by cops. "You turn to the left, there's another alley, cops are there. Then you turn to the right, there's another alley, cops are there," she said.[56] Some of the protesters danced to Michael Jackson. Some cops chased protesters inside homes, and tear gas flooded into at least one living room.

Months later, in September 2020, Glover was standing outside of DC mayor Muriel Bowser's home, where a small number of protesters had gathered.

"Jason, go home," Glover said to one man. "Glover, fuck off," the man, who was covered in black, responded.[57] Glover's intelligence here was good. He was, in fact, talking to Jason Charter, an antifascist activist who had been arrested in July for attempting to tear down a statue of former president Andrew Jackson in Lafayette Square, the park near the White House.[58] That effort was unsuccessful, but antifascists had better luck with a statue of Confederate general Albert Pike,[59] which DC's local government had been trying to get removed since 1992.[60] Charter was seen pouring a flammable liquid on the statue, setting it on fire, and then lighting a cigarette with the flame.

Glover had dealt with the Proud Boys in November and December too, when there was a stabbing after a Proud Boys member initiated a confrontation with a counterprotester, purposefully running into him and then punching him.

"They'd go back to the hotels and were drinking heavily. They would come out in front. They would challenge law enforcement. Antifa was looking for them," Glover said. It took a lot of resources.[61]

The law enforcement posture with groups like the Proud Boys looked much different than it did with loosely affiliated left-wing groups. The Proud Boys "always want to make it look like they're law enforcement's friends," while antifa says "Fuck the police," Glover said. The Right was "law and order," while the Left would "be that antigovernment side."

In the lead-up to Jan. 6, Glover was on email chains with organizers of the Trump rally as they coordinated with law enforcement.[62] One of the organizers, Cindy Chafian of Women for America, was hesitant to provide lists of speakers at the events in November and December, and Glover thought she was hiding something.

"I think she knew full well what was going on. But I think she was hiding a lot of the information of who was really funding things and who was really organizing things," he said. By Jan. 6, other organizers had pushed Chafian aside.[63]

When Jan. 6 arrived, MPD appeared to be more prepared than anyone. They took their investigative units "down to the bare bones" and put as many people as they could in uniform. They activated the MPD processing center at the police academy, so they had the ability to process mass arrests. They used their recruits, the ones not yet out of the academy, for prisoner control. Their cadet corps—a program for seventeen- to twenty-four-year-old DC residents, who earn a small salary and attend DC community college as they work at MPD part-time[64]—took over administrative functions in some MPD stations to "free up all our sworn bodies" and get them out on the streets.

Still, Glover found himself wondering if it had been enough after they got the call about the uppers and the crowd reacted with hostility. The dynamics "were much different" than what happened in November and December. "It was definitely not something that we were expecting," he said.

Soon, he got a call about the pipe bombs at the RNC and DNC buildings. Then he got another call, saying the Capitol was in trouble. They organized near the Botanical Gardens on the Capitol grounds and walked into a full-scale riot.

Bear spray. Gas masks. Pieces of bike racks. Flag poles. Fire extinguishers. "They were using pretty much anything they could as a weapon at that point," Glover said.

Glover walked through blood on the inauguration platform and later found out that an officer lost a piece of his hand when it got caught in a bike rack. Officers had orange dye on their faces from being bear sprayed. They had cuts from being hit. They were exhausted. And they were getting outflanked.

"It was the first time in my career I didn't think we were going to go home," Glover said, knowing that officers had dealt with individuals with guns earlier in the day. "I was concerned that this was actually going to become a gun battle."

Glover took to the radio. "We lost the line! We've lost the line! All MPD fall back!" Glover said. "10-33. I repeat, 10-33, West Front of the Capitol. We've been flanked and we lost the line."[65]

To police officers, 10-33 means this: Stop whatever the fuck it is you're doing and go. "No matter where you're at, if you're hearing that, you're going to help," Glover said. It's an emergency declaration, a signal that a cop's life, or cops' lives, are in danger. Glover, in the course of nearly thirty years in law enforcement, had only used it four or five times.

"Never in my life have I heard that, a police commander telling us to retreat," recalled MPD officer Sara Beaver, who'd soon find herself stuck in the "tunnel of doom," unable to breathe.[66]

MPD sergeant Jason Mastony went into the western tunnel, not knowing where it led, other than into the Capitol building. Shit hit the fan, and a "ramshackle" group of officers assembled to make their final stand. "Nobody knows each other's names. There's not a lot of commands," Mastony said.

There is a set of two gold doors in the tunnel. Police had managed to close the first one, and now a mob on the other side of the door was bashing it open.

"He's got a weapon in his hand," an officer said. The glass shattered. "Shields up front!"

They closed the second set of doors, but they soon swung open, as if by ghosts, as men stood behind them to let the mob in. A man in a gas mask and a protective vest peeked his head out from the door on the right and then came at the cops. It was hard to make out what he said, his yelling muffled by his gas mask, but he pointed behind them and advanced.

"Get the fuck out of here!" a cop yelled. The battle was on. The man in the gas mask—identified by sleuths in 2021, but not yet arrested as of this writing—grabbed an officer's shield and then pushed against it. Other members of the mob swarmed into the tunnel.

"Hold the line!" a cop shouted. "We don't wanna hurt you, man," a rioter in a sweatshirt insisted. "1776!" someone yelled from the back. "USA! USA! USA!" the mob cheered.

———— ◆ ————

Reps. Stephanie Murphy and Kathleen Rice were about forty paces away from the tunnel, listening to the mob. The duo of Democratic moderates had made their way to the hideaway office of Sen. Kyrsten Sinema, one of Murphy's closest friends in Congress, who had called Murphy that morning offering up her pink-walled hideaway office in the Capitol basement—decorated with pictures from and books about Arizona and mini cacti—to use on what Sinema thought "could be a dangerous day."[67]

Murphy and Rice were roommates, having signed a lease together about three days after they met, after Murphy won her swing district in 2016. They held a housewarming party, burned food in the oven, and then jokingly dubbed their rental the

"Smokehouse" after filling the place with smoke just as their guests arrived. Until COVID-19 hit, they hosted rare bipartisan events with their colleagues, playing Jenga and card games with a group that included McCarthy, whose security team would sweep the "Smokehouse" before he entered.[68] McCarthy even joined them for a 7:00 a.m. workout class at a Solidcore location in DC's Navy Yard.[69]

Both were concerned about the events of Jan. 6, bringing their work clothes into the office the night before and arriving at the Capitol that morning "incognito," in their workout clothes, no member pins.[70] Neither of the members were necessarily household names nationally, but there were plenty of Trump supporters from both Murphy's home state of Florida and Rice's home state of New York who might recognize them. Better to play it safe.

In a town where government workers, contractors, politicians, and journalists alike sometimes boost their status with a bit of unnecessary mystery, Capitol hideaways are an alluring tale. For decades, the press has written fun features about the existence of the offices, titillating readers about being let in on the secret lives of the country's most powerful people.[71] Sen. Barbara Mikulski's hideaway was once described as "amid staircases that wind and corridors that disappear in a space reminiscent of Hogwarts."[72] Sen. Patrick Leahy thought his hideaway was occupied by ghosts.[73] Decades earlier, Robert Parker, the son of a sharecropper who rose to become the maître d' of the Senate Dining Room the year before the passage of the Voting Rights Act, recalled in his memoir how Lyndon Johnson had acquired a total of seven hideaways spread out over two floors that the Texas senator dubbed "Johnson Ranch East." Parker said three or four white women and at least three Black women used to visit Johnson's "love nest" regularly, and Johnson would invite female employees into his hideaways to "take dictation."[74]

By 2020—with booze legal and sexually harassing employees and parading sexual conquests through the halls of Congress slightly less in vogue—hideaways were a bit more mundane. Senators would sometimes give reporters video tours of the spaces, showing off the little items that reminded them of home, the private spaces where they had a bit of time to relax.

In the QAnon and Pizzagate era, when delusional Americans believed that a pizza shop had a basement full of sex-trafficked children groomed for elite pedophiles, and with election deniers sharing maps of the tunnels that connect the Capitol to the surrounding office buildings, the playful coverage of these "undisclosed locations" suddenly seemed a lot less cute.

Murphy and Rice had barricaded the door with a chair and kept the television on with no sound. They didn't want to make any noise. Instead they listened to the sounds of rioters yelling and of officers struggling, coughing from tear gas and wincing in pain as it stung their eyes and officers made their way to a nearby bathroom.

"We had taken refuge in that office because we thought for sure being in the basement at the heart of the Capitol was the safest place we could be," Murphy said. "And it turned out we ended up at the center of the storm."

"For about an hour, it sounded like there was a herd of elephants smashing through windows and making all this noise," Rice said. "We spent the next four or five hours in that room, just waiting for them to get control of that Capitol."[75]

"I couldn't believe that here I was, a refugee immigrant who had fled authoritarianism, now trapped in the basement of the Capitol, the place where I thought I would be safest," said Murphy. "The heart of our democracy."[76]

———————— ◆ ————————

Standing at the door to the House floor was Damon Michael Berkley, who had some experience in storming capitols. Back in

November, he'd been at the Georgia State Capitol for a Stop the Steal event featuring Sept. 11 and Sandy Hook truther Alex Jones, white nationalist Nick Fuentes, and far-right provocateur Ali Alexander. Enrique Tarrio showed up.

"Who is going to be ready to storm the Capitol with us in a couple of minutes?" Alexander asked that day. "Peacefully!" he added with a smile. He said they wanted to call a special session.[77] Alexander said they were going to leave their signs outside. Vernon Jones led the way. "No knives, no guns," said one person. "Nobody armed, no arms, it slows everything up."[78] They appeared to go through security. Inside, Jones—a former Georgia Democrat who endorsed Trump in 2020 and would announce his plan to change his party affiliation to Republican during his speech at the Ellipse on Jan. 6—sung an off-tune rendition of "Georgia on My Mind," soon admitting that he couldn't sing.

A few weeks after chatting with Jones in Georgia, Berkley was ready for violence.

"It's not a slap in the face; it's a DARE for hardened patriots to Charlie Hebdo these traitors and end their audacious tyranny," Berkley wrote on social media in December, referring to the terrorist attack that killed twelve at the offices of a French satirical newspaper in 2015.

"Coups demand revolutions. Congress should be pulled into the DC streets Khaddafi style!" he wrote in late December, referring to the death of Colonel Muammar Gaddafi, the powerful Libyan dictator who was found hiding in a sewer and then was placed on the hood of a truck, bloodied, before dying at the hands of a mob.[79]

Now Berkley was at the front of a mob that had shattered the window. Inside the chamber, someone told lawmakers that the backs of their chairs were bulletproof. They were instructed on how to use their escape hoods, emergency breathing devices that made a whirling noise when opened.[80]

A rioter in ski googles was trying to win over an officer outside the door, saying he should join their cause. "If you let us in there, you deserve a medal of honor," he said.[81]

"I don't wanna get shot man, really," Berkley said. "We, we want, we just want our grievances redressed like the Constitution."

Eventually police forced him out, and he addressed a cameraman on the stairs of the Capitol. "Vice President Pence, my name's Damon Michael Berkley and I do not appreciate one bit this situation *you've* caused here, sir," he said. "All this violation and everything was 100 percent unnecessary, okay?"

The blood of Ashli Babbitt, the Jan. 6 rioter who'd been shot after she entered a broken window leading to the House Speaker's Lobby, was on Pence's hands, he said.

"We're not putting up with this tyrannical rule. If we gotta come back here and start a revolution and take all of these traitors out, which is what should be done, then we will."

Over on the Senate side, there was mass confusion as rioters took over the building. They didn't know where to begin, what the plan was to clear the building. "Dude, I don't fucking know," said one Capitol Police officer. "You want to talk about getting caught with your pants down?"

"Wow, if only we had some warning that this was gonna happen!" said another officer, sarcastically.[82]

———— ✦ ————

The battle had raged for nearly two hours on Jan. 6 when rioters dragged the body of Rosanne Boyland, a Trump supporter who'd been crushed by the mob, to the front of the West Terrace doors. The Trump supporter had gotten sucked into QAnon conspiracies in the summer of 2020 and had traveled to DC with Justin Winchell. "My family thinks I'm crazy, but I'm heading up from ATL to be shoulder to shoulder with my true brothers and sisters," Boyland wrote on Parler, before attaching a QAnon hashtag.

"My president has asked me to come support him. And I've done so many stupid things in my life that I'm going to do something that I really believe in," she told her mother, Cheryl.

Her family saw the chaos unfolding at the Capitol on the news. "You all good?" her sister Lonna texted her. No response.[83]

Inside the Capitol, officers and medical personnel were working, desperately, to save Boyland's life. As officers choked back gas, someone had lifted Boyland's shirt up, and they were pumping her chest again and again. As other officers were battling it out in the tunnel, another team worked for at least ten minutes straight to revive Boyland.

Officers lined up, taking turns pumping her chest, trying to get her breathing again. "No pulse," someone said. They counted as they pushed on her chest.

*1, 2, 3, 4, 5, 6, 7, 8, 9, 10, 11, 12, 13, 14, 15, 16, 17, 18, 19, 20, 21, 22, 23, 24, 25, 26, 27, 28, 29, 30.*

Finally, they lifted Boyland's body onto a stretcher, strapped her body in, and wheeled her body out.

"Protester?" asked an officer, coming upon the scene. "Got trampled?"

Seconds after they wheeled Boyland's body out, there was a commotion down the hallway, a new call for backup.

The rioters had broken a window and were inside the building, inside a suite of senators' hideaway offices, right down the hallway from where two members of Congress, Murphy and Rice, had been holed up in Sinema's office.

A bit over a year earlier, the year Sinema joined the Senate, Murkowski welcomed Sinema, Murphy, and Rep. Susan Brooks of Indiana to her own hideaway office in the Senate basement. The hideaway had not only windows but a nice view too. You just had to know where to find it. Sinema joked she had "no idea where we are" as they headed to the squirreled-away location. They walked through one door, went down a set of stairs, and went to her office.

"It's a rabbit warren, but it's our rabbit warren," Murkowski said, describing the suite of connected offices, before opening the doors to her office, which was flooded with sunshine and looked out at the National Mall and Washington Monument.[84]

Scaffolding for the inauguration was blocking Murkowski's view on Jan. 6 when she stopped by her hideaway before heading back to the Senate floor.[85] Suddenly she heard someone inside the office suite, coming down the stairs, and then the "gut-wrenching" sounds of heaving. An officer was washing his face in the sink, his eyes red and swollen shut. She asked if she could help. "No, I've got to get out there," he replied. "They need my help."[86]

Now, with Murkowski with the other senators in a secure location in a Senate office building, the battle over the tunnel raged on, and rioters were inside her office suite. Officers scrambled to the other side of the door, stacking everything they could—a table, a moving cart, some sort of plaque, one of those black food carts. MPD cops, already unclear where they were, got the rundown on what was inside. There were offices in there, suites, hideaways, a Capitol Police officer told them. They lined up, waiting for the rioters to bust through. An MPD officer went off, searching for a drill to make sure that the door stayed closed. They couldn't get outflanked again.

———————— ◆ ————————

"You are not going to take away our Trumpy bear!" yelled Gina Bisignano, a Beverly Hills cosmetologist wearing a Louis Vuitton sweater and holding a bullhorn outside the broken window, as rioters flooded inside. "You are not going to take away our votes!"

"George Soros, you can go to hell!" she said.[87]

Through the shattered window, inside room ST2M of the US Capitol, rioters had overturned a table and barricaded a door with a blue couch. A man with a pink respirator and an American-themed hat held a broken table leg.

"I need to know right now who's following me," said the man later known as #SwedishScarf, who had taken on a leadership role. "Who's with me?"

"We need more bodies in here!" said another man, luring people in through the window.

#SwedishScarf wanted music, suggesting "Fortunate Son" by Creedence Clearwater Revival. Trump, who inherited the equivalent of at least $413 million from his father,[88] had unironically placed the song in regular rotation on his campaign playlist, but #SwedishScarf had a suggested adaptation. "It ain't me, it ain't me, I ain't Joe Biden's son," he sang.

"We need a cameraman! We need the press! Where's the press?" #SwedishScarf soon asked, after climbing out the window and summoning over a videographer. "This is the story of your life right now, brother. This is the story of the century."

Inside, the videographer found a remarkable scene. A table was overturned, and a blue couch had been barricaded against a door.

In front of an overturned table stood two friends from Newport Beach, California, Melanie Belger and Michelle Estey. Belger, brown hair, sunglasses on her head, black puffer jacket on, held a bottle of Captain Morgan's. Estey, in a pom-pom Trump winter hat and with a Trump flag draped on her back, was on her phone, having set her gloves aside on an overturned piece of office furniture.

Belger had kept her Facebook friends updated on their trip. "I am at the Capitol doors been [pepper] sprayed. Round 2 coming," wrote Belger, who'd worked at an investment firm. "Back at hotel! Was in the Capitol im safe. Pence=NOT SAFE," Belger wrote.[89] Months later someone would tip off the FBI when Estey revealed she'd entered the Capitol during a dinner with friends, and the FBI would go back through their massive database of tips, finding Belger's Facebook posts.

#SwedishScarf made preparations for the push. Danny Rodriguez, the MAGA fanatic from California who went to DC predicting a bloody revolution, stood up on a chair nearby to address the mob. "They have shot and killed a girl," he said. Some of the rioters left, but many didn't seem fazed.

They went through one door. To the right, through the door to the suite, officers were waiting.

A woman in a pink hat and sunglasses poked through the broken window. Striking the tone of chaperone leading a group of eighth graders on a tour of the Capitol, she laid down the law.

"People should probably coordinate together if you're going to take this building," she said. "We're in, we've got another window to break to make in-and-out easy, and this window here in the other room needs to be broken."

People started handing things out the window—a tabletop, furniture legs—that were used as weapons against officers in the tunnel.[90]

"Going in the Capitol," said Imelda Acosta, who went by Mariposa Castro online. Back in 2006, Donald Trump almost hit her with a golf ball at a course at Pebble Beach, and they ended up having a conversation.[91] Now she was storming through a broken window on his behalf. "We're in!" she said, before someone asked her to take a photo for them. Later, as she left the grounds of the Capitol, she said this was the beginning of war. "It's just the beginning," Castro said. "As Trump says, 'The best is yet to come.'"

Bryan Betancur, a self-professed white supremacist who was wearing a Proud Boys shirt and a GPS monitoring device because of a 2019 burglary conviction, was inside too, and he had someone snap his photo holding a Trump flag. Earlier, he'd posed with a Confederate flag.[92] Later, he told the FBI that his only regret about Jan. 6 was that it would likely prevent him from joining the US military. He then flashed the "okay" hand gesture to the agent who interviewed him.

The rioters inside the conference room grabbed a table and moved into the hallway to protect themselves in case the cops on the other side busted through the door.[93] They broke the door off its hinges, and someone handed it out the window. Then they started trying to break into a locked room ahead.

#SwedishScarf and a man wearing a football helmet didn't have much success kicking in the next door. Then Nicholas James Brockhoff, a twenty-year-old from Covington, Kentucky, who traveled more than eight hours and five hundred miles to be in DC on Jan. 6, had an idea.[94] The high school graduate had one semester of college under his belt when he took off for DC, and he was living at his childhood home with his brother and father.[95] He busted out the bottom section of the door and reached through and up, unlocking it from the inside. He was disappointed with what he found.

"Aw, it's just a fucking room!" he said. It was a dead end, another office that rioters soon began ransacking.

They busted into the door to the left too. That was assigned to Sen. James Risch, a Republican from Idaho. He was lucky to get a hideaway office in late 2009, when he was a freshman. Until that year, there weren't enough hideaways for all one hundred senators, but the opening of the Capitol Visitor Center on the east side of the Capitol had freed up some space after the Capitol Police locker rooms that had been crammed into the Capitol basement were moved to the new space. A Risch spokesperson said the senator's hideaway was medium-sized, with enough room for a couch and a desk.[96]

During the 2016 campaign, Risch supported his fellow senator Marco Rubio, saying that he believed the Republican Party should "move away from Donald Trump" and that the GOP could do much better.

"This is not my Republican Party that I've been a member of half a century," Risch said at the time. "These dustups he'd had with women, and with minorities, and with different religions, and handicap people, I mean, this is nuts! This is not the Republican Party."[97]

How things had changed. One month before rioters trashed his office, the *Washington Post* ran a story running down what every member of Congress had said about the presidential election, whether they had conceded the obvious fact that Biden had won.[98] Risch's office didn't initially respond, but the paper cited a *Spokane Spokesman-Review* piece in which Risch acknowledged that there would be a transition. Trump was upset with the piece.

"I am surprised there are so many. We have just begun to fight," he tweeted. "Please send me a list of the 25 RINOS."[99]

Risch's office soon contacted the paper and "asked that he not be classified as saying Biden won as he was in the initial story." In a new statement, Risch said he continued "to support the President's efforts to ensure an accurate vote count in the Presidential election."[100] He said the new deadline was Dec. 14, when the electoral college met. That's when this thing would be over. But when that date passed, he didn't make any announcements on Twitter. Every tweet he did send, including those wishing others happy holidays or showing off the Idaho Christmas tree at the White House, was answered with demands that he overturn the will of the American people because of bogus claims about fraud.

On Jan. 5, back in his home state, hundreds of Trump supporters gathered outside his office, holding signs reading "Stop the Steal," "Election Fraud," and "Election Integrity Matters."[101]

Finally, after months of nonsense and around the time rioters were trashing his office and smashing out his window, Risch decided it was time to speak up.

"This nonsense and violence needs to stop now," he tweeted at 4:23 p.m.[102]

In the office next door, one of the looters found some document about the Green New Energy Deal.[103]

"Green New Energy Deal, get that!" Rodriguez said. "Get all that!" Rodriguez yelled as he made his way through someone's backpack, which had a lock he couldn't seem to open. He asked around if anyone had scissors or a knife, maybe a letter opener he could use to pry it open. He found some scissors in a desk.

"We're going to open this up looking for intel, we've got Green New Deal stuff, we've got emergency prep class, open these up, what's in here guys?" Rodriguez said. "Open these up! Open these up! Emergency escape hood! It says emergency escape hood! . . . Who needs to get out of here alive?"

"Burn this motherfucker down!" someone said as Rodriguez emptied out a bag. Still wearing the MPD helmet he'd stolen earlier in the day, Brockhoff walked over to the watercooler and poured himself a glass. But the end was nearing. Reinforcements had arrived from Virginia, with more people and more munitions, and rioters scattered off, back through the windows.

There was a jam-up as rioters filtered their way off the inauguration platform. Many of the rioters were still dealing with the effects of tear gas, and a police officer offered up advice: rubbing would only make it worse.

"You said don't rub just blink? I've been rubbing for the last ten minutes, fuck my life," one man said.

The officer approached the man. "Be careful when you take a shower, because it reactivates down here," the officer advised, gesturing toward his groin. "Be careful!"

The man, a twenty-six-year-old Proud Boy from North Carolina named Ryan Barry who had been arrested the previous night for assault, smiled.[104]

"I'm good down there, I'm straight," he said. "It's just the face."

Barry assured the man that the mob liked police, that it was the politicians they disliked. "You can only fuck people over for so long," Barry said. "We gotta make sure we are not being fucked over anymore."[105]

Soon, officers were finding their colleagues, surveying the damage, and exchanging war stories. Dozens of officers were injured, but backup had arrived. They had the numbers on their side. At the US Attorney's Office, it was all hands on deck, with authorities preparing to process mass arrests and churn the cases through the court system. That never happened. Very few rioters were arrested on Jan. 6, and many who'd been detained had been released because law enforcement didn't have the manpower to process them. One man who was arrested late in the evening, Christopher Alberts, was covered in body armor, and investigators would later learn that he'd been at the front of the mob and charged at the police line with a wooden pallet.[106] A concealed handgun had been tucked away, fully loaded, inside Alberts's waistband. Another Jan. 6 defendant brought two guns to the Capitol; dropped one of them in the melee; and then filed a false police report in Indiana, claiming the gun he'd lost had been stolen.[107] Still, some kept insisting, the mob was unarmed.

———— ◦ ————

By the time the grounds had cleared, the rumors had spread. *It wasn't Trump supporters who had stormed the Capitol; it was antifa.*

Just as happened with voter fraud disinformation about the election, there were two camps: those who believed the bullshit and those who found the bullshit useful. Once again, distinguishing who fell into which camp was difficult.

"This has all the hallmarks of Antifa provocation," tweeted far-right Rep. Paul A. Gosar when police officers were still trying to get protesters off the West Terrace.[108]

Within twenty-four hours, there were at least 411,099 online mentions of the lie, which spread with the help of election denier Lin Wood, who spread the conspiracy theory on Parler. A far-right televangelist tweeted that the Capitol siege was a "staged #Antifa attack" and gained a like from the president's son Eric Trump. Soon Sarah Palin backed the theory on Fox News.[109]

The conservative *Washington Times* ran a story claiming that a facial recognition company identified some rioters as members of antifa, a piece they had to retract when the company pointed out that it wasn't true and that their analysis had identified two neo-Nazis and a QAnon promotor.[110]

Still, the story was up long enough to give Trump ally Rep. Matt Gaetz something to distract with when he took to the floor once the House reconvened. Citing "reports," he said that "some of the people who breached the Capitol today were not Trump supporters. They were masquerading as Trump supporters and in fact, were members of the violent terrorist group antifa."

Inside the White House, Chief of Staff Mark Meadows initially thought that blaming antifa was a good strategy and was part of what White House aide Cassidy Hutchinson would later describe as the "deflect and blame" caucus. "Let's blame antifa," she said, describing their thinking. "These aren't our people."[111]

Trump himself, sitting in the White House dining room during the attack, was on the phone with Kevin McCarthy, the Republican House leader and future House Speaker, who was urging the president to get on television or Twitter and call off the mob.

"Well Kevin, these aren't my people," Trump told him. "These are Antifa." McCarthy later recalled to others that he tried to introduce reality, saying that these were Trump supporters, that staffers were running for their lives, and that the president needed to call off the mob.

"Well Kevin," Trump said, "I guess they're just more upset about the election theft than you are."[112]

Stewart Rhodes, the Oath Keepers' founder, knew it was bullshit. "Look, I WAS THERE. I WAS RIGHT OUTSIDE. Patriots stormed in. Not Antifa," he wrote in a message at 6:22 p.m. "And I don't blame them. They were justifiably pissed off."[113]

On the grounds of the Capitol, there were plenty of true believers, as rioters turned on one another and called one another antifa.

One Trump supporter whom others accused of being antifa was ripped away from a window he was smashing to get into the

Capitol, despite having Trump and "Make Liberals Cry Again" stickers on the back of his helmet. He'd previously been arrested for threatening a Democrat in his home state over COVID-19 restrictions. After storming the Capitol, he showed up on Jan. 7 to memorialize Ashli Babbitt, who was already being turned into a martyr. He appeared to be wearing the same sweatshirt. As of this writing, in May 2023, many Jan. 6 participants and Trump supporters still think he's antifa, and he has not yet been arrested by the FBI. "UNAPOLOGETIC, MAGA 4 EVER GOD SAVE OUR REPUBLIC 🦅," his Facebook page reads.

Another group that grabbed the attention of Trump supporters was a trio referred to as the "quick changers," who swapped items of clothing underneath a tree. A bystander thought it was suspicious and took video, walking by the group very quickly. "Delete that," the young woman said when she saw them recording.

Minutes earlier, the woman had rushed away from the police line in a cloud of tear gas, along with her two companions. "I had to change; it's fucking all over me," she explained to one bystander. "They thought she was spraying them; they tried to arrest her," one of the men said. Online sleuths were later able to identify all three, including the young woman, who was a specialist in the US Army Reserves. A least one member of the group may have committed an assault, but none of them entered the building. No members of the Ohio trio have been arrested as of this writing.

A Vice News reporter's tweet featuring a photo of three men wearing matching "MAGA CIVIL WAR" T-shirts became a meme, with some of the Right claiming they were federal agents who were only pretending to be Trump supporters. In reality, online sleuths later told me, the men were all blue-collar workers in Kentucky, and two of them were related. I checked their public records, and each was registered as a Republican.[114]

By Jan. 7, it was practically canon. *Of course* it was antifa who stormed the Capitol.

"Please, don't be like #FakeNewsMedia, don't rush to judgment on assault on Capitol. Wait for investigation. All may not be (and likely is not) what appears," Rep. Mo Brooks had tweeted by Thursday morning. "Evidence growing that fascist ANTIFA orchestrated Capitol attack with clever mob control tactics."[115]

The whiplash could make your neck snap. One rioter had posted the day before the attack that Trump supporters "need to take over the [Capitol] building tomorrow" and that Trump supporters should rip the doors down and rush the Capitol because authorities "can't stop a million people." He quickly pivoted after the attack, writing that "antifa . . . was let into the capital by the crooked police."[116]

Online, much of the discussion centered on John Sullivan, a self-described activist, reporter, and businessman who filmed high-quality footage of the Capitol riot and licensed off clips of it to networks, which were scrambling for footage that showed what had happened in the mob.[117] Much of the live coverage of Jan. 6 didn't reflect the brutal reality of what had happened: many reporters and videographers inside the Capitol had to flee for safety, and rioters targeted reporters outside too, assaulting members of the media and gathering around for the modern-day equivalent of a book burning, as Trump supporters smashed media equipment and, unsuccessfully, attempted to set metal ablaze without an accelerant. Sullivan's compelling video filled the demand, and he knew it. "Everybody's gonna want this. Nobody has it. I'm sell it, I could make millions of dollars," he said upon emerging from the Capitol. The networks were in fact willing to pay. The government later determined that Sullivan "received at least $90,875 in payments from at least six companies for the rights to use his video footage of the events at the U.S. Capitol" and seized $62,813.76 in funds from his account.[118]

"Where a criminal defendant profits to the tune of $90,875 from his charged crime—proceeds that, based on the totality of facts and evidence specific to this particular case and this particular defendant, would not have obtained but for the defendant's obstructive acts on January 6, 2021 at the U.S. Capitol—there is a strong governmental interest in taking the profits out of the crime, and removing the financial incentives for such behavior going forward," federal prosecutors would argue, supporting the seizure of Sullivan's funds. US District Judge Emmet G. Sullivan denied Sullivan's attempt to have the funds returned ahead of his trial.[119]

Sullivan's video showing the shooting death of Ashli Babbitt did get a lot of coverage. His actual ideology, so much as he had one, was tough to pin down. He opposed Trump and said he supported Black Lives Matter but was disowned by BLM activists back in his home state of Utah, who said he would often provoke violence at their events too.[120]

Sullivan made provocative comments in the video he filmed on Jan. 6 too, illustrating that he wasn't so much an observer of the scene as he was a participant.

"There are so many people. Let's go. This shit is ours! Fuck yeah," he said in the video. "We accomplished this shit. We did this together. Fuck yeah! We are all a part of this history. . . . Let's burn this shit down."[121]

Sullivan spoke with the FBI on Jan. 7. Within a week, he'd be charged.[122] Later, the government would find footage of him stating that he came to "instigate shite" at the Capitol. "I was like, guys we're going inside, we're fucking shit up," Sullivan said. "I'm gonna make these Trump supporters f— all this shit up."[123]

———— ◆ ————

The conspiracy quickly evolved. *Not only was antifa involved, but so was the FBI, or maybe the FBI and antifa conspired together.*

When an FBI undercover employee later met with Jan. 6 participant and navy reservist Hatchet Speed, he explained that Jews give out scholarships to teach "how to be a good antifa person" and that both Biden and antifa were controlled by a sinister cabal.[124]

The very first person to smash out a window of the Capitol was Edward Kelley of Tennessee, who then jumped through the broken glass. He soon became a target of antifa conspiracies too. He wore a green tactical helmet and was covered up well, wearing a gas mask, goggles, gloves, and a backpack. It was the black sweatshirt with the letters "TCAPP" that eventually gave him away: Kelley had been a protester at "the Church at Planned Parenthood" in Tennessee. The FBI found his bank records showing that he had bought the sweatshirt for forty dollars a few days before the riot, on Dec. 29, 2020. By the time Kelley was arrested—first in May 2022 for his actions at the Capitol,[125] then again in December 2022, for allegedly conspiring to murder FBI employees in retaliation for his first arrest[126]—convincing Trump supporters that the black-clad rioter who helped lead the initial breach of the Capitol was one of them seemed like a long lost cause.

Later, a right-wing attorney for Jan. 6 defendants would declare on Tucker Carlson's Fox News show that a man with red face paint and a "Keep America Great" hat who pushed against police in the tunnel was "clearly a law enforcement officer" and an "agent provocateur." Nope! He was a St. Louis man who went by the name Rally Runner and did loops around the Cardinals' stadium wearing that same red face paint. He'd even posted Facebook videos bragging about his actions and posted about getting a visit from the FBI.[127]

Two years later, Rally Runner told me, he hadn't heard back from the FBI. "I only heard from them the day they came over back then," he said. "I just went to stand up for Trump. To run for him and America. But everyone was going to the Capitol so it was natural to go."

The lawyer who appeared on *Tucker Carlson Tonight* didn't care much what the truth was. "If I'm wrong, so be it, bro. I don't care," the attorney, Joseph McBride, told me when I called him to break the news. "I don't give a shit about being wrong."

In real time, many of the conspiracy theorists who stormed the Capitol on the basis of delusions of mass voter fraud were a bit frustrated with their fellow conspiracy theorists for stealing their thunder, believing that they should get credit for their work and that it wasn't something to be ashamed about. Jonathan Mellis— later dubbed "Cowboy Screech" for his resemblance to the *Saved by the Bell* character and his ten-gallon hat—took to Facebook to declare that antifa and Black Lives Matter supporters were "too pussy" to have done what Trump supporters had. Mellis, who used a stick to swing and stab at police officers, said Trump supporters "proudly take responsibility for storming the Castle." Ryan Nichols said that "every single person who believes that narrative have been DUPED AGAIN!" And Brandon Straka, the Stop the Steal organizer and online influencer, said it wasn't antifa but rather "freedom loving Patriots who were DESPERATE to fight for the final hope of our Republic because literally nobody cares about them."[128]

That the false claim fell apart with just a bit of critical thinking hardly mattered. It was spoken long enough to take hold. By June 2022, more than half of Republicans believed a claim with zero basis in reality: that left-wingers were responsible for Trump's riot.[129]

The reality on the ground was that Trump supporters, believing that the country was at risk and that Pence had sold them out, engaged in activity that, for many of them, was out of character. It became a trope during Jan. 6 cases that defendants would say they were "caught up" in the chaos, that they'd been swept up in mob mentality. In their circles, acting like antifa was just about the worst thing you could do, and they'd done it. A friend of Joshua Dillon Haynes, a Jan. 6 defendant who'd invaded the

Capitol through a window and smashed media equipment out-side, texted him during the attack. "Dont be like antifa tearing stuff up," they wrote. "Ahhhh tooooo late," Haynes replied. "Broke lotsa stuff . . . lol."

"I was going to be super hard and go punch an antifa terrorist in the face," said Eric Barber, a West Virginia councilman later arrested by the FBI. "And I end up being the terrorist. Plot twist, huh?"

———————◆———————

QAnon Shaman, standing shirtless in his horned headdress in front of the Columbus Doors on the east side of the Capitol, was trying to reason with a woman he thought was a bit batty. Hold-ing his spear with an American flag, he told his fellow Trump supporters it was time to go home. Donald Trump said so.

"We're gonna pull up the tweet; Donald Trump has asked everybody to go home," Jacob Chansley said. "I'm here delivering the president's message. Donald Trump has asked everybody to go home."

They played the president's video.

"I know your pain; I know your hurt. We had an election that was stolen from us. It was a landslide election, and everyone knows it, especially the other side. But you have to go home now," Trump said. "This was a fraudulent election, but we can't play into the hands of these people. We have to have peace. So go home; we love you; you're very special."[130]

The woman in the crazy hat didn't think it was real.

"That doesn't sound like him," she told QAnon Shaman.

"That is him," he said. "It's his fucking video!"

"That looks prerecorded!" she yelled.

"Chill the fuck out!" QAnon Shaman said.

"That didn't sound like his voice," she said.

He'd sent a tweet too, one of his last on the platform.

"These are the things and events that happen when a sacred landslide election victory is so unceremoniously & viciously stripped away from great patriots who have been badly & unfairly treated for so long," Trump tweeted before he was kicked off for nearly two years. "Go home with love & in peace. Remember this day forever!"[131]

———— ♦ ————

Sergeant Aquilino Gonell of the US Capitol Police seemed to be everywhere on Jan. 6. He was on the inauguration platform, taking hits from the mob, and in the tunnel, taking hits there too, where a rioter yelled, "You're gonna die tonight" and kicked a shield.[132] "I could feel myself losing oxygen and recall thinking to myself, 'This is how I'm going to die,'" Gonell would later testify.[133]

Around 11:00 p.m., the Dominican Republic immigrant and Iraq War veteran headed back to the tunnel. He'd come down earlier too, in search of an ID belonging to the dead rioter police had tried, unsuccessfully, to revive. He wanted to give it another go, and he started poking around where he'd been fighting rioters hours earlier.

In a few hours, a scaffolding crew would find the inauguration platform a "complete disaster," full of trash, weapons, hats and even abandoned shoes. Workers used leaf blowers to blow away evidence. Later, an official would tell me it was the modern-day equivalent of putting John F. Kennedy's limo through a carwash after his assassination. "This place was a war zone last night," said Steve Ott of Scaffolding Solutions as the leaf blowers recirculated tear gas. "Unbelievable."[134]

That night, Gonell found law enforcement officers from the FBI and Capitol Police examining the scene, lining up items that had been used as weapons. He talked with them, let them know what he was doing, and began to search.

About three steps down, he found it: Rosanne Boyland's identification card.

Boyland's mother had been calling, and she got the call around midnight. "I believe we have your daughter," they said. Cheryl Boyland thought Rosanne had been arrested, but then the officer asked if she had a tattoo. "Then I knew she was dead," her mother said. "Right then."

The next day, Boyland's brother-in-law Justin Cave made a statement to the media on his family's behalf. Then he went off script a bit.

"It's my own personal belief that the president's words incited a riot that killed four of his biggest fans last night, and I believe that we should invoke the Twenty-Fifth Amendment at this time," he said.

———— ◆ ————

Sen. Murkowski went back to her hideaway days later, seeing glass and water bottles along the way. "They cleaned up some of the mess, but it was still just kind of really eerie and very, very unsettling. The plexiglass riot shields were still out in the hallway there," she said.

When she got into her hideaway, she couldn't close the door. "I was unsettled being there," she said. She ultimately moved out of the location, giving up the view.

"I really liked it, but it was just too much déjà vu. That memory is still there. That little public bathroom right across from my hideaway—I can just still hear the awful sound of the officer as he was trying to rid himself of whatever the spray was," Murkowski said. She became one of just a handful of Republicans to vote to convict Trump in his impeachment trial.[135]

"If months of lies, organizing a rally of supporters in an effort to thwart the work of Congress, encouraging a crowd to march on the Capitol, and then taking no meaningful action to stop the violence once it began is not worthy of impeachment, conviction, and disqualification from holding office in the United States, I cannot imagine what is," Murkowski would say.

The night of Jan. 6, after his office was trashed, Risch would condemn the attack on the Capitol as "unpatriotic and un-American in the extreme" and vote to certify the 2020 presidential election results. He tossed election truthers a bone, saying that the events showed there was a "deep distrust in the integrity and veracity of our elections."[136] He'd vote against Trump's impeachment, saying it was "time we stop the political hate and vitriol and move forward wiser and stronger just as America has countless times before."[137] Risch kept silent about what happened to his office for two years. Once court documents revealed that his office had been trashed, my colleague Sahil Kapur approached him in a hallway of the Capitol and asked for comment. "I don't do interviews on Jan. 6, but thanks," Risch said.[138]

As for Braun, who'd met with future rioters that morning and assured them he had their best interests in mind, he flip-flopped yet again on Jan. 6.

"Today's events changed things drastically," he tweeted. "Though I will continue to push for a thorough investigation into the election irregularities many Hoosiers are concerned with as my objection was intended, I have withdrawn that objection and will vote to get this ugly day behind us."[139]

He'd also vote against convicting Trump at his impeachment trial, but he took the easy way out. Instead of judging the facts on their merits, he claimed he believed it was unconstitutional to hold a trial to bar a president from holding office.[140]

Earlier in the day, after he met with constituents who'd later storm the Capitol, Braun posted an image on his social media accounts. A Facebook user made a prescient comment.

"Perfect," they wrote, "now facial recognition technology can cross check these people with ones who crossed the barricades!"

# PART III

# CHAPTER 7

# "We Do Big"

## Jan. 8, 2021

One of the first people[1] to breach the Capitol on Jan. 6 showed up unannounced at the Des Moines Police Department about thirty-six hours after and one thousand miles away from where Vice President Mike Pence gaveled the Senate back to order after what he called "a dark day in the history of the United States Capitol."[2]

Doug Jensen, sitting under the fluorescent lights of a police interview room on Friday at 8:39 a.m., had had a wild seventy-two hours. He'd woken up at 6:00 a.m. on Tuesday and worked a full day at a construction site until 4:00 p.m. Then he rushed home, took a shower, and tossed on his QAnon gear. His truck wasn't up to snuff, so he hopped in his wife's car and drove all night, aided by Red Bull, along with his close friend who wasn't that into politics but was willing to join him on the trip. The GPS showed them getting to DC about thirty minutes before Trump's rally on Wednesday morning. They pulled up to the Capitol Hilton hotel near the White House, dropped the car off with the valet, and headed over to Trump's rally after a grueling drive of more than one thousand miles. Trump was running late, so they made it in time for the speech to begin at noon.

When the rally wrapped up, Jensen's friend peeled off and headed back to the hotel, but Jensen headed over to the Capitol. He grabbed his phone and filmed a selfie video.

"This is me, touching the fucking White House," he said, with his hand on the retaining wall of the US Capitol. Then he scaled the twenty-foot wall and filmed another video. "Storm the White House! That's what we do!"

Congressional reporter Igor Bobic was sipping coffee in the Senate press gallery, watching a video feed of the electoral college certification, and working on a story about what he thought would be the biggest news of the day: Sen. Mitch McConnell's speech urging his colleagues to ignore the "sweeping conspiracy theories" and avoid going down a path that "would damage our republic forever" and send democracy into a "death spiral."[3] A security alert went off, advising those inside the Capitol to stay away from the windows. Minutes later, a colleague emerged from inside the chamber, shouting that Pence had been pulled from his chair on the Senate dais.

Bobic went down to the second floor to see where Pence was headed and heard a commotion one floor below. Reporters are typically banned from recording video near the Senate chamber, but the thought hadn't crossed Bobic's mind when he instinctively pulled out his phone as he descended the stairs and arrived on the ground floor, filming what would become one of the most iconic moments of the day. Bobic, who'd arrived in America as a refugee of civil war in Bosnia, rounded the corner to find a lone Black police officer facing down an angry mob. A man had just used a Confederate battle flag to jab at Officer Eugene Goodman, a DC native who had served his country in Iraq as a member of the US Army.[4] "I'm not leaving!" Kevin Seefried, the man with the flag, said. "Where are the members at?"[5]

Jensen was now at the front of the pack next to Seefried, who was confronting Goodman, having breached the building through a broken window. Bobic's video shows Goodman with his right hand on his service weapon as Jensen continued pressing forward.[6] Jensen just "kept coming," Goodman recalled.[7]

Bobic's video shows Jensen at the front of the crowd, chasing Goodman up an ornate marble staircase leading to the entrance to the US Senate, one floor above. "Keep running, motherfucker," someone in the mob behind him shouted. "He's one person; we're thousands!"

"They breached the Senate!" Goodman yelled over the radio, dashing up the stairs, with Jensen in hot pursuit. "Second floor!" Goodman reached the landing, turned around to see Jensen in his face, and took out his extendable baton.

Because of the chaos of the attack, it would take a few days to realize the significance of what Goodman did next, at 2:14 p.m., when he gave Jensen a light shove and lured him to make a left, diverting the mob away from the Senate's most heavily trafficked entrance. Inside the chamber, as members huddled, authorities were able to lock down the doors at 2:15 p.m. Had Jensen made a right at the top of the stairs, and not been baited by Goodman to turn left, rioters could have reached the senators on the floor. Vice President Mike Pence and his family were still nearby off the floor of the Senate waiting to escape, and Goodman's ploy gave the Secret Service time to evacuate Pence, whom the mob saw as a traitor who should hang.[8]

Goodman's heroics would soon be widely recognized: He escorted Vice President Kamala Harris at the inauguration, and, in a rare moment of bipartisanship, the entire US Senate rose to cheer for Goodman during Trump's second impeachment trial in February 2021. He'd later receive the Congressional Gold Medal.[9] For many Black Americans, seeing Goodman use himself as bait against a majority-white mob to protect senators just six months after George Floyd's murder brought them to the edge of tears. "Black bodies have just had to do so much heavy lifting in this country—from slavery til now—and all black people want is the chance and opportunity to be treated like everyone else," one friend texted me after viewing Bobic's viral video.

Bobic went into the Senate gallery to see that rioters had taken over the dais. Soon, realizing the danger of the situation, he went into hiding with fellow reporters in a tucked-away media room, where they turned off the lights, locked the doors, and covered up the sign indicating it was a media room. As for Jensen, he continued confronting officers at the Capitol, was escorted out of the building, and then made his way back inside through another entrance. Eventually, after the citywide curfew went into effect, he went back to his hotel room, so exhausted that he slept for about twelve hours. When he woke up Thursday morning, he and his friend drove the more than one thousand miles back to Iowa, which took the whole day.

He didn't know it at the time, but several people who knew Jensen had already called up the FBI and reported his name. Some of those tipsters reached out to local media too. Jensen got a Facebook message on Thursday morning from a reporter with Des Moines station KCCI, which had received tips about his presence at the Capitol. "You are FAKE news," Jensen told reporter Todd Magel.[10] Jensen's phone kept pinging along the way with updates from those who had seen photos of him in the Capitol or the story on KCCI that identified him, so he decided to delete all the apps on his phone and shut it off.

"I deleted all my Twitters, my Facebooks because I'm paranoid thinking Mark Zuckerberg and his henchmen or somebody's going to try to locate me through one of these apps, so deleted every freaking app on my phone and shut my phone off, you know, because I was afraid of being killed on the way home being a poster child. I feared for my family, you know," the QAnon conspiracy theorist said. "There are lots of people that are completely brainwashed by the media."

Jensen got back early in the morning of Jan. 8, pulled into his garage, and then walked over to a friend's house, where his wife, April, was sleeping on a couch because she was scared to be at home. She yelled at him to leave, acquiesced to a hug, and told him to go take care of the problem.

"She was just like, *What the hell?* You know? *You're a terrorist, Doug*," Jensen said. "I'm like *Baby, I am not a terrorist*."

By the time Jensen got to the police department the morning of Jan. 8, he was once again running on little sleep after a long drive. He sat in an interview room with a bottle of water for about twenty minutes, waiting for the FBI to arrive. When they did, he was ready to spill.

Jensen went through his life story: how he'd been a construction worker for twenty years; how he'd been molested as a child; how he'd gotten into meth; how this one time he went to Las Vegas with his wife and they won $1,500 on a slot machine, got super drunk before a Gwen Stefani concert, and then woke up the next morning broke and tattooed.

He told them how he'd voted for Obama and thought he was going to vote for Hillary Clinton too, but then started reading things online. How QAnon came out and he started following it "religiously," checking for "Q drops" every day. How he'd lost friends and family who thought he was "insane," how coworkers thought he was a "wacko," but how he thought he was "the only one that sees it, and everybody's sheep that doesn't know." How he thought his brain was overloaded with "all this information I've stored in my head for two years."

Jensen told the FBI that he thought Jan. 6 was "show time" and a bunch of arrests were about to begin. He told them how excited he was to see the Washington Monument because of his background in masonry. He told them that he honestly thought he was "at the White House at first" and that he was glad he didn't post the video to Facebook because he looked "like a complete idiot." He said he knew that what he did "was stupid," but he thought it had meaning. He thought they were going to save the country and still thought Mike Pence would be arrested any day now.

"I consider myself a digital soldier," Jensen said. "I believe in Q 100%. I still believe that Trump's gonna be our President, and that there's some trick he has left, you know? And that all these arrests are gonna happen, and there's gonna be this emergency broadcast that's gonna, like, broadcast the videos of all their— you know, them admitting to all this stuff. And I'm still holding on to that, I guess. And then, I was kind of hoping General Flynn would become the Vice President, you know, because that's more realistic than JFK Jr., who probably passed away, you know?"

He rambled on like that for a while, making statements that would test even the most seasoned law enforcement officer's poker face. "I know I have to watch out for dis-info," said Jensen, who believed that the late Republican senator John McCain really didn't die of cancer but had been secretly executed over ties to ISIS. "Q always said watch out for dis-info."

Jensen told the FBI he did "what my President told me, and that was to go to D.C. on the 6th for a rally." He regurgitated Trump's rhetoric, saying he believed Trump was "draining the swamp," complaining about the social media fallout for the president, and questioning why the FBI hadn't done something on Trump's behalf and investigated all the conspiracy theories Jensen believed.

"They blocked Trump on Twitter. He has no way to speak to the people right now," Jensen said. "What I want to hear from you guys is if you guys are the FBI, why haven't you guys looked into this stuff? Why haven't you guys made a move, you know?"

When the FBI asked him whether he had any regrets about his actions on Jan. 6, Jensen said he wasn't sure yet.

"It depends on if the outcome I wanted happens; then it would have been worth it," Jensen said. "But if nothing happens except for negativity from this, and I'm a rioter, then, yeah, I completely regret it."

As they wrapped up the interview, FBI special agents let him know it might take a bit before he heard from the bureau again and asked him to be patient.

"It does take time for the wheels to grind, so don't feel like we're giving you the silent treatment if things, you know, take a little while before, you know, the next steps," one agent said.

Reality still didn't click for Jensen. He told them he hoped this would all disappear in the next ten days, believing that they were at the start of "our 10 days in darkness" before "the storm" and that Trump was going to use the emergency broadcast system to speak to the people, to show video confessions, and to send his enemies to Guantanamo, which Jensen falsely believed had been expanded over the course of the Trump administration. He had so much riding on it, and he hoped the FBI special agents would let him in on the secret.

"Seriously, can you guys just let me in on that if you know? Like are these arrests really, or am I just completely following a whole shit thing?" Jensen asked. "Because I completely believe this with my heart, and it'd be the biggest let down for me, you know?"

The agents deflected, with one saying there was "a lot of information" and "a lot of different things" out there and that he was going to go back and try to "really understand" the "full picture that you're talking about." But when Jensen expressed outrage over being portrayed as a terrorist, one of the special agents did reassure him that they didn't think of him as one.

"We get paid to separate the sheep from the goats when it comes to real terrorists and not," one of the FBI special agents told him. "And you are—you do not fit our definition either, my friend."

———◆———

What were we calling this, exactly? An attempted self-coup, or failed autocoup? An insurrection? A terrorist attack? The attack did, as federal officials would later concede, meet the federal definition of an act of domestic terrorism. But the feds avoided the strict labels, initially, with FBI director Chris Wray referring to the "siege of the Capitol"[11] and Acting Attorney General Jeffrey Rosen condemning the "intolerable attack on a fundamental institution of our democracy." Down the line, when it came time for judges to determine prison sentences, federal prosecutors would seek sentencing enhancements in several Jan. 6 cases. It was tricky. The Capitol attack itself was clearly an act of domestic terrorism, but individual defendants had a wide range of conduct that day. A defendant who'd plotted to overthrow the government and spent hours battling police on the west side of the Capitol was on a much different footing than someone who followed the mob inside the building for a few minutes, snapping photos of the building or parading around with a protest sign.

Initially, there was some Republican support for the description of Jan. 6 as a terrorist attack. Sen. Ted Cruz (R-Texas), upon learning about the death of Capitol Police officer Brian Sicknick, called the "horrific assault on our democracy" a "terrorist attack" and said that every "terrorist needs to be fully prosecuted."[12] At the time, calling Jan. 6 an act of terrorism wasn't inherently sacrilegious within the GOP conference. But when Cruz used the same language on the one-year anniversary of the attack, he came under swift attack from many on the right and went on *Tucker Carlson Tonight* to call his own comments "frankly dumb."[13] By February 2022, the Republican National Committee had declared the attack on the Capitol and the events that preceded it "legitimate political discourse."[14]

President-elect Joe Biden thought it was terrorism. During Trump's speech, before the Capitol attack began, word emerged that Biden planned to nominate Judge Merrick Garland as his attorney general, along with Lisa Monaco as deputy attorney general and Vanita Gupta as DOJ's no. 3 official.[15] That Garland had seen the Oklahoma City bombing investigation in 1995, in addition to the prosecution of the Unabomber and the Atlanta Olympics bombing, made him seem like the man for the moment. "How much more experience could you possibly have in domestic terrorism?" asked one former prosecutor.[16]

Biden's announcement had been set for Thursday, at the Queen theater in Wilmington, Delaware, and it went on despite the Capitol attack. Garland, who'd been on the bench for years and knew the importance of the Justice Department's independence, had to choose his words carefully. Nothing about this situation was normal, but Garland would soon be overseeing what would grow into the largest investigation in FBI history, one that would result in charges against hundreds of Trump supporters and potentially affect the outgoing president himself. He had a much different role to play than Biden, who could attack Trump as he pleased.

"What we witnessed yesterday is not dissent. It was not disorder. It was not protest. It was chaos. They weren't protesters—don't dare call them protesters—they were a riotous mob. Insurrectionists. Domestic terrorists. It's that basic, it's that simple," Biden said.

"I wish we could say we couldn't see it coming. But that isn't true. We could see it coming," Biden said, calling Jan. 6 the culmination of Trump's "unrelenting attack" on the institutions of our democracy.

Biden, going off script, recalled texts he had exchanged with his granddaughter Finnegan Biden as the riot unfolded. The college senior had texted Biden a photo that was going viral again online, showing the dystopian scene of a contingent of masked members of the National Guard lining the steps of the Lincoln Memorial during protests after George Floyd's death months earlier.[17]

"No one can tell me that if it had been a group of Black Lives Matter protesting yesterday, they wouldn't have been treated very, very differently from the mob of thugs that stormed the

Capitol," Biden said. "We all know that's true. And it's unacceptable. Totally unacceptable."[18]

Garland came to the podium next. He emphasized the importance of norms, of the independence of the Justice Department. He made only passing reference to the Capitol attack, a comment that wouldn't give defense attorneys in any future cases anything to complain about.

"As everyone who watched yesterday's events in Washington now understands—if they did not understand before—the rule of law is not just some lawyer's turn of phrase," Garland said. "It is the very foundation of our democracy."

———— ◆ ————

Back in Des Moines, the rule of law was coming for Jensen. The FBI's regional Joint Terrorism Task Force had begun investigating Jensen as soon as local station KCCI ran a story on him on Thursday. One of the agents who interviewed Jensen on Friday primarily worked domestic and international terrorism, and as they spoke with him, JTTF officials were in communication with federal prosecutors, working out the logistics of Jensen's arrest.

"JTTF would like to know how quickly do we want to move on this guy," read an email from a Justice Department official as the interview unfolded.[19]

Pretty quickly, it turned out. After the FBI special agents gave Jensen a ride home, and after he handed over his phone, a DC federal magistrate judge signed off on a warrant at 6:36 p.m. central time, and the FBI knocked on his door to arrest him around 8:00 p.m. on Friday night.

Justice Department officials had spent Thursday scrambling to put together fifteen federal cases, mostly against people who were already in custody. A handful of rioters had been arrested on-site at the Capitol, and even two video journalists from the *Washington Post* had been temporarily detained.[20]

At the time Jensen walked into the Des Moines Police Department on Friday morning, federal authorities had arrested just one defendant outside of DC: a Proud Boy from Hawaii, whom the FBI rushed to arrest when they learned his flight was arriving at the Honolulu airport on Thursday evening.[21] (Federal prosecutors moved so quickly there that they filed language in a motion that described an entirely different investigation, a scheme involving an attorney who used a law firm credit card for personal expenses.)

Jensen wasn't the only one who got a home visit that day. Some of his fellow viral stars of the Capitol attack were arrested the same day he was.

In West Virginia, FBI special agents walked state senator Derrick Evans out of his home in sweats and into the back of a waiting SUV.[22] A woman who identified herself as Evans's grandmother walked out, told a reporter he was a "fine man," and blamed the president for her grandson's predicament. "Thank you, Mr. Trump, for invoking a riot at the White House," she said. Evans had live streamed himself storming the Capitol on Jan. 6 while yelling "Derrick Evans is in the Capitol!" Before he entered the building, he told his audience that he "bet Trump would pardon anybody who gets arrested for goin' in there."[23]

In Florida, law enforcement authorities arrested[24] Adam Johnson, who bragged to friends that he "broke the internet" when he was photographed carrying Pelosi's lectern through the rotunda.[25]

In DC, a federal magistrate judge signed off on an arrest warrant for the "QAnon Shaman," who had stormed the building in a horned headdress and stood near Jensen. (Jensen, having yet again been duped by internet disinformation, now believed the QAnon Shaman was actually affiliated with Black Lives Matter, maybe antifa, and was possibly a pedophile because, Jensen believed, his tattoo was a secret code. "I would have turned around and beat the living crap out of that guy, you know, just for having that tattoo," Jensen said.)

A lawyer for Albert Ciarpelli, who was photographed next to Jensen, reached out to the US Marshals and the FBI to see if there was a warrant out for his client's arrest.[26] He'd be arrested soon too.

In Bentonville, Arkansas, the FBI arrested Richard Barnett, who stormed the Capitol carrying a stun gun and put his feet up on Pelosi's desk. "Nancy, Bigo was here, bi-otch," he wrote in a note he left in her office, before eventually heading outside and telling his story to anyone who would listen. During his FBI interview, Barnett placed his feet up on the interview room table as he explained how he drove home overnight, turned his cell phone off, and used cash because he "knew this was coming." In a jailhouse phone call, he suggested his "Bigo was here" phrase was copyrighted.[27]

The attorney general doesn't issue a statement for just any arrest, but he did for this one. It was important to drive home the message that the FBI and the Justice Department were on the case and would take this seriously.

"The shocking images of Mr. Barnett with his boots up on a desk in the Speaker of the House's office on Wednesday was repulsive," Rosen, the acting attorney general, said in a statement after Barnett's arrest on Friday. "Those who are proven to have committed criminal acts during the storming of the Capitol will face justice."[28]

The FBI's "Bigo" investigation already showed signs of trouble. After Special Agent Johnathan Willett brought Barnett out to a vehicle following his interview, Barnett later said, he had a question for him. "Can I take a selfie with you for my family?" Barnett said Willett asked him. Two years later, at Barnett's trial, Willett would deny that ever happened. "I did not take a selfie with Mr. Barnett," he said. Then he got to thinking about it overnight and said he wasn't sure if he had. He wanted to change his testimony. The jury was told that Willett did "not have a memory either way."[29]

Lamond, the DC police lieutenant who tipped Tarrio off about his arrest before Jan. 6, was back in communication with Tarrio after the attack, after the Proud Boys leader got a new cell phone. Tarrio told Lamond that he thought he "could have stopped this whole thing," but Lamond wasn't so sure, mentioning that the founder of InfoWars couldn't calm the crowd.

"You know it's fucking bad when [Alex Jones] was the voice of reason and they wouldn't listen to him," Lamond wrote. ("We're not antifa, we're not BLM," Jones had said at the Capitol. "Let's not fight the police and give the system what they want."[30])

On Jan. 8, Lamond checked in with Tarrio again.

"Looks like the feds are locking people up for rioting at the Capitol. I hope none of your guys were among them," Lamond wrote.

"So far from what I'm seeing and hearing we're good," Tarrio replied.

"Great to hear. Of course I can't say it officially, but personally I support you all and don't want to see your group's name or reputation dragged through the mud," Lamond wrote.

"Thanks," Tarrio wrote to Lamond, but things were probably past that point. "The brunt of this will come down on us."[31]

———————— ◆ ————————

Jan. 8 saw a flurry of law enforcement activity around the country, as tips flooded into the bureau at an unprecedented pace.

At the Ronald Reagan National Airport, just outside of DC, an airport police officer was scrolling through his Instagram feed around 4:15 p.m. when he spotted someone who looked familiar. About forty-five minutes earlier, a Delta flight had come back to the gate because a passenger was continuously yelling "Trump 2020!" and disturbing other fliers. The officer had kept an eye on the passenger, John Lolos, who had been booked on another Delta flight and was waiting by the gate.

A video popped up on the officer's phone showing rioters streaming out of the Capitol. Lolos was there, wearing the same shirt he had on at the airport, yelling, "We did it, yeah!" as he left the building. The airport officer quickly told the US Capitol Police Dignitary Protection Division officers stationed at the airport, who went to the gate and detained Lolos. FBI special agents soon arrived, attempted to interview Lolos, and then placed him under arrest on the basis of the video evidence. They found his "Trump 2020 Keep America Great" flag among his belongings.[32]

In New Jersey, Mick Chan got an FBI special agent's phone number from a mutual friend and shot over a text message for advice. "Any fallout u think I should be aware of?" he texted. In a phone call the next day, Chan said that he "broke into, well air quotes, broke into" the Capitol.[33] Footage showed him removing a barricade and flipping off a surveillance camera as he entered the building.

In New York, the Fashion Institute of Technology's campus police sent in a tip about a student who went into the Capitol, and the FBI quickly obtained a warrant for his social media records.[34]

Jan. 8 is also the day that the FBI started making house calls. In Kentucky, agents showed up to the workplace of Bobby Bauer, who a tipster said had posted photos of himself giving the finger inside the US Capitol. He claimed not to know that Congress was even in session but said that they wanted to "occupy the space" because they were angry, among other things, about pedophiles. He and his wife missed their 5:00 p.m. flight back.[35]

In Chicago, the FBI showed up at the home of Kevin Lyons, who'd posted on Instagram before Jan. 6 indicating that he was making an eleven-hour, twenty-two-minute drive from Chicago to DC. "I refuse to tell my children that I sat back and did nothing," he wrote. "I'm heading to DC to STOP THE STEAL!"

When he reached the Capitol, he commented that he loved "the smell of teargas in the mid-afternoon," that the mob was "storming the Capitol building," and that he guessed they were all "going to jail." Climbing the stairs, he said he was part of "a fucking revolution." He followed the mob into the Capitol, calling the police officers he encountered along the way "oath breakers," "fucking Nazi bastards," and "traitors."[36]

Lyons soon made his way to Nancy Pelosi's office, with the mob. "Let's find these fucks!" one man yelled. "Nancy! Nancy! Nancy!" rioters chanted. Inside Pelosi's office, Lyons recognized Elijah Schaffer,[37] a far-right provocateur who worked for Glenn Beck's Blaze until he was fired for reportedly drunkenly groping a colleague's breasts during the premiere of *Uncle Tom II*,[38] a right-wing movie he coproduced that purported to depict "the gradual demoralization of America through Marxist infiltration of its institutions."

Schaffer had infiltrated an American institution himself, dashing off tweets that identified himself as ideologically simpatico with the mob raiding the House Speaker's office.

"BREAKING: I am inside Nancy Pelosi's office with the thousands of revolutionaries who have stormed the building," Schaffer wrote. He later deleted that tweet after backlash, claiming that he only used the words "patriots" and "revolutionaries" because that was how the rioters self-identified. His employer defended him too, saying Schaffer was under unfair attack by big tech and the media. "He was there as a journalist, and he was there reporting," Beck said. "This is something *The New York Times* should have done."[39]

Inside Pelosi's office, while wearing a press credential on his black protective vest, Schaffer wasn't acting like much of a journalist. When Lyons walked up to introduce himself, as a man fumbled with a laptop near a set of overturned chairs and rioters spread out in the room, Schaffer didn't say he was just there to document but expressed solidarity with the rioter. "I'm with you bro, let's do this!"

Schaffer walked up to a gold-plated mirror to film a video, one that he—wisely—never ended up posting online. "*We* are in

Nancy Pelosi's office right now," Schaffer said to his reflection. "*We* are occupying the Capitol building."

Lyons walked away and continued filming the criminal activity unfolding around him. "Watch your, uh, fingerprints," Lyons said to another rioter scribbling a grammatically challenged "THIS OUR HOUSE" note on a manila folder, as he used his smartphone to film himself committing several crimes. Lyons grabbed a coat off the rack and reached into a pocket, pulling out a brown wallet that contained about fifty dollars in cash, a TSA precheck card, two bank cards, and a driver's license. Lyons stuffed the wallet into his pocket and walked out, heading for Pelosi's conference room, where another rioter had just stolen a laptop, and then into Pelosi's personal office, where another rioter who'd stormed the Capitol with his mother was complaining about Pelosi's chocolate selection.

"They don't have any caramel! That's the problem, they have no caramel," said Rafael Rondon, who moments earlier had helped disconnect cables from a laptop in Pelosi's conference room, stuffing the computer into another man's backpack.[40]

"They got no fucking taste," Lyons replied.

Lyons, stolen wallet in his pocket, grabbed a pair of pink boxing gloves from a nearby table. "Anybody want Nancy's fucking boxing gloves?" he asked. "You want Nancy's pink boxing gloves?"

On the fireplace mantel, which was covered in shattered glass from the mirror rioters had smashed above it, was a framed photo of Pelosi with the late civil rights hero Rep. John Lewis, who'd died of pancreatic cancer less than six months earlier.[41] Lewis was one of the Freedom Riders, he coordinated sit-ins and lunch counters in the South, and he helped organize the 1963 March on Washington, where Martin Luther King Jr. gave his "I Have a Dream" speech at the Lincoln Memorial. In 1965, as he tried to lead a march for voting rights across a bridge in Selma named after a Confederate general and Ku Klux Klan leader, an Alabama state trooper cracked Lewis's skull with a nightstick. When he died in 2020, he became the first Black lawmaker to lie in state in the Capitol rotunda, his casket resting upon the Abraham Lincoln catafalque,[42] a set of rough pine boards covered in a black cloth that was built after the president's assassination by a Confederate spy toward the end of the Civil War.[43] Thousands of Americans had lined up on the east side of the Capitol just to get a glimpse of Lewis's casket when it was displayed outside amid the COVID-19 pandemic, just as they had in April 1965, when "one continuous stream of human beings" made their way "up the steps of the East portico and to the rotunda"[44] to pay their respects to Lincoln.

Now furious mobs of Trump supporters were streaming up the same east stairs, through an ornate set of bronze doors sculpted before Lincoln's death,[45] and into the rotunda to try to disenfranchise Americans yet again, battling cops who got in their way. Inside the Speaker's office, Lyons eyed the photo of Pelosi and Lewis. It showed the pair during a trip to Ghana to mark "the Year of the Return" on the four hundredth anniversary of when the first ships full of enslaved Africans made the brutal journey across the sea. Lewis stood and took a moment of silence at the "Door of No Return" at Elmina Castle,[46] where human beings, as Pelosi said, "caught their last glimpse of Africa before they were shipped to a life of enslavement." The congressional delegation viewed one of the dungeons, viewing the markings on the wall that had been made by enslaved people before they were shipped off. Pelosi and Lewis posed for a photo at the "Door of Return" after a reentering ceremony. Pelosi called the trip, which took place less than a year before Lewis's death, deeply transformative.[47]

Lyons moved toward the fireplace mantel.

Back at home in Chicago on Friday, Lyons was impressed with the FBI's work. He was evasive at first, not wanting to say whether he had entered the Capitol, but he said he "100 percent guaranteed" that he had seen nothing damaged. He spoke hypothetically, talking about a dream he had in which he was in a mob of people banging on doors and throwing paper and explaining how in the dream people didn't really have much of a choice but to enter. If he *were* inside, Lyons said, it would have been for about forty-five minutes.

The FBI showed him a copy of the Instagram post he'd made that showed the plaque outside of Pelosi's office. "Wow, you are pretty good," Lyons said. "That was up for only an hour."

Lyons opened up a bit more, admitting that he went into the "big boss" office and that there were about twenty or thirty people inside. He said that he saw a broken mirror but that it had already been smashed before he entered. He claimed he left the building after a Capitol Police officer entered with his pistol drawn and ordered him to leave, and he said he passed officers with AR-15 rifles on his way out, through clouds of tear gas. He said he went right to his vehicle and drove home to Chicago. He said that he was willing to share videos he'd filmed but that the file would be too big and that he'd upload the videos to YouTube and send agents a link. An email arrived the next day.

"Hello Nice FBI Lady, Here are the links to the videos. Looks like Podium Guy is in one of them, less the podium," he said, referencing Johnson, who'd grabbed Pelosi's lectern. "Let me know if you need anything else." He included three YouTube links: a ninety-second video showing people outside the Capitol; a three-minute, fifteen-second clip showing people inside a hallway; and a forty-eight-second clip of people in the rotunda.[48]

What it left out is a longer video that Lyons filmed inside Pelosi's office, the one that showed him meeting Schaffer and stealing the wallet. He also didn't turn over the photo he'd had his Uber driver snap, the one that showed him sitting in the back seat, displaying his souvenir: the framed picture of Pelosi and Lewis that he had smuggled out of the Capitol under his sweatshirt.

"I took this off Pelosi's fucking desk!" Lyons texted in a message the FBI would later recover.[49]

---◆---

The flurry of law enforcement activity continued in the coming days. Those arrested within a week of the attack included Larry Brock, a military veteran who went to the floor of the Senate with protective gear and flex cuffs and who gave an interview to the *New Yorker*;[50] Eric Munchel, who brought a Taser to DC and stormed the Capitol with his mom, stealing zip ties along the way;[51] Thomas Baranyi, who emerged from the Capitol with Ashli Babbitt's blood on his hands and said "it could have been me, but she went in first";[52] and Josiah Colt, who scaled the stairs and rappelled onto the floor of the US Senate, thinking he was in the House and bragging about how he took Nancy Pelosi's seat.[53] The feds also got Robert Keith Packer, the guy who wore a Camp Auschwitz sweatshirt,[54] and—just shy of a week after the attack—obtained an arrest warrant for Kevin Seefried, who'd carried a Confederate flag inside the US Capitol.[55]

As Acting US Attorney Mike Sherwin later explained, the FBI and law enforcement had done an initial "surge" of investigations and arrests that mostly netted obvious suspects. Nearly every arrest in the first week involved someone whose name was widely known or whose photo had been seen nationally.

"Social media absolutely helped, where we picked off the Internet stars; you know, the rebel flag guy; you know, Camp Auschwitz; the individuals in Pelosi's office," he said. "The easily identifiable individuals that we were able to quickly find and charge."[56]

Plenty of Jan. 6 participants went mini-viral in their own social circles and ended up getting turned in by people they knew. While Steve D'Antuono of the FBI's Washington Field Office thanked friends and family members for making the "painful" decision to turn in loved ones, there were plenty of people who were clearly eager to turn in those who were only their "friends" in the social media sense of the term.[57]

Pat Stedman, a self-described "Dating & Relationship Coach for Men" from New Jersey, predicted before the Capitol attack that Jan. 6 would "be a national holiday akin to the 4th of July."[58] After he posted videos from inside the Capitol, he was turned in by Witness 1, a college classmate, and Witness 2, a high school classmate. Another man who'd stormed the US Capitol while wearing his nearly decade-old high school varsity jacket was turned in by many who knew him. "We all stormed the us capital and tried to take over the government," Brian Gundersen wrote. "We failed but fuck it."[59]

Some people were so sucked into conspiracy theories that they ended up incriminating themselves online. A man named Albuquerque Head, who dragged Officer Mike Fanone into the mob while declaring "I got one!" before Fanone had a stun gun driven into his neck, chose to spread claims that the QAnon Shaman had been with "several know[n] antifa members." In a tweet on Jan. 8, Head also expressed skepticism that a man with a Confederate flag was truly there, explaining that he was at the Capitol on Jan. 6 and "went to all confederate flags I seen because I'm son of Confederate veterans" and that he'd "never seen this guy."[60] The next day, he retweeted a post about the importance of encrypted messaging. Head was arrested in April and was eventually sentenced to seven and a half years in federal prison.[61]

In addition to bragging about their actions at the Capitol, some rioters used social media to inform their friends and followers that they believed they were wanted and would be taking precautions. "They call us terrorists, we are patriots. Each one a George Washington! Most of them are Benedict Arnolds! Sold out to China and Satan!" wrote Jordon Stotts on Jan. 8. "Peace Out Facebook! Apparently I'm a wanted man and will be going off the grid for a while!" He was arrested in March.[62]

The most consequential development of Jan. 8, though, didn't come with a press release. It came when DC residents who'd been randomly selected from lists of registered voters made their way downtown, just outside of the perimeter of newly erected eight-foot-tall black fences manned by the National Guard,[63] and entered the federal courthouse. There was a grand jury to be sworn in.[64]

With extra precautions in place—a larger space for grand jury meetings, video conferencing in witnesses when possible—grand jurors met to keep up with an unprecedented number of indictments. While masked, federal grand jurors returned indictment after indictment against defendants, many of whom made the grand jury's jobs easier by refusing to cover their own faces as they committed their crimes. By the end of the month, the bureau had opened over 400 case files, prosecutors had charged over 150 defendants in federal cases, and the Justice Department had gotten more than 500 grand jury subpoenas and search warrants.

The tougher investigations and the more complex cases loomed, and the task ahead was daunting. "Marathon," an FBI official wrote at 10:31 p.m. on Jan. 6, "not a sprint."[65]

———◆———

Andy Kim, a South Jersey native and son of immigrants from South Korea, had endured racist attacks in his first run for office in 2018, when the New Jersey GOP ran a photo of him next to a photo of dead fish on ice along with a caption—printed in the Chop Suey font—that called him "REAL FISHY." A Republican super PAC ad warned voters in Obama-Trump swing districts that the former State Department aide was "not one of us."[66] Kim won anyway, and on Jan. 6 he was three days into the 117th Congress, serving his second term.[67]

In the early morning of Jan. 7, just before voting in favor of democracy by certifying Joe Biden's electoral college victory,[68] he walked through the rotunda, which he thought of as a "physical manifestation of our Constitution" and the "most beautiful room in the most beautiful building" in the country.[69] Seeing the mess left by the rioters, he grabbed trash bags and spent nearly two hours cleaning up.

Photos of Kim's actions gave the country some hope after the attack, and Kim was flooded with notes from Americans across the country who appreciated his work. He later donated the blue J. Crew suit he was wearing to the Smithsonian.[70] Some of what Kim and others picked up, though, wasn't just trash. It was evidence. One item he picked up looked to be a stick wrapped in tape that may have been used as a weapon. Across the Capitol, custodial workers also tossed items into the trash, including a large red banner reading "TREASON," which one reporter retrieved from the trash, believing it to be of historical value.

In the rush to get back to business, the building was cleaned up as quickly as possible. Two years later, during the Proud Boys trial, prosecutors would have to rely on the video shot by Scaffolding Solutions to illustrate the crime scene that was the inaugural stage. When the FBI tried to determine what kind of explosive device a rioter had thrown in the western tunnel, an explosives and hazardous devices examiner at the FBI laboratory in Huntsville "was not able to conclusively identify the precise dimensions, charge size, or whether the explosive device thrown was improvised or commercially manufactured," though they said it was capable of causing bodily injury.[71]

Some Jan. 6 defendants were doing cleanup too, deleting incriminating evidence and scrubbing their social media profiles. Inside the encrypted Proud Boys chat, one member predicted on Jan. 7 that the previous day's events "are gonna be used to smash us in ways we couldn't even imagine." He was later hit with a search warrant.[72] Stewart Rhodes, the head of the Oath Keepers, sent a message though Kellye SoRelle, an attorney working with the Oath Keepers. "ALLCON FROM STEWART," it read, using the military lingo for "all concerned." "DO NOT

chatter about any OK members doing anything at [Capitol]. Stop
the chatter. I told you before that anything you say can and will
be used against you. Apparently that wasn't strongly worded
enough for some to get the message. So let me say it like this:
CLAM UP. DO NOT SAY A DAMN THING," it read. "Let me
put it in infantry speak: SHUT THE FUCK UP!"

Others, like Garrett Miller, didn't seem so worried. After being
detained on Jan. 6 and then released by officers because they didn't
have enough personnel to process him amid the chaos, Miller
tweeted his desire to "assassinate AOC"—referring to Rep. Alexandria Ocasio-Cortez—along with other members of Congress
and to lynch the Black officer who'd shot and killed Ashli Babbitt
after she jumped through a broken window leading into the
Speaker's Lobby. On top of "becoming a keyboard warrior," the
feds said, he fancied himself "a revolutionary" and brought a rope,
grappling hook, mouth guard, and bump cap to DC.[73] When federal agents came to pick him up on Inauguration Day, he was
wearing a T-shirt that read "I was there, Washington, D.C.,
January 6, 2021."

One Florida man was back to living his life in Clearwater, and
he was back on Facebook, sharing memes about election fraud.
"In the event of a Civil War, I'm not afraid of the 81 Million Biden
Voters," read a meme he posted on Jan. 12. "Half are dead and
don't exist!" By Inauguration Day, he was in mourning. "America Born July 4th 1776 Died January 20th 2021 Rip," he wrote in
one Facebook post. On a trip to the gas station, he snapped a photo
of the prices at the pump and then opined in a Facebook post that
an administration that had yet to take office was to blame for the
cost of a gallon of regular. "The Biden effect is already taking
place at the pump and Mr. Forgetful and his Commie VP are not
even in office yet!" he wrote. "GOD WERE [sic] DOOMED!" His
post received two likes.

The Florida man didn't know it yet, but gas prices were the
least of his concerns. Thousands of miles across the country, a

woman the Florida man never met had come down with COVID-19 and was stuck isolated at home. That spelled trouble for the Florida man.

Amy, a woman living in a rural area out west and identified here by a pseudonym, was quarantining away from her job as a federal employee. She'd grown increasingly angry watching footage of the deadly attack on the US Capitol.

"I was in quarantine, glued to the television," Amy told me. "I just started spending my evenings and weekends kind of glued to watching the videos because it was so horrific. The more that I watched, the more that I felt like I had lost control over what this country was supposed to be."

But she soon found a way to regain control and lose her sense of hopelessness. She became a "Sedition Hunter."

Amy started scouring Twitter, where a growing community of citizen sleuths was using open-source crowdsourcing to find and identify Capitol attack suspects. As she watched video after video, one suspect stuck with her: that man in the star-spangled hoodie emblazed with Trump's name who was wearing a "FLORIDA FOR TRUMP" hat. Target acquired.

Four years earlier, Trump had emerged from the tunnel for his inauguration, where he vowed to end "American carnage." In the final gasps of his presidency, the man in an American-flag jacket emblazoned with "TRUMP" on both sides had emptied a fire extinguisher on officers and then tossed the canister at the police line.

"Check out this older Floridian who also likes fire extinguishers. He clearly attacked the officers but something (tear gas?) turned him around. Good face shot," she tweeted days after the attack, attaching a photo of the Florida man in the American-flag jacket retreating from the police line.

By a week after the insurrection, she's settled on a nickname: Florida Flag Jacket. On the FBI's Capitol Attack Wanted list, this man eventually became "246—AFO," the acronym-loving bureau's shorthand for "Assault on Federal Officer." It would be over a month—long after Amy had already tipped off the FBI about Florida Flag Jacket's identity—before the swamped FBI would finally post photos asking for information on the man in broad stripes and bright stars.

The identification of Robert Scott Palmer wasn't particularly difficult; it just took some crowdsourcing. Once he had a nickname, someone stumbled upon a video live stream from later that night in which he identified himself by name and hometown. He'd identified himself in two separate live streams, actually, including one in which he displayed his bare belly after he got shot with a less-lethal munition. "I was just yelling at them," he claimed, falsely, in the Breitbart version.

Amy felt as if her messages were being buried at the bureau, and she eventually contacted me with the completely solid identification. When my colleague Jesselyn Cook called Palmer for comment on our story, he confirmed he was at the Capitol, complained that the Biden administration was trying to "vilify the patriots" involved in the riot, and once again claimed it was justified.

"I'm just going about it and letting them make the mistakes that they want and ruin the country as they want, and I'm just trying to live my life right now," Palmer said. "I'm not getting myself any—not deeper, 'cause I didn't do anything wrong—but I'm not involving myself anymore." He hung up when Jesselyn mentioned the fire extinguisher.[74] Palmer sought out an attorney, Bjorn Brunvand, when our story ran, walking into Brunvand's office and saying he considered himself a patriot. Brunvand responded that he considered himself one too and pointed to the picture of himself with former president Barack Obama that he had on display in his office. Brunvand contacted the FBI, and Palmer was arrested twelve days after our story ran.[75] Palmer was eventually sentenced to what was, for eight months, the longest sentence in any Jan. 6 case: more than five years in federal prison.[76] "He's kind of a true believer, you know?" Brunvand later told me. "I don't think he traveled to Washington with the intent of doing the things that he did; I think it sort of spun out of control when he was there."

It would be more than a week after Palmer's arrest that someone from the FBI finally reached Amy on her phone. She was at her daughter's softball game at the time, and she stepped away when she saw there was no caller ID. The other parents didn't know about her side gig chasing insurrectionists.

The FBI special agent she spoke with seemed to just be making his way through calls. He didn't know about the arrest, the information she'd sent in, or the publicity that our coverage of Palmer's identity and Amy's sleuthing work had received.

"He was arrested; he's already appeared in court and is out on bail," Amy told the agent. Amy didn't walk away from the call with much, but she at least had a direct contact she could reach if she came up with another good tip.

"The whole experience was kind of funny," Amy said. "Why did it take six weeks—actually closer to two months? Why did it take so long?"

Amy worked for a large federal agency and understood how slowly bureaucracies could churn. *We're all understaffed and we're all doing multiple jobs*, she thought. *Why would the FBI be any different?* Still, a nearly two-month delay to get a call back on a great tip about a violent insurrectionist was tough to understand. She realized she might need to reevaluate her impression of the FBI.

"We've all watched too many movies," Amy said. "Do we all just overestimate their capacity based on watching too much TV?"

———— ◆ ————

Founded in 1908, the Federal Bureau of Investigation was controversial from the start. One congressman labeled it a "bureaucratic bastard" because President Theodore Roosevelt's administration created what was at first called the Bureau of Investigation (BOI) without congressional oversight. The mass-produced automobile—the Model T debuted that same year—and the bureau were tied together from the start as the new method of transportation began creating whole new categories and methods of crime and breaking down city borders. More coordination was needed. But there was still sharp resistance to getting the federal government involved in law enforcement, which was (and still largely remains) a local and state matter.

After growing in both prominence and power during the infamous Lindbergh baby case, when the bureau adopted an early version of what might be called crowdsourcing, BOI became the FBI in 1935. J. Edgar Hoover was named director and spent the next thirty-seven years until his death building the bureau's pop culture reputation while abusing Americans' rights and wielding secret power over presidents. Through books, radio, movies, television, and the news media, the bureau became "a remarkably successful public relations pioneer," bombarding the public with "a never-ending deluge of positive portrayals of the FBI" that "trumpeted the law enforcement high points and obscured the domestic surveillance low points of the J. Edgar Hoover era."[77] Hoover was just twenty-nine when he was named acting director of the BOI in 1924, and headed what would become the FBI until he died in his sleep forty-eight years later, in 1972. Today, more than a half-century after his death, Hoover still looms large over the FBI, and the bureau's headquarters along Pennsylvania Avenue bears his name.

The brutalist Hoover Building has been crumbling for years: part of the facade is swaddled by netting to prevent cascading chunks of the concrete from killing passersby. Most DC residents don't agree with Trump on much, but they'd concede he had a point when he called FBI headquarters "one of the ugliest buildings in the city." The Hoover Building got some upgrades over the years—*A Starbucks! Picnic tables in the courtyard!*—but there's only so much you can do to renovate a building made of raw concrete. So FBI headquarters is now a heavy-handed metaphor for the bureau as an institution: big, imposing, a bit stuck in the past, and immutable even when things are falling apart at the edges.

The FBI is still the nation's premier law enforcement organization. But it's also a behemoth bureaucracy that doesn't match up to the Hollywood hype. Stepping inside the FBI building for the first time as a cub reporter, I found decor that evoked *The Office* rather than *CSI*. The FBI had extraordinary technological resources at its disposal, yet a top FBI spokesman was mystified by the existence of a digital-only news organization when I became the first "new media" reporter to attend the bureau's reporter roundtables. When a college buddy became a spokesman for one of the bureau's field offices, it was yet another reminder that the bureau isn't full of superhumans, just regular people (some of whom, presumably, also threw killer parties back in the day and now dominated the dance floor at their friends' weddings).

To keep up with twenty-first-century threats, the FBI needs to beef up its tech teams. But as a government employer, they can't compete with the salaries that tech companies can offer. The private sector has generous benefits, flexible schedules, and incredible office campuses in popular cities and Silicon Valley. For FBI special agents, the bureau has a rough training session in Quantico, a ban on the use of marijuana, and possible assignment in Anchorage or Omaha. It's not necessarily the first job that computer whizzes seek out.

The cadre of FBI intelligence analysts hired in the aftermath of the Sept. 11 attack, meanwhile, found themselves treated as second-class citizens, given assignments like taking out the trash, answering phones and radios, operating switchboards, and escorting cleaning personnel or repair workers. A 2005 report from the Justice Department's internal watchdog, produced when the first iPhone was still in the early development stages, recalled that agents gave intelligence analysts "the administrative work the agents prefer not to do, such as Internet searches," with one analyst saying that much of their job didn't require a college education. "Special agents hate to do their own research, even if it means finding out who the new [special agent in charge] in Kansas City is," one FBI analyst told investigators. Such mundane assignments, along with a muddled career path that left little room for promotion and growth, sent the FBI's newest and most highly educated recruits scampering for the exits.[78]

"It wasn't the photocopying or the lack of promotion potential that compelled me to leave my job as an FBI analyst," one former analyst wrote in 2005. "It was the frustration of working in a system that does not yet recognize analysis as a full partner in the FBI's national security mission."[79]

Given the bureau's historic struggles with technology, it wasn't necessarily shocking when the first images of Capitol insurrectionists wanted by the FBI—released around midnight on Jan. 7—came out as PDFs, with tiny photos of ten different suspects jammed onto a sheet. That strategy didn't make a lot of sense in 2021. There would be hundreds of suspects, and dozens of physical "Wanted" posters wouldn't work well when suspects were spread out all over the country. A month later the bureau redesigned their website, allowing them to post several high-quality photos of Capitol suspects with permanent individual links.

Slowly, the bureau began improving their social media presence, resurfacing images to expose them to new audiences that might not make a regular habit of checking the FBI's website. It was new territory for the FBI, which was used to calling attention to a Most Wanted list of high-profile targets as well as targeting bank robbers in certain regions. Seeking hundreds of suspects across the United States was just unprecedented. "We haven't had an investigation to this scale and scope where we needed to identify this many individuals," an FBI official told me. "Because all these individuals came to DC and then dispersed across the country, that adds to the complexity of it."

As the FBI published single still images, online investigators were compiling multimedia databases that tracked suspects' every move. With their open-source database, a suspect known only as a trespasser—a "MAGA Tourist," as they were called—could, with a newly uncovered video, turn into an assailant. The power of the crowd was apparent.

Capitol attack crowdsourcers also deployed a longtime FBI strategy that the bureau itself had yet to use against unknown insurrectionists: nicknames. The FBI regularly nicknames bank robbers, using names like "the Geezer Bandit," "the Grandma Bandit," "the Pink Lady Bandit," "the Fake Hair Don't Care Bandit," or "the Spelling Bee Bandit" for the bank robber who spelled "robbery" wrong. The FBI's catchy names offer a major publicity boost, spreading photos of suspects further because the nicknames can generate headlines and social media discussion.

For the Sedition Hunters network, nicknames were useful as both publicity and crowdsourcing tools. Volunteers were staring at hundreds of suspects, and it was a lot easier to remember someone with a catchy nickname that referenced their appearance rather than a random number. One Sedition Hunter helping generate the nicknames explained to me that a good nickname called up a mental image and entertained at the same time. Funny nicknames helped break apart the grueling process of looking at video after video of Capitol violence, they said. There was "Bald Eagle," the guy wearing an American-flag suit and an eagle mask who, when he took the mask off after joining the mob trying to push through a police line, revealed his bald head. "Tricorn Traitor" for the guy in the colonial hat. "Pippi Long Scarf" for the guy with the long scarf. "Pinky N the Brainless" for a woman with pink hair and her partner.

The Sedition Hunter network was chugging along, churning up plenty of solid leads on suspects, and they were submitting them to the FBI. Then those tips fell into a black hole at the bureau.

Sammy, identified by a pseudonym, was a Sedition Hunter who started hunting for insurrectionists as she recovered from cancer surgery. Sammy found herself "immersed" in the work, which she found to be a wonderful escape from reality. "It's a great distraction from my own worries," she told me. "I feel like I'm doing something useful when I can't do much of anything else."

Yet Sammy was also growing flustered not knowing whether the tips she submitted actually ended up in the right hands. She wished the FBI would take a page from the Domino's Pizza Tracker app, imagining getting alerts assuring her that her tips weren't just buried in an inbox somewhere. *1 p.m.: Agent Smith has received your tip*, she imagined they'd say. *3:45 p.m.: Agent Smith is cross-referencing your tip.* "Then when they actually go and get the guy, they show like a Google Map in real time," she joked.

There are good reasons that the FBI can't go into that much detail about an unfolding investigation, of course, but tipsters felt as if the feds weren't paying attention. The bureau had to sort through hundreds of thousands of tips, and solid leads on violent insurrectionists were being overlooked. Frustrated tipsters had no clue what was happening behind the scenes and were left thinking that the solid information they submitted to the bureau was just buried.

Often, their suspicions were correct. Just as they had been during the Lindbergh investigation eighty-nine years prior, the bureau was swamped, chasing leads in too many different directions. The FBI was also still reeling from the murder of two special agents who were serving a warrant in a child exploitation case in Florida on Feb. 2. It was extremely rare for FBI agents to be killed in the line of duty: the last death had taken place nearly thirteen years before, in 2008. The FBI typically overwhelms suspects with a show of force and brings enough manpower to suppress any fight-or-flight instinct. I couldn't help but wonder if the unprecedented, countrywide Capitol manhunt that was

stretching the bureau's resources had negatively affected preparations for the raid.

Even as the FBI community mourned the deaths, the bureau
was in what a source later described as a "balls to the wall" posture. Nearly every field office in the country was involved in an
investigation. They were still drowning. FBI agents who typically specialized in areas like child pornography and gang crime
were instead popping up in Capitol case affidavits.

"This is definitely an all hands on deck no fail moment for the
FBI," Alan Kohler, the assistant director of the FBI's Counterintelligence Division wrote on Jan. 10. "Whatever we need to do to
support we will do."[80]

Many of the FBI employees who ended up investigating the
Capitol attack had some skin in the game. Special Agent Carlos
Fontanez, who worked cyber matters and drug-trafficking cases
and was assigned to the criminal squad that investigates violent
crime and gangs, arrived to a chaotic scene.

"It was quite something when I got there. There was smoke
everywhere, broken doors and windows. I even saw certain parts
there was blood on the floor," Fontanez said. "It was definitely
more like a crime scene rather than the U.S. Capitol."[81]

FBI special agent Sylvia Hilgeman, who later worked the case
against the Oath Keepers, had been at the bureau for about
twelve years. Hilgeman didn't come to the bureau from the path
most of the public thinks about. She was a certified public accountant, or CPA, and worked in public accounting for about five years
before joining the bureau. She spent eight years in New York for
the FBI, working bank fraud and accounting fraud cases from
beginning to end: document review, bank statements, business
emails, you name it. Then she came to FBI headquarters and
spent two and a half years working terrorism-financing cases.

Hilgeman was out at her office in Manassas, Virginia, when
she heard the Capitol had been breached. Her supervisor told her
to respond, and she and her squad drove into DC, where she and

at least one other CPA assisted the Capitol Police and Metropolitan Police Department in clearing the Capitol grounds.[82]

It was a different story across the country, where some FBI employees had no personal connection to the Capitol and didn't realize the whole scope of what took place. One email laid out some issues the bureau would face in its investigation, including a lack of enthusiasm from some in the FBI, who were politically aligned with the president and didn't see what the big deal was about Jan. 6.

"There's no good way to say it, so I'll just be direct: from my first-hand and second-hand information from conversations since January 6th there is, at best, a sizable percentage of the employee population that felt sympathetic to the group that stormed the Capitol," the former FBI employee wrote to Paul Abbate, who'd soon be the bureau's deputy director.[83] "Several also lamented that the only reason this violent activity is getting more attention is because of 'political correctness.'" The FBI, the person wrote, had a problem on its hands:

> *I literally had to explain to an agent from a "blue state" office the difference between opportunists burning & looting during protests that stemmed [from] legitimate grievance to police brutality vs. an insurgent mob whose purpose was to prevent the execution of democratic processes at the behest of a sitting president. One is a smattering of criminals, the other is an organized group of domestic terrorists.*
>
> *I was talking to [a supervisory special agent] in a "red state" office who was telling me that over 70% of his [counterterrorism] squad + roughly 75% of the agent population in his office disagreed with the violence "but could understand where the frustration was coming from" which led to the "protesters getting carried away."*
>
> *An analyst in a "purple state" described watching horrified as the events were unfolding on the news, while several coworkers chalked up the insurgency as a "response to*

*everyone being quarantined at home for months" and more*
*"on edge, because so many lost jobs and lack steady income*
*because of COVID."*

The email mentioned that the squad's televisions were tuned to the right-wing propaganda network known as Newsmax after the election because Fox News was "playing to the left" and reporting "fake news." A former FBI senior analyst, less than two years off the job, had also fallen into conspiracy theory land and had a "Facebook page full of #StoptheSteal content," the email said.[84]

One FBI special agent, who was at a shooting range with other members of law enforcement on Jan. 6, laughed about the attack and the "goofballs" behind it. "We were literally laughing, people were cracking up, you know, somebody has Nancy Pelosi's podium."[85] An FBI supervisory intelligence analyst in the FBI Boston Field Office, one who supervised analysts who assisted a few Jan. 6 cases, would soon retire from the bureau and would start spouting conspiracy theories about how Jan. 6 was "a setup" and call the FBI "the Brown Shirt enforcers" of the Democratic National Committee.[86]

The FBI's Jan. 6 investigation was facing major challenges: dissent from within the bureau, a seemingly insurmountable digital pile of evidence that grew each day, a database of tips that would grow into the hundreds of thousands, and a long list of criminal suspects spread out all over the country.

The day after the attack, Sherwin said he wasn't "going to play Monday morning quarterback" and question why rioters "weren't zip tied as they were leaving the building by Capitol Police" but said the lack of information had given the FBI a long task ahead.

"The scenario has made our job difficult because, look, now we have to go through, process, cell site orders, video footage, to try to identify people and then charge them and then try to execute their arrest," Sherwin said.

The FBI was getting an unprecedented level of tips and had "literally hundreds of people vetting social media," Sherwin said. Soon they'd have the results of a search warrant that gave them info on anyone who'd live streamed within the Capitol on Facebook,[87] lists of Google accounts that were used on phones within the Capitol, and cell site returns that provided the phone numbers of most of those who went inside the building.

"The scope and scale of this are unprecedented, not only in FBI history but probably DOJ history," Sherwin said. "We're looking at everything from simple trespass, to theft of mail, to theft of digital devices inside the Capitol, to assault on local officers, federal officers both outside and inside the Capitol, to the theft of potential national security information, to felony murder. Just the gamut of cases and criminal conduct we're looking at is really mind-blowing."[88]

Nonetheless, the FBI, Washington Field Office assistant director in charge Steven M. D'Antuono told reporters, was "quite familiar with large-scale, complex, and fast-moving investigations" and was up to the challenge. "As Director Wray says, 'The FBI does not do easy,'" D'Antuono said, describing a "24/7, full-bore, extensive operation" being run by the FBI. "The men and women of the FBI will leave no stone unturned in this investigation."[89]

He said that the FBI was "in the first quarter" of the investigation and that the bureau was "methodically following all the leads to identify those responsible and hold them accountable."[90] He said "scores of dedicated agents, analysts, and other specialized personnel" were working behind the walls of every FBI field office throughout the country, "chasing down leads, reviewing evidence, and combing through digital media to identify, charge, and arrest anyone who was behind the siege we all saw on the sixth."

D'Antuono encouraged rioters to self-report, saying rioters weren't in the clear just because they went home. "Even if you left D.C., agents for our local field offices will be knocking on your door if we find out that you were part of the criminal activity at the Capitol," he said.

The FBI, he said, had "been here before," with the WFO leading "major complex investigations—including the terrorist attack on the Pentagon on 9/11, the D.C. sniper case, and the Navy Yard shooting." They were up for this "enormous endeavor," he said.

At the time, federal authorities had no idea just how enormous it was. The FBI didn't have any "hard and fast" figures on how many people entered the Capitol, a top FBI official said, but federal officials were estimating internally that eight hundred people entered.[91]

Later, as the Sedition Hunters organized and used open-source tools to examine the Capitol attack from every angle, they realized the FBI's initial estimates were way off. Within nine months, their database included more than two thousand Capitol "insiders," as they called them.[92] By the two-year anniversary, the number had surpassed three thousand.

"Everyone is all-in on these cases," Sherwin said at the time, not realizing the full scope of the task ahead. "There is no manpower issue. If a crime was committed, we are going to track that person down, and they will be charged." Sherwin predicted the cases would "continue over the next several weeks, several months, and probably even into the end of the year."[93]

D'Antuono was similarly insistent: the FBI was ready.

"It is complex and it is big—both in size and scope. But at the FBI, we do big, and we do challenging, and we do complex," D'Antuono said. "What happened at the Capitol on January 6 has not occurred in over 200 years. We owe it to the American people to find out how and why it did. We are committed to seeing this through—no matter how many people it takes, how many days it takes us, or the resources we'll need to get it done. We will get to the bottom of this. The American people, and this country, deserve no less."[94]

# CHAPTER 8

# "Those Meddling Sleuths"

The night of Jan. 6, as police helicopters swirled in the sky above her apartment in DC's Navy Yard neighborhood and Trump supporters streamed back to the nearby hotel, a twentysomething professional in the nation's capital felt that she needed to do something. Claire, identified here by pseudonym, watched news coverage of the Capitol attack with anger. The FBI needed help. They wanted the public to identify rioters.

*Okay, fine*, Claire thought, *I will.*

Claire's roommate was out of town, and she was sitting alone on her couch when she pulled out her phone and opened Bumble, the dating app where women make the first move. She had an "on-and-off relationship" with Bumble and hadn't done much online dating in the middle of the COVID-19 pandemic. Now it was time for a profile overhaul. First, she took down photos of herself doing "very non-MAGA things," like attending the Women's March that immediately followed outgoing President Trump's 2017 inauguration. She used a photo of herself doing some sort of "vanilla recreational activity," maybe on a boat, maybe at a brewery;

244 | Ryan J. Reilly

she can't quite recall. Next, she swapped her political beliefs to conservative.

Then she started swiping. For democracy.[1]

Unlike Hinge, the dating app Claire used more regularly, Bumble didn't limit the number of people she could match with, and it wasn't limited to her extended social media circle. Claire could just "endlessly scroll through these guys," looking for Jan. 6 suspects.

Claire sought out conservative men who were visiting from out of town, who would have almost certainly been there for Trump's speech and likely for the march to the Capitol. She took on an alias, pretending to be a woman who'd fallen for or chosen to believe Trump's lies about voter fraud and who was also easily impressed by self-proclaimed "patriots" who'd brag about their actions to a girl they'd only just met on the internet. Endlessly flattering men wasn't her typical approach to online dating, but it was effective here. She strategically regressed back to a type of "teenage flirting" that felt extremely high school.

The men she spoke with were "very on-brand for a MAGA rally," parroting the long-debunked election fraud talking points they'd heard from prominent Republicans, and they couldn't see Claire roll her eyes on the other side of the phone. "They just wanted to regurgitate a lot of these ideas to somebody, and it seemed like I was a willing participant," Claire said. "It definitely didn't take a lot of arm twisting to get them to start talking about it. Basically me being like *Wow, so cool—then what? What else?* Was pretty much all it took."

Part of her wanted to break character, tell the men, "This is terrible and you're a terrible person," but that wasn't her mission. Claire had launched a modern-day honeytrap, a classic intelligence strategy in which targets are lured into disclosing information with the hope of romance.

"I felt a bit of 'civic duty' I guess, but truthfully, I was mostly just mad and thinking, *Fuck these guys*," she later told me.

Claire ended up sending messages to about a dozen men. Plenty took the bait. One was a man named Andrew, who was in the area from Texas and whose photos included an image of himself riding a motorcycle. Like many date-app users who live in DC, Claire "almost always" screens out potential romantic partners located across the river in Northern Virginia. Andrew was in Alexandria, about eight miles away. In real life, dating someone from the Virginia suburbs would've been very inconvenient and sent Claire's monthly Uber expenditures skyrocketing. But Claire's MAGA alter ego didn't have any reservations about crossing the river.

She swiped far right.

"Hey how's it going?" Claire wrote.

They got to chatting. She made small talk and basically tried to get Andrew thinking they were on the same page. Andrew submitted to the charm offensive, telling Claire he was at the Capitol.

"Were you near all the action?" she asked.

"Yes," he replied. "From the very beginning." He sent her a selfie he said he took about thirty minutes after being pepper sprayed.

"I was [pepper] sprayed, tear gassed, had flash bangs thrown at me, and hit with batons for peacefully standing there," he claimed. "Safe to say I was the very first person to be sprayed that day . . . all while just standing there." (Spoiler alert: he was doing much more than that.)

Andrew conceded he was "fucking fed up" with the government when he arrived at the Capitol, but he started telling Claire that the people who did the damage were *obviously* antifa.

*Who do you think incited this violence? That's what they did all summer with BLM stuff. Only takes a few to get a crowd going. Majority of the people attacking police and smashing windows were all antifa. They just threw on a Trump hat or shirt they bought on the street. They have had this all planned out for weeks. The media is doing what they always do and lying to the people. The mayor and democrat leaders pur- posefully made sure they did not put up the tall fencing they have up now, and kept the police and security forces limited. They wanted this to happen.*

In the days following the attack, Andrew was back in Texas and wanted to video chat, but Claire begged off. Sitting on her couch, with the city on virtual lockdown, she told him that she was out at a beer garden with friends. (That was suspicious, given it was the dead of winter and COVID-19 vaccines were largely still unavailable, but Andrew didn't seem to catch on.[2]) Trying to gather more intel, she asked him if he'd be back in DC anytime soon. Perhaps, she suggested, they could check out the beer gar- den she was pretending to be at.

"Maybe depending on what happens with election," Andrew wrote. "Biden still isn't in office . . . and there is too much crimi- nal stuff to come out. There are many many Patriots ready and willing to head back depending what happens."

Using the skills she'd developed while using dating apps for non-insurrection purposes—finding a guy's real identity with just a first name and a few clues—she got to work.

"Having been on dating apps in the past, I'm good at finding guys' last names before I go meet up with them in a strange place," Claire said. "I feel like that's a skill you inherently pick up on dating apps that ends up being very helpful in this context."

She narrowed down the scope of her search using Facebook filters since she knew he went to Houston Community College and owned a business called Hi-Flow Houston, and she found his full name. Then she sent the information to the FBI, along with the names of two other men she'd chatted with. None of the men came right out and said that they'd assaulted cops, but Claire thought her tips could help put the puzzle together. Maybe they were lying and had gone inside.

"If they have pictures of these guys, now they had a name," she said. "I think I figured it was basically just to help with legwork on the backend."

Claire was amazed that Andrew and the other men, thanks to her "comically minimal ego-stroking," had given up so much info. Her strategy was basically saying, "Wow, crazy, tell me more" to guys on repeat "until they gave me enough," she said.

"One of my friends was like, 'You basically got all these confessions just being, like, *Haha! Then what?*'"

Once it was clear that the conversation wasn't going to result in any evidence, or when things got "a little too kooky," with MAGA fans sending her conspiracy theories, Claire would unmatch with them and block them. "I basically didn't want to have any matches currently living on anybody's profile to be like *Hey, this girl might have been the one to turn me in*," she said.

Three months later, Claire heard from an FBI special agent in the Houston Field Office in Texas. The female special agent ("Let the women do the work," Claire joked) had some questions to ask. Bumble stings weren't typical FBI territory. "It was clear we were both doing this for the first time," Claire said. "Clearly, it's not a common scenario, because I've never met this person, I don't know him. I just got his name."

Many of her friends who lived in DC were "emotionally wrecked" after Jan. 6, so Claire did not end up sharing her sleuthing work with them, but she did talk to friends who didn't live in the area about what she did. Bumble stings proliferated in the days after Claire sent her information to the FBI, and the service even temporarily disabled its political filters "to prevent misuse" a bit over a week after the attack. They soon restored it after an outcry, though by that point the secret was out.[3] That frustrated Claire.

"We're targeting insurrectionists; it doesn't seem like that goes against your code of conduct," she said. "Meanwhile, people get dick pics all the time, and I was like, *That seems a bigger problem for you to handle.* If you're just about protecting people, maybe protect women who are getting unsolicited pornography, but that's a bigger fish to fry."

In late May, the FBI showed up to Andrew Taake's home and talked with a FedEx driver who had just delivered a package to Taake, and they got the driver to positively ID the man.[4] They arrested him in July.

Although her discussions with Andrew had been the most troubling of any she had had with Jan. 6 participants, Claire was still shocked when she saw images of what Taake was accused of: pepper spraying officers and then attacking them with a metal whip before storming the building, weapon in hand.

"I assumed they mostly just needed a name to match with the face but I had no idea all the images they had of him at the riot/all the shit he got into," she wrote me.

Taake was a self-employed handyman and the owner of a pressure-washing business who had a felony record but kept multiple firearms at home anyway. He was on bond when he stormed the Capitol and assaulted law enforcement, with a pending charge for soliciting a minor online.[5] After the FBI came to arrest him, a federal magistrate judge ordered him held until trial.

"Overall, Mr. Taake's willingness to physically assault officers with bear spray and a metal whip, **while on bond and conditions of release for the felony offense of solicitation of a minor,** demonstrate by clear and convincing evidence that he poses a concrete threat both to the community and to specific individuals," US Magistrate Judge Christina A. Bryan wrote, bolding part of her order for emphasis.

When Taake was arrested, and after Claire shared the story of her Bumble operation with me using a pseudonym, praise rolled in.

"Give this woman the Presidential Medal of Freedom," wrote one Twitter user. "Claire is a hero and these men are so stupid oh my god," tweeted another. "Not all heroes wear capes," added one. "Catfish for the cause," said another. "Ask not what your country can do for you, ask what you can do for your country," tweeted one. "*Bonk* go to horny jail," wrote another. Even a Trump-appointed federal prosecutor was impressed.

"Just in case there are still folks out there not fully convinced that the criminals on 1/6 were less than brilliant," wrote former US attorney Jay Towns, "'Claire' helped bring them to justice, which is what patriots do."

In the days after Jan. 6, there were a lot of Claires. Organizing them—so much as you can organize a mostly anonymous group of online sleuths—would take time. Regular citizens would become the most effective tool in the FBI's Jan. 6 investigation.

"If normal people have to do the work, we will, but it's annoying," Claire said. "It definitely does feel like it's kind of like a team sport . . . it feels like kind of a drop in a larger bucket."

---

Joan (also identified by a pseudonym), a mother and a former teacher from south central Pennsylvania, has a "weird hobby." She's a Facebook detective.

"Some people crochet. Some people paint. I look up people," said Joan. "I'm the go-to person for all my friends if they meet a new man. They're like 'Hey, look him up, give me all the details.'"

Joan was at her home in Hershey on Jan. 6, and she found herself in tears as she watched what was happening in DC. "I was so upset," she later told me. "This is not who we are as a country, or it's not who we're supposed to be." So just over a week after the deadly attack, Joan was thrilled to have a new opportunity for Facebook investigative work land in her inbox.

Someone in the budding Sedition Hunters community reached out through a regional political Facebook group Joan helped administer. The internet detectives were trying to identify the man who invaded the floor of the US Senate wearing a "Hershey Christian Academy" sweatshirt, and they thought she could help spread the word.

Joan got to work. She went to Hershey Christian Academy's Facebook page and started digging, learning everything she could about the small school with barely a dozen staff members. It'd only opened in 2019, and now—because someone had the bright idea of storming the floor of the US Senate while wearing school swag—it was at the center of the Capitol insurrection.

"I went and started looking at all the likes, all the comments. Then on every single like and comment I would go look on their profile and snoop around and say, 'Oh, that guy's too fat, that guy's too bald, that guy's too bearded, it's not him," she said.

Soon she stumbled on a "pretty vanilla" Facebook profile of a man named Zeeker whose profile photo only featured a snowman. When she dug around a bit more and plugged Zeeker's name into Facebook's search bar, she turned up photos that he'd been tagged in. She realized she might just have a match. She took some screenshots, poked around his Instagram, and did a bit of Googling.

"Zeeker Bozell," she learned, was Brent Bozell IV. He was the son of Brent Bozell III, the high-profile conservative activist, founder of the Media Research Center, and grandson of Brent Bozell Jr., the ghostwriter for Barry Goldwater and supporter of Joseph McCarthy.[6]

*Woah*, she thought as she went down a Google rabbit hole and read his grandfather's Wikipedia page laying out his ties to William F. Buckley, the founder of the *National Review* and a man considered the intellectual godfather of the conservative movement.[7] *I've really stumbled onto something.*

"I mean, I'm just, like, a mom," Joan later told me. "It was kind of a holy shit moment."

Joan dialed up the FBI's National Threat Operations Center just after midnight. The bureau was completely overwhelmed in the aftermath of Jan. 6, so she was stuck on hold for about forty-five minutes. She gave her info and waited for the FBI to get in touch.

Ten days later, an FBI agent gave her a ring. Joan laid everything out. What Joan called her "weird little detective hobby" had left the FBI special agent impressed. *How did you get all of this just from seeing his picture?* he asked her. *You got his whole life story!*

But the FBI was swamped. Thousands of tips were pouring in every day, and sorting through the chaos was a logistical nightmare. The charges against Bozell wouldn't come through for weeks. Joan had at least gotten a call back—a real, live FBI special agent was on the case. Still, like thousands of other FBI tipsters, Joan grew a bit impatient as time dragged on. *Come on,* she thought. *I handed this guy to you on a silver platter! Where is his arrest?*

Then it happened: Leo Brent Bozell IV—a.k.a. "Zeeker"—was arrested, and the charges against him unsealed. It turned out that two people who knew Zeeker had turned him in: someone affiliated with the school and someone who knew him as a girls' basketball coach.

Zeeker's dad, Brent Bozell III—who had gone from calling Trump "the greatest charlatan of them all"[8] to signing an open letter calling for fake electors to be appointed and claiming there was "no doubt President Donald J. Trump is the lawful winner of the presidential election"[9]—didn't speak out about the charges against his son. He had already condemned the riot the day it happened, although he tossed the rioters a bone.

"Look, they are furious that they believe this election was stolen," he said. "I agree with them."[10]

He called the attack disturbing. "You can never countenance police being attacked. You cannot countenance our national Capitol being breached like this. I think it is absolutely wrong." He also called for a "thorough investigation" and raised the possibility that maybe "the other side" was involved. "It's just a hunch," Bozell III said.

Joan kept going, taking to the broader hunt for insurrectionists with zeal as she scoured videos and followed suspects known only by their hashtag as part of the crowdsourced efforts. She kept an eye out for Bozell too, and she eventually spotted that sweatshirt again. Twice, in fact. In one video, Bozell is on the front line of the battle with police officers, attempting to rip down a tarp and let the mob through. In another, Bozell was smashing a Capitol window. So weeks after she helped the FBI identify Bozell, she pointed federal authorities to evidence they seemed to have overlooked.

"Thank you very much for this new information," the FBI special agent wrote in an email after Joan passed along links. "I will make sure I share this with the prosecutors in the case."

The next day, prosecutors presented their evidence before a federal grand jury in Washington, which indicted Bozell on seven counts. Among the new felony charges: destruction of government property for breaking a window.

Joan was blown away, and a bit intimidated, by the role she'd played in identifying a member of a conservative political dynasty as a Capitol insurrectionist and helping the FBI build the case against him. A hobby Joan sneaked in while homeschooling her kid could land Bozell IV—the namesake of men at the forefront of America's conservative movement for the better part of a century—in federal prison for years.

She remained astonished that Bozell wore a sweatshirt bearing the name of a tiny school his child attended as he stormed the US Capitol, smashed out a window, and took over the floor of the US Senate in an attempt to overturn the election on behalf of Donald Trump.

"He probably would've gotten away with it," Joan joked, "if it weren't for these meddling sleuths."[11]

———————◆———————

No one is quite sure where the photo came from at this point, but it grabs you. It is shot from above the "tunnel of doom" and centers on an officer, face down, being dragged down the stairs by a mob.

A Twitter user, under the vanity name "No Nazis for Me Thanks," posted the image on the evening of Jan. 9.[12] The news of the death of US Capitol Police officer Brian Sicknick was less than forty-eight hours old, and details were sparse. Authorities had only announced that Sicknick "was injured while physically engaging with protesters," had collapsed at his division office, and had died at the hospital at 9:30 p.m. on Jan. 7.[13] The videos that played on cable news were mostly outside shots filmed from a distance, showing rowdy demonstrations on the east steps of the Capitol, the side where networks had cameras set up before the media pen was ransacked.[14] Social media, filled with images and videos from journalists who'd risked their safety entering into the mob and from rioters themselves, looked much different.

The split-screen experience began on Jan. 6 itself. While the folks at the FBI were watching cable news and trying to figure out what was happening, Donell Harvin was watching the riot unfold online. Donell, the executive director of the National Capital Region Threat Intelligence Center, was partially responsible for why the Metropolitan Police Department was more prepared than most law enforcement agencies on Jan. 6. He'd seen groups discussing occupying Congress to stop the vote, even if he didn't think it was possible. "I didn't think that they would be able to do it," he later said. "We knew what the intent was. No one figured that they'd be able to storm a federal facility, and now we all know otherwise."

Given his role, Harvin was chiefly concerned with the streets of DC, thinking the Capitol Police were prepared to defend the building. He had one of his analysts brief the DC medical examiner to prepare for a mass fatality event. "I was scared that there would be bloodshed," he said. His fears increased when he got stuck in traffic amid cars with out-of-state license plates. He saw Trump flags, Confederate flags, "Don't Tread on Me" flags, you name it. As chief of Homeland Security and Intelligence for DC, he arrived at the fusion center and began watching it all unfold.

He saw the mob move toward the Capitol. He watched the tugging and pulling on the "loose and ragged" defenses the police had set up: bike racks. He watched things deteriorate quickly. He texted someone that the mob was going to breach the scaffolding; then they did. He watched them storm into the building. What was happening was "not being carried in live media because the media aren't down range.

"I wasn't watching on television like everybody else was. We have other means to look at some of these things, most of them as social media. People are live streaming them. So we have access to that. It's OSINT."[15]

OSINT, or open-source intelligence, is a phrase self-described "gray haired grandma" Donna had to Google when she first started getting involved in online sleuthing efforts after Jan. 6. She watched the attack on two screens as well: on television and on her laptop, as she scrolled through Twitter.

"What I realized was it was a catastrophic, monumental failure of law enforcement," Donna said. "I just realized that if citizens did not get involved and start collecting evidence independent of them and with a tenacious amount of focus and steadfastness and determination, that these people would not be held accountable." She took an OSINT crash course and got involved.

One of her first points of entry was that photo of the officer being dragged, face down, on the stairs just outside the tunnel. Not much was known about the officer at that point: some initially believed it was Sicknick and later thought it was Michael Fanone, another police officer who'd been attacked in the crowd. The overhead photo showed one suspect with his hand on the back of the officer's neck, wearing what an early Sedition Hunter described as a "distinctive scalloped grey backpack." He got the nickname #Scallops, which was, admittedly, not the best one they'd generate.

Donna also focused on Individual4, who would go on to become the white whale of the Sedition Hunters community. Individual4 seemed to be everywhere at the Capitol but was frustratingly disciplined about keeping his (or her, perhaps?) face covered. They were part of the initial breach, streaking up the lawn waving a Trump flag; they cut the tarp on the scaffolding around the inauguration stage with a knife, went inside the Capitol, and were on the front lines of confrontations with law enforcement; and then they smashed media equipment outside the Capitol building after being pushed out. "Fight for Trump!" they yelled, while wearing a protective vest with zip tie handcuffs attached.

Donna whipped together a composite image that featured all the gear Individual4 was wearing, with hopes of identifying him. She worked on some other cases too. Soon, though, she ended up spending a lot of time trying to "herd the cats." With so many people helping, things could go haywire, with people duplicating work.

Donna wanted to eliminate inefficiencies. Once she had a direct contact with the FBI, she tried to convince them to give the sleuths some sort of feedback, a signal on whether they were spending their time in the right place. A common scenario was for sleuths to spend a lot of time working to identify a suspect, only to find out months later that the bureau had gotten a bunch of tips from people who personally knew a defendant. That could be time wasted. Donna said her bureau contacts wouldn't budge. "Keep doing what you're doing" was the extent of the feedback, and while she understood the underlying policy about not talking about ongoing investigations, she found it very frustrating, very antiquated.

There was also the problem of separate systems. The Metropolitan Police Department had put out lengthy PDFs featuring suspects with one set of numbers. The FBI had another set of numbers and those "Wanted"-style posters featuring multiple small images of suspects, as though rioters were going to be caught in 2021 on the off chance someone spotted them on a piece of paper hung up at their local post office.

The lack of a feedback loop was extremely frustrating. "They were behind the eight ball," Donna said. The same concerns that the sleuths had about their interactions with the FBI—the lack of even basic feedback—had been flagged in an inspector general report in 2015.

"Several private sector representatives told us that providing information to the FBI is akin to sending it into a black hole—the information goes in and the entities never hear any more about it," the report said.[16]

To many of the sleuths, the technological limitations were jaw-dropping. "The thing that kept me going was how slow they were and how awful they were at catching up to us on Twitter," Donna said.

"They're just slow. This many people committing crimes on the same day obviously just overwhelmed the system; it's not designed to handle that," she said.

———•◆•———

A young woman, let's call her Harriet, was working in tech on the West Coast, and like her colleagues, she wasn't getting much work done on Jan. 6. She soon joined the online manhunt. Everything she read online said she should separate her real identity from her online sleuth identity, just so she wouldn't become a target, so she fired up a new Twitter account. She's still not sure who first came up with the name Sedition Hunters. Harriet adopted it after someone added her old screen name to a Twitter list under that banner. She had no idea at the time it would land in FBI affidavits and become a major driver of the bureau's sprawling investigation into the Capitol attack.

"I didn't think it would be the name that would be used everywhere," Harriet later told me. "Sedition isn't quite the word for what most people are getting charged with. Hunters is very aggressive. I feel like I would probably change a lot about it, but at the time I was like, *Oh that's perfect.*"

Sitting at her kitchen table, she began tweeting.

> *Tips for #SeditionHunters!*
> *- Choose a person of interest*
> *- Make a thread*
> *- Collect sources*
> *- Describe actions and clothing*
> *- Research interesting details*
> *- Create a hashtag and use #SeditionHunters*
> *- Lead all information for them*
> *- Tag @SeditionHunters with important info* ☺

She wanted to be a resource for people. And while she couldn't take credit for the Sedition Hunters name, she did take credit for the poster style she developed, featuring an American flag in the background.

"I put that together in Canva"—the free graphic design tool—"in five seconds," Harriet said. "Red, white, and blue, and an American-flag background, because I knew they were going to call us traitors. . . . This is a very American view, to try and keep democracy alive, to encourage police officers not to be beaten to death, and so I used that imagery."

Harriet didn't stick with the account for long, handing it off shortly after it started. "It all felt really too big for me," Harriet said.

What she created from her kitchen table would become one of the most effective drivers of the FBI's Jan. 6 investigation.

———— • ————

After the attack, the FBI realized how outdated their technology was. They were "hamstrung," as one report put it, on doing basic searches through the data they'd collected. The FBI director described the system as "cumbersome" and "difficult," while a news story said the bureau's computer systems were in the "high-tech equivalent of the Stone Age."[17]

The year was 2002, two years past the Y2K panic, and the bureau was grappling with the signals it had missed in the lead-up to Sept. 11 the year before. "Traces of Terror: The F.B.I.; Computer System That Makes Data Secure, but Hard to Find," read a headline in the *New York Times*. It took a decade and $451 million, but the FBI finally deployed a new computer system in 2012, the year Barack Obama was reelected and the iPhone 5 debuted.[18]

Even now, twenty-two years after Sept. 11—enough time for an FBI special agent to have enrolled at Quantico in their midthirties and then left the bureau at the mandatory retirement age of fifty-seven—the bureau's technological capabilities are still frequently behind the curve.

One sleuth who worked closely with the bureau on Jan. 6 cases described the FBI computer system as "fucking stupid" and said it makes it "very difficult for them to use and find stuff." Multiple sleuths told me they had to format the reports they sent to the FBI to make sure they didn't surpass the email file size limit, since the bureau wasn't keen on using file-sharing programs. Once they got to the FBI, those reports had to be chopped down even more to be entered into their system, which could only accept files up to a few megabytes: roughly the size of an eight-second iPhone video or a series of high-quality photos. Sometimes the FBI would dispatch special agents to pick up USB drives.

In the early aughts, FBI email addresses were formatted as @ic.fbi.gov, which stood for "internet café," from the days of dial-up and AOL. Inside the FBI, it was a huge "pain in the ass" to move files from the "low side" (unclassified) to the "high side" (classified) and vice versa, which would require special permission and a thumb drive. A few years ago, even some news websites and Twitter would be blocked on the FBI's internal system.

It wasn't until 2014 that the bureau even began recording inter-views.[19] Then there were the phones. Oh god, the phones.

"The phones were terrible," said one former FBI official. They were consumer products but had been loaded up with software operating in the background that made them super slow. "They didn't really work as intended."

There were understandable reasons why the bureau needed to take extra precautions to make sure their network wasn't pene-trated. Bureau data was a huge target for outsiders, especially for-eign adversaries. Even so, the FBI's policies kept them cut off from what was happening online.

At the bureau, there was an attitude that agents were "only as good as your undercover sources and your covert sources," said Sherwin, the former US attorney. But in the past several years, he said, "with the proliferation of social media, it's critical to have a very robust intelligence directorate that mines open-source, social media because this is a new phenomenon."

Sherwin thinks the bureau is learning that. "I think maybe the Bureau or federal law enforcement looks down on a lot of social media type drivers or open-source drivers of informa-tion that maybe they're clownish, or, you know, it's all just unverified—and a lot of it is, okay? But I think there has to be a more robust attention given to data mining in the social media realm and really leveraging—maybe even building partnerships with some groups, you know, to better assess open-source intelli-gence," he said.

Old habits die hard, though, and the FBI culture is built much more around covert sources than OSINT. The bureau also had reason to be skeptical of the work of anonymous online investiga-tors. They'd been burned before. During the 2013 hunt for those behind the Boston Marathon bombing—an attack the FBI first learned about via tweet—Reddit exploded as a mass of amateur sleuths looked for the figures who they thought may have been responsible for the attack.

260 | Ryan J. Reilly

It was a disaster. Someone posted photos of two people they thought were bombers. They were not. They were sixteen-year-old Salaheddin Barhoum and twenty-four-year-old Yassine Zaimi, who attended the event as spectators and departed before the bombs went off. The *New York Post* splashed their image on the cover. "Bag Men," the headline screamed. The paper eventually settled a lawsuit on undisclosed terms.[20]

The Boston Marathon bombing was a cautionary tale about how online sleuthing could get out of control easily. It also changed the trajectory of the FBI investigation, with the false rumors popping up online influencing law enforcement's decision-making process.

"In addition to being almost universally wrong, the theories developed via social media complicated the official investigation, according to law enforcement officials," the *Washington Post* reported. The decision to release photos of the actual suspects "was meant in part to limit the damage being done to people who were wrongly being targeted as suspects in the news media and on the Internet," the paper reported.

"Investigators were concerned that if they didn't assert control over the release of the [suspects'] photos, their manhunt would become a chaotic free-for-all, with news media cars and helicopters, as well as online vigilante detectives, competing with police in the chase to find the suspects," the *Washington Post* said. "By stressing that all information had to flow to 911 and official investigators, the FBI hoped to cut off that freelance sleuthing and attend to public safety even as they searched for the brothers."[21]

That wasn't the end of it, though. When the FBI released the photos of the real suspects, Reddit began speculating that Sunil Tripathi, a missing Brown University student from suburban Philadelphia, was behind the attack. When his family shut down the Facebook page they'd set up to find him, the internet only saw that as confirmation of their theory.[22]

An accurate tip from a family member did come in. Eventually the FBI caught up to Dzhokhar Tsarnaev, the brother who survived a shootout with police and hid in the back of a boat on a trailer in a backyard in Watertown, Massachusetts.

After Tsarnaev's capture, law enforcement officials held a press conference in the parking lot of the Watertown mall. "This was truly an intense investigation, and I do emphasize a truly intense investigation," said Rick DesLauriers, the FBI special agent in charge, thanking the media for publicizing the photos after their release.[23]

Standing in the parking lot during the press conference that night after Dzhokhar's capture was a Boston K-9 officer. An ABC News camera on the scene picked up a nice, clean face shot, and a screenshot from the video ended up on a website.

Eight years later, that former K-9 officer would storm the US Capitol building on Trump's behalf, wearing a Boston sports-themed hat.

In the spring of 2022, online sleuths would send his name to the FBI after getting a facial recognition match. Twelve months later, there was movement. A decade after working the Boston Marathon bombing case, retired Boston Police officer Joseph Fisher was arrested in March 2023, charged with attacking a Capitol Police officer with a chair.[24]

This time, the sleuths had done it right.

———————— • • ————————

Alex was sitting alone in his garage somewhere in the South and feeling guilty. Here he was, a year into the pandemic, skipping out on a family vacation over spring break, when his family had booked a place on the coast. But he had an app to finish.

Behind the scenes, Alex had already played a major role in the Sedition Hunters community. Worried that the image composites Sedition Hunters was posting weren't getting enough traction—and knowing that his own kids primarily consumed content by scrolling through sharable video clips—he started producing short videos of some of the violent suspects Sedition Hunters and the FBI wanted to identify.

"It's one thing to see a face on a sheet and something else to see a video of that person beating a police officer," Alex said. "I still believe that video is a way to make these crimes connect with people."

Seemingly inspired by the Sedition Hunters videos, the FBI soon began producing a few of their own videos for dissemination to the public. Now it was mid-March 2021, and Alex had spent the last several weeks developing an application that had grown out of one of the roles he took on within the Sedition Hunters community: finding the best images of suspects' faces.

Alex's app would soon become one of the main tools of the manhunt: driving investigations, organizing information, generating new leads. The sleuths, using tools like Alex's app, would affect hundreds of investigations. There are people in prison right now who might not be there but for Alex's app.

Alex didn't know that at the time. He thought his app would be useful, but it grew beyond what he could have imagined when he decided to skip out on his family trip.

"It was painful because my wife and kids are everything to me. I want to be a good Dad and cherish the time we have together. But there's this thing I was working on that was so much bigger than any of us," Alex told me. "Here I was in a position to make a real difference and I needed to hurry because so many people were depending on me."

So he said goodbye to his family, sat in his garage, and got to work. To be clear, it's a nice garage with an office setup. There's a luxury vehicle on the car lift. Alex is a software engineer who sold a company years ago and worked as a consultant with a flexible schedule and financial stability.

Alex grew up in a red state and owns guns. Used to work with firearms, in fact. He's got a lot of Republican friends. "I definitely have friends that think the election was stolen," Alex said. One of them was in DC on Jan. 6, actually. Alex tuned into Fox News to see if he could spot them on television. That's how he watched the day unfold.

There's a common perception among supporters of Jan. 6 defendants—seemingly rooted in stale stereotypes about computer whizzes—that Sedition Hunters are washouts with too much time on their hands, perhaps living in their mom's basement. In fact, many of the sleuths I spoke to have rewarding family lives and thriving careers. One sleuth, in a moment of frustration, pointed out to an online troll that he worked on cancer cures, and probably made a lot more money than his Twitter nemesis too.

Aside from giving dated notions about the personal lives of computer programmers a twenty-first-century refresh, Alex also thought it was important for Americans to realize that the Sedition Hunters community isn't a monolith, that they aren't just partisans trying to take down Trump supporters.

"This is not some effort of Antifa, or even of Democrats," Alex said of the sleuthing effort. "This is the people that love their country, and who saw that happen that day at that building that means something to me, and that's what got us involved."

Working with sleuths, he'd sometimes joke he was "a Liz Cheney Republican," referring to the congresswoman who'd voted to impeach Trump and who'd led the congressional Jan. 6 committee. He did lean right, but he wasn't a hard-core partisan: he'd been one of those Obama-Trump swing voters you hear about. He faded away from the GOP, though, after the Capitol attack, frustrated at Republicans who refused to condemn Trump's actions that led to so much violence and continued gaslighting the country about what Alex knew, better than most

Americans, had happened on Jan. 6. "I'm done with the party," he told me in 2023.

Alex's app was an instant hit when it launched in the spring of 2021. The tool allowed online sleuths to quickly search all the archived videos and photos that Sedition Hunters had pulled from all over the web. Light bulbs were going off; new finds were discovered. The sleuths were now working with "a nuclear arsenal" instead of "a paintball gun," said one.

"After using your tool for two days . . . everyone should be focused on finding new content. Nothing else," wrote another sleuth, Josh, in a message to Alex in the early days. "All of the twitter sleuthing, going through sets of pictures, watching and annotating videos . . . they're all showing up to a gun fight with toothpicks. And people using your tool are showing up with a nuke."

"Also, can we teach the algorithm to recognize eagles?" he added, referencing a suspect nicknamed Bald Eagle who was involved in clashes with police near the tunnel while wearing an American-flag suit and an eagle mask.

Soon there was even a corporate-style welcome pack for users of the app, which told users the tool had "proved extremely helpful to us in our joint mission of tracking down the bad guys, and we are excited to have you join us."

One of their most significant early finds from the app was the man the sleuths had dubbed CatSweat, because he was wearing a sweatshirt branded with the Caterpillar Inc. logo as he dragged an officer into the mob.[25]

CatSweat was also wearing sunglasses, a hat, and often a mask during the assault, which meant facial recognition searches weren't turning up much. Now there was an app for that. Images they plugged into the system turned up other photos of CatSweat, including one from earlier in the day, when he was at Trump's rally near the White House and had his sunglasses hanging from his shirt collar. A Parler user had quickly panned past CatSweat's face. That was all it took. Sleuths entered his photo into a facial recognition website and got plenty of hits.

Logan Barnhart, a Michigan construction worker, had been a competitive bodybuilder. He had also been a cover model for romance novels with titles like *Stepbrother UnSEALed: A Bad Boy Military Romance* and *Lighter*, which included the slogan "Wrong never felt so right."

If they needed additional confirming evidence, on top of the right-wing memes he posted all over social media, Barnhart had also posted an image of himself wearing the same hat he wore to the Capitol and a workout video in which he punched a punching bag while wearing his "Cat" sweatshirt.

After a sleuth tipped me off about his identity, I went digging, found his Twitter page, and went to the Replies tab. He'd replied to a Trump tweet promoting Jan. 6.

"I'll be there," Barnhart wrote.

When I contacted the FBI, they requested that I hold off on naming Barnhart. They made a solid case, saying that naming him would complicate bringing him to justice. Two FBI special agents had been shot the prior month when they showed up to serve a search warrant in a child exploitation case, and I didn't want to do anything to give a violent criminal like Barnhart a heads-up that he was wanted, so I agreed. Barnhart was arrested a few months later, in August.

For a long time, the app stayed quiet. A core group of Sedition Hunters who joined the app agreed to keep mum about how they were organizing things behind the scenes. I knew about the spreadsheets and the tracking, but it became clear something even deeper was going on. When I found out months later what was happening, things started to make sense.

Alex's app and other open-source tools used by Sedition Hunters became critical resources for the investigation, allowing sleuths to organize and quickly point the FBI to open-source information on the internet that the bureau could check out and vet themselves.

"It ended up turning into something that is profound in the impact that it has," Alex said.

———————◆———————

Two years after Claire's Bumble sting, I went back and checked in on some of the other men she'd reported to the FBI. I wanted to see if the sleuths could figure out what exactly the men Claire spoke with had done on Jan. 6.

Not everyone had fallen for Claire's charm. One man told her he went to the "main, normal rally" and "left after the president gave his speech." He was on the metro back to Arlington, he wrote, when he started seeing all of the events unfold at the Capitol. "Quite wild," Greg wrote. "Also, for clarity up front, your profile says conservative, is that genuine? It's simply pretty rare in this area, that's why I ask 😊." He was onto her.

One of the men who did fall for it was named Sean. His profile said he had no pets, he didn't know what he was looking for yet, he wanted kids someday, he was conservative, and his love language was physical touch. "I'm a real nerd about . . . Computers 😁," his Bumble profile stated.

Figuring out his last name wasn't tough: he was wearing a military uniform in one of his photos, featuring his last name on his chest. His Instagram account shows that he'd snapped that photo at Fort Hood and that he'd once been stationed in Romania.

After Jan. 6, Sean told Claire he'd been protesting in DC but "didn't go inside the capitol," although he "was close" to all the action.

"At first it felt right, and then I realized it was wrong. . . . Once I read the president's tweet to go home I did," Sean wrote. "I feel ashamed for those who fought police. It's wrong 😔."

Sean didn't say exactly where he was outside the Capitol—whether he was on the west side, where the most violence took place, or on the east side, where Jan. 6 defendants who arrived late enough may not have realized the severity of what happened—but said it was Trump's tweet that caused him to leave. "Yea some old man said his wife texted him the tweet and I instantly knew it was time to go," Sean wrote.

"Yea . . . I was happy we stopped the count, but when the media and big tech are against you, there is no victory we could have had that day . . . so sad," Sean wrote. "As if another group didn't burn and riot all summer and now we're terrorists.

"They are labeling us terrorists so now they can come after anyone who supports DJT. It's coming!" he said. "One time the right fought back and now they are going to censor and crack down. Just have to listen to multiple sources for the news. Not legacy media."

Working with a sleuth, I dug into what Sean was up to and what he'd done that day. His LinkedIn profile said he had more than eight years' experience working with the Department of Defense in military intelligence and as a contractor "supporting many entities across the Intelligence Community through various industries." He held an "active TS/SCI w/ CI Poly clearance," meaning a top secret/sensitive compartmented information with counterintelligence scope polygraph clearance, which would be used to determine if the candidate had any knowledge of, among other things, sabotage or terrorism against the United States.

The app pulls up a lot of info on Sean. There he is, wearing his Trump beanie, standing behind a man with a head wound, in an Instagram photo. There he is again, in a Parler video, near the front lines as rioters rush police firing nonlethal rounds. There he is again, standing near a rioter holding a massive Confederate flag with an assault rifle superimposed on top. There he is again, standing at the top of a massive media tower that had been set up so photographers and videographers could capture Joe Biden's inauguration, which now gave him a bird's-eye view at that mob violence unfolding below him. There he is again, appearing to snap photos or record on his phone, standing just yards in front of the tunnel, where the most brutal assaults on law enforcement took place. There he is walking away from the Capitol, as a Trump supporter has a medical emergency on the lawn.

Sean also appears to have returned the next day, rolling up on a red rental e-bike as a middle-aged North Carolina woman with a mom mullet (the sleuths dubbed her "PoofTop") talked about election fraud. The woman was telling a man live streaming on Facebook that she had been back at her hotel watching on C-SPAN when she learned the building had been stormed. She talked about that batch of affidavits she heard about—"Karen was upset" and other complaints—and complained about communism. She complained that Black Lives Matter, regurgitating things she'd read on conservative websites, was a "Marxist organization."

She was accompanied by a man in a "Desert Storm Veteran" hat who'd been up by the inauguration platform the previous day. He insisted that antifa had stirred up things on the east side and that the west side—where most of the deadly violence took place—was "peaceful."

The streamer, a Black man, tried to have a reasonable discussion with the Trump-supporting election deniers, who had quickly transitioned into insurrection truthers.

"The rhetoric that Trump used from 2016 to today, what did y'all expect to happen?" he asked. "When has the Capitol ever been stormed?"

"It wasn't even stormed yesterday, those were paid antifa people," PoofTop insisted, with unshakable confidence. "It was not stormed."

Sean soon rolled up in the background, his backpack in the front basket. Authorities had quickly erected a tall black fence that wasn't as easy to toss aside. Someone was blasting Bob Marley's "One Love" on repeat.

"There wasn't massive violence, it was just a couple crazies," PoofTop said. "We felt punched in the stomach, because there's a lot of people—Democrats, Republicans, independents—who do believe there was fraud. So this was very convenient, as soon as a man stood up and raised his hand and said I am going to be a senator that's bold enough, then that happened. I think it was all planned."

Sean rolled off when PoofTop started ranting about Karl Marx and *The Communist Manifesto*.

It is difficult to say for sure whether Sean will be charged, but the odds are slim. Sure, no reasonable person could believe that it was lawful to join a mob and ascend a media tower and then the inauguration platform, watching from above as rioters assaulted police. But there were thousands like him that day. One of the only cases the government pursued against a person who committed conduct similar to what Sean did is Couy Griffin, the founder of Cowboys for Trump. Griffin was arrested early in the investigation and had hired a videographer to trail him throughout the day on Jan. 6. That meant a judge could watch, in high-definition video, as Griffin hopped over the stone wall, then climbed a bike rack that had been repurposed as a ladder, then walked up a broken piece of plywood, and eventually made it up to the inauguration platform while stating, "I love the smell of napalm in the air."[26] A Trump-appointed judge who heard his case and who criticized the Justice Department's handling of the Jan. 6 investigation found a way to acquit Griffin on one charge, but even he couldn't ignore the overwhelming evidence and found Griffin guilty of entering restricted grounds. He was sentenced to time served.[27]

Given that he doesn't appear to have participated in any violence, should his presence at the Capitol riot follow him for the rest of his life? Opinions vary. Whether he continues to hold a government clearance might ultimately be up to the government to decide.

"They call us terrorists and now they will come after us," Sean had written on Bumble, "making it harder for us to get jobs."

# PART IV

# CHAPTER 9

# "Give That Fan a Contract"

A few weeks after he got out of the hospital, DC Metropolitan Police officer Michael Fanone got a call from one of the federal prosecutors investigating the rioters who'd nearly killed him. Fanone wasn't feeling particularly trusting at that point. He had recently gotten word that his coworker Yari Babich, an MPD detective who'd been assigned to the task force propped up to help the FBI investigate assaults on MPD officers, had been shit-talking him behind his back.[1]

In the days after Jan. 6, Fanone's phone had blown up while he was at the hospital, as reporters called hoping to tell his story. His ex-wife—mother of the children who were in Fanone's mind when he told the rioters who'd abducted him that he had kids to remind them of his humanity—had posted about Fanone on Instagram[2] and Facebook. Fanone, she wrote, was not always her favorite person,[3] but he was a hero that day. "We are so proud of you. ♥ ," she wrote.

Fanone hadn't sought the spotlight, but it found him. Flooded with requests, MPD's Public Information Office eventually authorized Fanone and other officers to speak to Peter Hermann, the *Washington Post* reporter on the MPD beat,[4] and tell their stories to media crews from a rooftop overlooking the Capitol building. Fanone—with his neck tattoos, chiseled rough-and-tumble look, and profane-laden candor—became a quick media favorite.

"A lot of people have asked me, you know, my thoughts on the individuals in the crowd that helped me, or tried to offer some assistance," Fanone told the cameras. "The conclusion I've come to is, like, you know, *Thank you, but fuck you for being there.*"

It was TV gold. It also caused some tensions, jealousy, and anger within MPD. A former MPD employee got word that Babich was calling him a bullshit artist, an attention seeker, an embarrassment to the force. *But hey,* Babich wrote, *at least he got a new Tinder picture out of it.*[5] One of the recipients of such a text from Babich was also friends with Fanone's partner, Jimmy Albright, and gave him the heads-up.

"I mean, put this in context," Fanone later told me. "Imagine if you were a rape victim, and I was investigating your case, and I sent disparaging remarks about you and your credibility to an outside person. I mean, this was horrific."

Fanone also knew how cops were. This wasn't just Babich talking smack; this was a bigger thing within the department. Fanone was livid; he was in the throes of posttraumatic stress disorder and the trauma from the attack, and it was difficult for the hardened officer to admit just how much he was affected by his near-death experience. Now to be attacked by his own coworker, the guy who was supposed to help identify Fanone's assailants? This was a betrayal.

"They're charged with getting justice for me, and they're talking shit about me behind my back," Fanone said.

What strikes Fanone two years later is how early it started. "MPD started talking shit about me the week after I got injured." This was before his testimony to the Jan. 6 committee, before the CNN contributor contract that began when he left the department after they gave him a boring office job, before his book became an instant *New York Times* bestseller.

The gripes within MPD didn't let up. As the two-year anniversary approached, Fanone attended the Congressional Gold Medal ceremony in the Capitol rotunda. A few cops in the Special Operations Division, which Fanone said was known internally as "Meal Team Six," heckled him, calling him a piece of shit and mockingly calling him a "great fucking hero" while clapping. They said that he was a disgrace and that he didn't belong at the ceremony because he wasn't a cop anymore.

Two years after the attack, Fanone said it was increasingly clear to him that it wasn't just that MPD officers thought he was a showboat. It's that he made their preferred political party look bad.

"Having the benefit of two years of hindsight, a lot of it has to do with the fact that MPD officers were sympathetic to what happened, to the rioters that day," Fanone, who himself voted for Trump in 2016, later told me.

When he got wind of Babich's texts in early 2021, Fanone called Babich's boss, whom Fanone knew from their days in narcotics. "This would ruin Yari's career," Fanone said he was told. "Do you really want to ruin Yari's career? He's actually a decent detective."

"I lost my ever-loving mind," Fanone says. He went further up the chain, to Ramey Kyle, who was then commander of the Criminal Investigations Division. Kyle knew what Fanone had faced down in the tunnel. He'd lived it, telling officers to prepare to go "old-school CDU" when the vicious mob was about to breach the doors, referring to MPD's Civil Disturbance Unit. "We are not losing the U.S. Capitol today," he said.[6] Kyle took action. Babich was soon off the task force, and soon no MPD officers were assisting the FBI. Fanone heard that was a DOJ decision.

That day in early 2021, Fanone headed to the FBI Washington Field Office, located on Fourth Street NW, just north of the Judiciary Square metro stop, MPD headquarters, and the federal courthouse. FBI WFO was not unfamiliar territory for Fanone, who had been an FBI task force officer for more than five years and a DEA task force officer before that. (Later, when meeting with politicians, Fanone would adopt a method he'd deployed in his undercover days, flipping on a recorder and running tape on Kevin McCarthy.[7]) He didn't quite have walk-in rights or an office, but he "had a fuckin' FBI-issued radio" that was with him at the Capitol on Jan. 6, and he was at WFO regularly, usually on a weekly basis.

He headed up to the third floor, where the criminal squads sat. In his job, Fanone interacted most with CR-6 and CR-3, which did narcotics and violent crimes. This time they headed into the conference room belonging to CR-1, the unit that handles sex crimes. Fanone walked into a room with a long table, lots of chairs, and walls covered in printouts of Jan. 6 suspects. Some had been identified; most had not. They sat down, and Fanone began a process that he'd repeat again and again, in congressional testimony, in television interviews, and in sentencing hearings. He told his story.

Fanone reached down in the corner of the room and grabbed one of the anatomical dolls that FBI special agents would use when speaking to child sexual assault victims. Fanone recalls it being nonbinary, plush white cloth, having a little stuffed thing that's supposed to be a penis but also breasts and midlength hair.

"Well," Fanone said, pointing at parts on the doll, "I got touched here, and I got touched here, and I got touched here."

"They all started fucking laughing," Fanone said. His dark comedy hid a deeper issue, one that he's been working through ever since. This wasn't how he saw himself in the world, as the person being saved rather than the one doing the saving, as the victim rather than the investigator. "I was usually the person at the spearhead of these investigations, and now I'm the—quote, unquote—'victim,'" he says. "It's just an uncomfortable position to be in as a police officer, at least in my opinion. I don't get to know the details, and I understand that, but it doesn't make it any less frustrating."

Fanone, by his own admission, wasn't the most pleasant person to be around at that time. This was before he got help dealing with the rage he felt about what happened to him and the reaction from his brothers in blue. "I was angry all the time," Fanone said. "The whole thing was just really difficult to process. I hadn't even started therapy at this point." He'd start getting threats, but there wasn't much for anyone to do about those. Like all those pre–Jan. 6 warnings, they weren't actionable.

There was also something else happening in Fanone's head. He was putting himself in the shoes of FBI employees who'd been told to stop what they were doing and work Capitol cases, who'd been ripped off their assignments to work Jan. 6 cases.

"How would I feel if I was an investigator and I got re-tasked to deal with this bullshit?" Fanone said. "I would've probably been pretty pissed off."

———◆———

In the weeks after the Capitol attack, many of the arrests focused on low-hanging fruit: the rioters who went viral online because they were already semi-famous, were wearing a ridiculous outfit, or did something to stand out. Lots of other viral Capitol rioters were arrested quickly in the first few weeks of the investigation, including far-right media personality Baked Alaska; and Texas Realtor Jenna Ryan, who posed for a photo in front of a broken window and promoted her business as she filmed herself storming the building.[8] "Y'all know who to hire for your Realtor. Jenna Ryan for your Realtor," Ryan said in the video. (Ryan, who tweeted that she was "definitely not going to jail," citing her "white skin" and "great job," would later be sentenced to sixty days in federal prison.[9]) QAnon Shaman, Jacob Chansley, who stormed the building shirtless and wearing his signature bullhorn headdress, called into the FBI on his car ride back to Arizona, was transferred, and then was told he would get a call back from the bureau. The FBI called him after his appearance on *The Alex Jones Show*. "I was wondering if I was going to hear from you," Chansley said. "I'm the guy that was in the Senate; I'm the guy in the horns and the face paint."[10] He was willing to talk.

"My only thing is, I don't want me talking to the FBI ending up coming back and biting me in the rear end, because, you know, I said something that legally you guys try to use against me," Chansley said. "As long as you're cool, as long as you're cool, I have no problem talking to the FBI in Phoenix."

Chansley said that he liked the FBI special agent he spoke with, a former marine. "It just sounded to me like he was a patriot doing his job," Chansley said. "I'll tell you this much, FBI will let you talk." Chansley was arrested on Jan. 9 when he went in for the meeting.[11]

About a week in, after those initial high-profile arrests, FBI National Security Branch executive assistant director Jill Sanborn went to rally the troops. She knew the FBI was going to take a pounding in the press over the intelligence failures, or the failures to act on intelligence. She also knew that the investigation was going to be one of the biggest challenges the bureau had ever faced. She told FBI employees that she needed their best.

"I don't need to tell you the gravity of the work that we are doing, but please know all of your hard work to hold accountable those who were responsible for the breach and subsequent violence and criminal activity inside the Capitol has not gone unnoticed—not by me, not by the executives . . . and certainly not by the country," she wrote. "Each intelligence product, subject identification, and subsequent arrest represents another milestone in this race. But we aren't done. With the tips and leads still coming in as well as the upcoming inaugural activities and State of the Union, we have much left to do and we need you.

"I need your continued energy, passion, and focus in the days and weeks to come. The significance of this investigation will be a milestone in the history of the FBI and I couldn't be more proud to stand with you as we do this work together," she wrote.[12]

The more serious investigations, the cases against rioters who hadn't just handed the government the entire case against them on a silver platter, would soon be under way. Those cases would require some legwork, which some FBI special agents wouldn't feel like doing. It wasn't necessarily always purely political: some FBI special agents just didn't think they should be working misdemeanor cases. But politics was undeniably a part of it.

Despite his half-decade-long campaign against the FBI, Donald Trump has said most FBI special agents "like (love!)" him "a lot."[13] His son Donald Trump Jr. has said that the FBI "door kickers love us" even if the "upper levels" don't.[14] They weren't wrong. Ron Hosko, a conservative former FBI official, laughed when I asked him about the notion that the bureau is some sort of hotbed of liberalism.

"Nothing could be further from the truth," he said. "At its core, the FBI is still a pretty conservative, right-leaning organization that tries to divorce itself of politics."[15]

One employee contended in 2016 that the FBI "is Trumpland."[16] There's a very strong case to be made that the bureau is directly responsible for Trump's presidency. Comey decided to send a campaign-shaking message to Congress in the days before the 2016 election that damaged Hillary Clinton, in part because he was worried about leaks from conservatives in the New York Field Office. (They *really* didn't like Hillary Clinton.) That helped put Trump in office.[17]

Some FBI special agents would regurgitate right-wing talking points about the Jan. 6 investigation. In a survey inside the WFO, some employees made "less than professional" comments about Jan. 6, according to an email sent in April. One FBI employee reportedly referred to the Jan. 6 investigation as "Crossfire Hurricane 2: the Revenge," a nod to the Russia investigation run by the bureau between July 2016 and early 2017, when James Comey was fired and the Mueller probe began.[18]

We'll probably never have a comprehensive portrait of how many Jan. 6 cases FBI special agents killed, or half-assed, or purposefully undermined. But it is not zero. In the early days of the investigation, cases that had been sent out to field offices were unilaterally closed without further action, even when FBI special agents obtained evidence of crimes or when suspects admitted that they had been in the Capitol. Some FBI agents chose to accept explanations from suspects that strained credulity: That Jan. 6 participants had *no idea* they couldn't enter the building, despite the broken windows or the blaring alarm. One special agent from Florida talked about interviewing a Jan. 6 suspect who claimed that a Capitol Police officer granted him explicit permission to enter the building and that he only entered because he was a nerd and simply wanted to view the architecture. The FBI special agent believed, or at least pretended to believe, him.[19] Another FBI special agent spoke about his desire to purposefully undermine Jan. 6 interviews. Prosecutorial decisions weren't theirs to make, but some FBI special agents were determined to step into that role.

Eventually, DC caught on and set up systems to make sure that field offices couldn't just kill investigations on their own, accounting for what a source close to the investigation diplomatically referred to as the "varying degrees of enthusiasm" from FBI field offices across the country." Still, they couldn't catch everything, and leaned on special agents in the field to bring cases to fruition.

---

"JOHN RICHTER! JOHN RICHTER!" the man on the floor of the US Senate said, calling out for his friend. "THIS IS US! THIS

IS OUR RIGHT HERE! THIS IS OUR HOUSE! THIS IS WHAT YOU DO WHEN YOU TAKE IT! DON'T GIVE IT BACK TO THEM NOW!"[20]

As rioters looted the desks on the Senate floor on Jan. 6, a man named Joseph Irwin[21] roamed around, shooting a selfie video. The former deputy sheriff in Hardin County, Kentucky—birthplace of Abraham Lincoln, neighbor of Fort Knox—began forming a plan with his old army buddy after Donald Trump lost the election, but it crystallized after Trump sent his "Will be wild" tweet on Dec. 19. "Muster for liberty," Richter had texted. "Let's show these liberal soy boys what the sleeping giant looks like when it's up, angry, and ready to fight."

"Are we going open militia or innocent/ready bystander?" Irwin asked in a text message.

"I think we are gonna be in a huge crowd mostly," Richter replied. "So we will have to be opportunists most likely. Gnome sayin? I like the ready bystander wildcard approach myself."

Irwin and Richter attended Trump's speech and then headed to the Capitol. They saw someone bust in the window of an emergency fire exit, reach the magnetic lock bar on the inside, and open the door. Violent rioters flooded in. The duo scurried right over; stormed the building, chanting "USA"; and made it onto the Senate floor, where they occupied Senate desks. They were some of the last rioters to leave when police finally had enough manpower.

About two months later, two men knocked on John Richter's door with some questions. He didn't answer at first: he was on a work call and thought it was probably just an Amazon delivery. They were persistent.

He glanced at his doorbell cam. It was just two dudes in jeans and polos, but they weren't leaving. *What are these, Realtors trying to get me to sell my house or something?* he thought. He had no idea they'd want to question him about storming the Capitol.

Why would he? The FBI had the wrong John Richter.

This John Richter worked for Joe Biden's campaign. Hillary Clinton's too. He'd also worked at the Capitol for the better part of a decade and had friends who'd hidden in their offices during the attack.

This John Richter had been in Pennsylvania during the 2020 campaign, helping Biden win the state. He favored button-downs, not camouflage pants. He'd returned to DC after the campaign, when he was based out of Scranton. When he popped open the door to his home on Capitol Hill and the special agents introduced themselves, this John Richter figured somebody must be getting a security clearance.

"It's DC; everyone's trying to get background checks, so like, *Oh, they're here trying to, like, confirm one of my neighbors lives here.* I don't really talk to my neighbors that much, so I was like, *Oh, I really don't have any time; I'm on a work call,*" he later told me.

They said it was important. They wanted to ask him about Jan. 5, where he'd been between the hours of 7:00 p.m. and 10:00 p.m. "When the FBI's at your door on a workday, and they ask you where you were on Jan. 5, you don't remember where you were on Jan. 5," he said.

"I have no idea," Richter said. The special agents questioned why he remembered what he was doing on Jan. 6 but not the previous day.

"It was a violent insurrection, like, five blocks from my house. I worked there for 8.5 years, I have friends that were trapped in their office, of course I remember Jan. 6," he said. "I don't remember Jan. 5."

His puppy was going crazy in his crate. He started going through his phone. John's Instagram was full of pictures of the Washington Monument, flowers on the National Mall, etc. If he had gone for a walk on Jan. 5, maybe he'd snapped a photo that night. As he scrolled back, one of the FBI special agents pulled out a manila envelope. They'd blown up his passport photo and had a photo of a masked figure in a gray sweatshirt.

"Do you know what this is a picture of?" they asked.

"Yes," Richter replied, increasingly nervous.

"How do you know what this is a picture of?"

"It's the person who dropped the bombs, it's on bus stops, it's all over Twitter, it's everywhere," he said. "What do you mean how do I know?"

Did he know who the bomber was? No!

Suddenly it all clicked. He'd adopted his eight-week-old puppy on Jan. 2. "I wasn't even leaving the living room without him whining," John said. Jan. 5 had been the first night that he'd given the pup a bath in the sink, along with his roommate and his roommate's girlfriend.

"I'm showing the FBI agents me playing with my dog, giving my dog a bath, and at the same time he's going crazy in his crate, so clearly they know I have a dog," Richter said.

If he did know who the bomber was, would he tell them?

"Yes! I worked for Joe Biden, I worked for Hillary Clinton, like, I wasn't trying to 'stop the steal,'" Richter said. The only time he'd even left his place on Jan. 6 was to take the bumper magnet off the back of his car. He didn't want rioters to burn his vehicle upon seeing Biden's name.

After Jan. 5, he even pored over photos of rioters on the FBI's website, hoping he'd recognize someone he could turn in himself. "I was like, *I want to find somebody I know that I could turn in. I'm from Scranton; there has to be somebody I know that is in these photos*," Richter said. "I told the FBI that! *Every time you put out a new photo, I'm searching through like Where's Waldo to see if I can find someone!*"

Richter had no idea why the FBI would question someone who worked in Democratic politics for fifteen years about leaving bombs near the Capitol to stop the electoral college certification. Did his Pennsylvania phone number pop up on some cell phone list? There certainly were plenty of Pennsylvanians who'd stormed the Capitol.

Asking Jan. 6 suspects if they knew anything about the pipe bombs was good practice and also routine. The FBI did so when questioning suspects that ranged from serious violent offenders to low-level defendants charged with petty misdemeanors.[22] FBI special agents showed up to the doorsteps of self-identified journalists who were at the Capitol looking for footage too.

Still, showing up to the home of a random person—a Biden campaign official, in fact—to see if they were the bomber seemed like something the FBI could've ruled out with a bit of research.

Richter's run-in with the feds became an ongoing joke among his friends. "Even last weekend, my friend introduced me to her new boyfriend, and I had to tell him the FBI story," he told me in 2023. There were jokes about him being on the no-fly list. Whenever the FBI raised the reward for information on the bombs placed outside the Republican National Committee and Democratic National Committee on the night of Jan. 5—they bumped it up to $500,000 in January 2023, signaling they were still desperate for leads[23]—DC John Richter would receive texts from friends joking about how the money was too good and they were going to turn him in and collect.

"Sorry Richter but the reward is up to $500K. I think we have to turn you in," one wrote in a group text.

"Maybe we should wait till the reward climbs to $1M," suggested another.

Months after visiting the wrong John Richter in 2021, the FBI arrested former police officer Irwin and obtained a warrant for his cell phone.[24] Then, two years after the FBI visited DC John Richter, the FBI arrested the *right* John Richter in March 2023.

DC John Richter saw the news, and suddenly things made sense. "I was always curious why they were asking me questions," Richter said. "This might make sense now. It's literally one of the craziest stories I have."

The group text blew up when John Richter's friends found out about the other John Richter.

"OMGGGGG," wrote one friend. "Omg I'm crying," wrote another.

———————◆———————

Thousands of miles away from DC, on April 28, 2021, a dozen FBI special agents raided the Homer Inn & Spa, home of a conservative Alaska couple who'd been at the Capitol in DC on Jan. 6.[25]

Marilyn Hueper and Paul Hueper, the FBI said, had been banned from Alaska Airlines for refusing to obey mask regulations. Days after they were identified as noncompliant passengers, someone contacted the FBI. They'd found an Instagram post on Paul Hueper's page. "Marilyn approaching the [Capitol]. As Patriots, there is a righteous revolution to take back our country," he wrote. "Keep praying . . . we are only getting stronger and will not quit until our country is restored. To be there was a once-in-a-lifetime experience. To be surrounded by a million Patriots who love this country is something I will never forget. STOP THE STEAL!"

"The FBI confirmed the woman in photograph's 225 A and B was MARILYN HUEPER by comparing MARILYN HUEPER's Driver License photograph," the FBI special agent wrote. They also said someone who knew Hueper personally "confirmed she is the person in photographs 225 A and B."

FBI employees surveilled the Hueper home.[26] Then they showed up for a search.[27] The couple was handcuffed for nearly three hours after agents busted through their door.

"We're here for Nancy Pelosi's laptop," the FBI special agents said, according to Paul Hueper. Neither of them knew what they were talking about. They showed them a photo of a woman on the FBI's Capitol Violence website, who bore a casual resemblance to Hueper. It just wasn't her.

Marilyn Hueper noted—correctly—that she has attached earlobes and the woman in the photo did not. Ears are one of the first things online sleuths know to look at. Plus, she said, the woman in the photo was wearing a sweater "you couldn't pay me to wear," said Marilyn.[28]

"Her hair parts on the opposite side. She has darker hair, big earrings, heavy makeup, and an ugly sweater that I'd never wear!! She has a black coat like mine and nappy hair like mine. And she's white. So some similarities," she wrote in a caption of the photo on Facebook.

"My wife is much better looking than that," Paul Hueper said.

The online sleuths saw the search of the Alaska couple's home as "an embarrassing fuckup" by the FBI. "I just can't believe the FBI would fuck up like this," they said. Sleuths never would have sent this tip in; it didn't look like her, and there were no forms of secondary confirmation. But initially they were skeptical that she was telling the truth.

"I wasn't sure if I believed Hueper's story because I didn't believe the FBI could mess up that badly," said one sleuth, whom we'll call Taylor. They started trying to solve the mystery.

Taylor had started running every face he saw posted on the Sedition Hunters account through facial recognition, and it was like "shooting fish in a barrel." He didn't say much about what he was doing, worried that others would misuse the tool and that innocent people would end up being wrongly accused because someone—an anonymous internet user, not the Federal Bureau of Investigation, the thinking went—failed to check their work. Soon the sleuths would come to "live off of face shots," and finding a person with their face uncovered became a key focus of the sleuthing efforts.

Taylor found every photo he could of the woman who'd been inside the Capitol. Finally, one shot of the woman known as #AirheadLady—she wore an emergency escape hood when she left the Capitol with a younger male—triggered a facial recognition hit. It came back as Maryann Mooney-Rondon, a Trump supporter from New York. She sat on the board of an organization affiliated with hospitals and nursing homes. That led to her Facebook page, where sleuths quickly realized that the young man she was accompanied by at the Capitol was her son. They even matched rings that the woman had been wearing with rings she was wearing in photos on Facebook.

It took a grand total of about thirty minutes to identify the correct woman. When I spoke with someone in law enforcement about the story, they asked me to leave out the fact that the correct suspects had been identified, since they were hoping to effectuate an arrest. When the FBI had searched the family home, agents found the cut barrel of a 12-gauge shotgun, and Rondon's son admitted he had an unregistered sawed-off shotgun stored at his relative's property.[29] The FBI arrested the duo in September.[30]

Taylor has been involved with a lot of finds, but the Rondon find remains at the top of the list.

"This was one of my most proud moments as a Sedition Hunter because it not only IDed people who were part of an important event on J6 (Pelosi laptop theft), it also exonerated an innocent woman who was being targeted by the FBI because they didn't fully vet a tip before acting on it," the sleuth, Taylor, later told me.

———————— ◆ ————————

Alise Cua felt "just stupid," she told the judge. The Georgia mother, a former veterinarian who stopped working to homeschool her kids, believed her president. She thought the election was stolen. Her husband and teenage son thought so too. So when the president called on his supporters to come to DC on Jan. 6, they decided to make a family trip out of it. They listened to Trump's speech and then headed to the Capitol.

Two months later, her eighteen-year-old son was in federal custody. Bruno Cua had boasted on the right-leaning social media website Parler about the need to take the Capitol by force. "If they don't do what's right MILLIONS of us will BREAK

DOWN THEIR DOORS!" he wrote on Parler on Dec. 19.[31] On Jan. 6, he helped do just that.

Bruno Cua left his parents outside the Capitol as he joined the mob storming the building. As his parents tried to reach him on his cell phone, Bruno Cua walked through the halls of Congress with a baton—which his parents knew he had with him—and shoved an officer to get inside the Senate chamber. Then he went back on Parler and bragged about it.

"The tree of liberty often has to be watered from the blood of tyrants. And the tree is thirsty," he wrote on Jan. 7. "What happened at the capital was a constitutionally protected right. WE THE PEOPLE have a right to rise up and overthrow a tyrannical government. . . . THEY HEAR US NOW!"[32]

A federal magistrate judge in Georgia was appalled by the Cua family's behavior. The day before the Senate acquitted Trump at his second impeachment trial, the judge ordered the teen held until trial, finding that the Cuas were improper guardians for the eighteen-year-old.

"We have a lot of really serious, a lot of really, really dangerous people who show up here in federal court," Chief Magistrate Judge Alan J. Baverman said. "But I have to say, this is the first time in a number of years [outside of a drug case] where his parents were—maybe not instigators—but aiders and abettors and didn't take steps to stop their child from going off the rails."[33]

Cua was in US Marshals' custody en route to DC, which with the Marshals' fractured transportation network meant a pit stop in Oklahoma. From jail, he joined the video hearing before a DC judge his family hoped would send Cua back home to Georgia while his case was resolved.

Alise Cua told US District Judge Randolph Moss that she'd been wrong. The election wasn't stolen after all. She said that she felt "ridiculous" about having believed Trump and that she "really should've known better." Her voice broke as she talked about her son's experience behind bars.

"We are completely broken and just honestly and truly remorseful to the core of our beings, and we're asking for a chance," she testified.

Her testimony echoed that of her husband, Joseph Cua, who told the Georgia judge that he was "pretty embarrassed" that he believed the president's lies. "I feel like I maybe should've known a little bit better at my age," he'd said earlier. "At some point, you're like, 'I think this is a bunch of BS.'"

Back home in Milton, Georgia, there was a lot of interest in the case. One resident I spoke with, who sent me photos of the "We ♥ You Bruno" display with the American flag outside the teenage insurrectionist's home, called Bruno Cua "a terror." The claims of Cua's parents, they thought, were just for show.

Assistant US Attorney Kimberly Paschall, the federal prosecutor arguing for Bruno Cua to be locked up until what seemed like his inevitable conviction, seemed to think so too. She told the court the actions of Cua's parents illustrated that they were "inappropriate" guardians for their son.

"His parents were fully aware that he was in possession of a weapon, fully aware that he was inside the building, fully aware that there was an altercation with a plainclothes officer, and they did nothing about it," Paschall said at the hearing. "They did nothing."[34]

Paschall had picked up some of the most attention-grabbing Capitol cases. There was the "QAnon Shaman" case against Chansley, who invaded the Senate floor with a spear and horn helmet and demanded organic food in jail. There was Beverly Hills salon owner Gina Bisignano, who shouted her name over a bullhorn as she urged rioters to fight the police at the Capitol. Paschall pitched in on cases against a man charged with beating an officer on the steps of the Capitol and a case against Jenna Ryan, the Texas Realtor who flew to DC in a private jet and promoted her real estate business as she live streamed herself breaching the Capitol.

Bruno's case, Paschall argued, was "one of the most terrifying" of the roughly three hundred Capitol cases the government had brought at that point. His posts on Parler eerily predicted what exactly was going to happen on Jan. 6, even if his violent, macho rhetoric left out that his parents were going to give him a ride.

"There are few other defendants who have stated their intentions so clearly and so knowingly on social media before showing up on Jan. 6," Paschall argued. "He knew exactly what was going to happen when the rest of us did not."

———————— ◆ ————————

A judge was about to toss the conviction of a protester who had interrupted a congressional proceeding, and Paschall wanted him to reconsider.

On Jan. 10, 2017, ten days before Donald Trump would walk through the upper West Terrace tunnel, take the oath of office, and vow to end "American carnage," a sixty-one-year-old woman named Desiree Fairooz was arrested at a congressional hearing. Sen. Richard Shelby (R-Ala.) was introducing Trump's nominee for attorney general, Sen. Jeff Sessions (R-Ala.), and claimed his then colleague had a "clear and well-documented" record of "treating all Americans equally under the law." In fact, the Senate had rejected Sessions's nomination to be a federal judge in the 1980s over concerns about his attitudes on race.[35] Fairooz, a seasoned protester with the left-wing group Code Pink, found Shelby's claim laughable, and she laughed.

A rookie Capitol Police officer in just her second week on the job, working a congressional hearing for the very first time, decided that was enough to make an arrest, also her first.[36] Fairooz's loud chuckle hadn't disturbed the hearing, but her arrest brought it to a stop. "Why am I being taken out of here?" Fairooz asked as officers moved her out of the room. "This man is evil, pure evil. Do not vote for Jeff Sessions! I was going to be quiet and now you're going to have me arrested? For what? For what? He said something ridiculous! His voting record is evil!"[37]

I was in the room covering Sessions's confirmation hearing when the arrest took place, and had tweeted out a video I'd recorded of the incident. Then I forgot about it. (The FBI director had been fired, and Robert Mueller was investigating Russian interference in the 2016 election, after all.) I'd assumed the case had been dropped, or maybe resulted in a small fine or something, the equivalent of a parking ticket. Months later, I found out that my video was being used as evidence at trial.

Federal prosecutors very rarely get to try a criminal case these days: just 2 percent of federal criminal defendants go to trial.[38] Fairooz's case, prosecuted by the Justice Department in DC Superior Court, was a perfect opportunity for lawyers without trial experience to find some. An attorney from the fraud section and an attorney from the money-laundering and asset recovery section at the Justice Department were brought in as special assistant US attorneys on the case. The fraud attorney grilled the rookie officer on whether the laughter was "loud enough to draw your attention" or if she recalled "seeing other people turning around." Money-laundering attorney David Stier, who told jurors this was his first jury trial, said during closing arguments that laughing at a moment when others did not laugh was enough to find Fairooz guilty.

"I would submit that laughter is enough, standing alone" to merit a conviction, Stier argued.[39]

To my surprise, the jury convicted Fairooz. They acquitted two other protesters, who'd dressed up as members of the KKK and pretended to support Sessions, on one of the three charges they'd faced, finding that because they'd put on their white robes and faux southern accents before the hearing began, they hadn't actually interrupted anything. Fairooz *had* interrupted the hearing, they said, by protesting her arrest and inserting political commentary. "She did not get convicted for laughing. It was her actions as she was being asked to leave," the jury foreperson told me.[40]

The judge, however, didn't like that argument about the laughter. At this point, the trial and Fairooz's conviction had become fodder for late-night comedians. It also drew the attention of Rep. John Conyers (D-Mich.), the ranking member on the House Judiciary Committee, who had "substantial questions" about the application of the law in Fairooz's case.

Paschall was brought in to play cleanup, to defend the convictions. Chief Judge Robert E. Morin of the Superior Court of the District of Columbia called the government's argument about the laugh "disconcerting" and said he was "concerned about the government's theory." Laughter alone "would not be sufficient" to submit the case to the jury, he said, and the government had not made clear that it intended to make that argument. Fairooz, Paschall told the judge, "wasn't just merely responding, she was voicing an opinion."[41]

"She did not merely laugh," Paschall had argued in a filing. "Just as a defendant cannot resist arrest when the arrest is unlawful, a defendant should not be able to disrupt a hearing before Congress based on her opinion that police action was unlawful." In court, Paschall also noted that jurors had told me they didn't convict because of the laughter, but that wasn't admissible. Morin ordered a new trial,[42] but the government ultimately backed off the case before the new trial date.[43]

I thought I recognized Paschall at the hearing, but I just couldn't place her. I did some research when I got back, and it finally clicked: we were in the same dorm freshman year, before she'd transferred. In the intervening years, I ran into her once or twice around Judiciary Square. Then she started picking up major Jan. 6 cases in early 2021. *Jackpot*, I thought. I had an in.

Unfortunately for me, just as I vaguely remembered from freshman year, Paschall is a stickler for the rules. She was unflappable. One time, I got word that she'd flown out to California for witness prep ahead of a bench trial. "How was LA?" I asked. She just smiled, refusing to bite. "Sorry about the Eagles," she said, referencing their Super Bowl loss. All that told me was that she'd probably been following my Twitter feed. I'd struck out again. The most I managed to get out of her came in response to a casual inquiry about how she was doing. "Tired," she replied. In another decade, she once told me in an elevator, maybe we'd grab a beer and talk this all over. For now, she was a lockbox.

Paschall would speak in court, though, and became practiced at reminding judges of the seriousness of the offenses, of what Jan. 6 meant for American democracy. Paschall—and I didn't learn this until more than two years later, when she mentioned it to a defense attorney—had responded to the US Capitol herself on Jan. 6. The prosecutorial response began that very day, and Paschall was a part of it.

"None of what happened on Jan. 6, 2021, was peaceful," she said during Chansley's sentencing in November 2021, which was among the first. She told the court to send a message to the "flag-bearer" of Jan. 6 that, regardless of your political beliefs, there are consequences when you try to obstruct American democracy.[44]

At the start of the investigation, the US Attorney's Office (USAO) for the District of Columbia spread the cases out among its staff. Section assignment didn't matter; everyone was getting one of these cases. But ultimately Paschall and her colleagues at the DC USAO couldn't handle this caseload on their own. The Justice Department called in help from around the country. Soon federal prosecutors from USAOs all around America stepped up to the plate, dialing in remotely for many hearings and eventually flying in for trials.

"What time is it in Alaska?" Chief US District Court Judge Beryl Howell asked at one hearing, after seeing the darkness outside Assistant US Attorney Adam Alexander's window. "It is 5:03 a.m.," he replied.[45]

There were plenty of missteps. The Justice Department violated the rights of one defendant, losing track of him in the system.[46] Overworked prosecutors missed court deadlines. Even if you managed to avoid the internet threats, it could be grinding, stressful, consuming work. "I've thought more about those four hours than I have about my wedding day," joked one.

Meanwhile, Justice Department leaders were assuring the public that the investigation had the necessary resources. For the line prosecutors working the cases, though, there was a huge disconnect between what Justice Department leadership was saying and their daily reality. The email system was a constant source of frustration, as inboxes would frequently reach their limits. The sleuths, on the other hand, had twenty-first-century technology.

———— ◆ ————

On the morning of August 18, 2021, Alex was back in his garage again. His app had been a tremendous success, and it was still a secret to the public. The rules the community had established had worked. Alex saw that a Jan. 6 defendant was set to be sentenced. There hadn't been many sentences at this point, barely half a dozen. This wasn't a particularly big case, but it was worth looking into.

Robert Reeder had been identified with the help of facial recognition from an official source: the Office of the State's Attorney for Harford County in Bel Air, Maryland, sent in a tip to the FBI, saying that facial recognition software had produced a hit off a photo they ran of him from the FBI website.[47]

Online sleuths had nicknamed Reeder "Chin Diaper" because he wore his blue medical mask under his jaw in most of the photos they'd found of him.

Alex ran his photo through the database and found new images they didn't know about, videos that showed him in a scuffle with officers on the stairs of the Capitol. He alerted the other sleuths, and someone quickly got the info to prosecutors. A Sedition Hunters account also broke the news about the last-minute finding.

"This truly is a massive undertaking," an account posted. "Did we make it in time?"

I hadn't planned on attending the hearing, but I rushed over to the courthouse, which was still relatively quiet. It would be months before jury trials would resume after the COVID-19 pandemic. Most hearings took place over the phone.

The elevator doors near the courtroom opened, and Reeder approached the courtroom with his lawyers. On my phone, I tried to show him one of the clearer videos that showed the attack, but his lawyers intervened.[48]

Once the courtroom doors opened, prosecutors told the judge that they had initially planned on bumping up their sentencing request to the maximum of six months because of the video. But they decided to ask for more time to review the footage, to see if additional charges were warranted. Reeder's sentencing would have to wait.

Later that night I tried to get a handle on how the sleuths had found that moment so quickly. The story didn't quite add up. At this point, I thought that sleuths were doing most of their work manually, watching videos and tagging certain moments. Certainly if someone came across a charged defendant doing something violent, they would have run that up the chain, made sure it got into the right hand sooner.

"Was that moment tagged somewhere in the database, or was it a last minute Where's Waldo thing?" I asked one of the sleuths.

"I'm not sure I should reveal detailed sources and methods at this point 😊," they wrote.

The sleuth, whom I'll call Liam, got sucked into the Sedition Hunters community to get some accountability, sure, but also because they loved the challenge of the hunt.

"I like solving puzzles, figuring things out, and there's a certain appeal to that as well," Liam said. "But there are a lot of puzzles in the world, and I haven't chosen to solve most of them, but this one I did. This was unprecedented and dangerous, and it needed attention."

Within the government, that near miss jump-started a process for making sure that they wouldn't miss something big again. They'd check with the sleuths.

The government had spent millions on a discovery system to deal with an unprecedented amount of material. Here's the thing: it was crap at searching videos. Paperwork? Sure, it could do that, scan for words. *Whoop-de-do*. But at the end of the day, what do the juries want to see? They want to see videos. That's what this investigation was built on. The system couldn't handle it.

"Relativity is a cloud-based eDiscovery platform that offers functionalities including document organization, review, production, and analytics within a single environment, and is an industry leader in eDiscovery hosting," the government wrote in a filing in September 2021, explaining where things stood in the discovery process. "Defense teams will be able to perform key term searches and metadata searches across hundreds of thousands of documents in the defense workspace."

There's that word again. *Documents*. This wasn't an investigation about documents. This was a case about video. Video after video after video. The system "was shit," said one law enforcement official. Defense lawyers agreed. "Relativity is a pile of shit. 1990's technology. Government paid a zillion bucks for it and got raped," said a defense lawyer. Well, $6.1 million. Deloitte could earn up to $26 million on the deal, which would last for years.[49]

The bureau had no way of independently combing through the video to find the most relevant material. They had to outsource it to the sleuths.

Think of a friend of yours. Open up your iPhone, and try to find a picture of them, but imagine that you don't know the date it was taken and that it's not geotagged so you can't use the map if you remember where the photo was taken. Your iPhone, luckily, has internal facial recognition, and it scanned all of the photos in your iPhoto library. All you've got to do is find the tag, and there are a bunch of photos of your friend. That's what the sleuths could do, but with a database of every video of Jan. 6 that went public.

The sleuths were also just better at collaboration. For the bureau, anything and everything has to be kept for discovery, so FBI special agents are very wary of putting things into writing. They do have Microsoft Lync, a messaging platform that FBI special agents and analysts can use to communicate.[50] Think of it like Slack, but shittier. Kind of like an old-school AOL chatroom.[51] Still, it was something. Two FBI employees on the Proud Boys case used it to gossip about what they thought was a far-right love triangle while investigating the group October 2021.

> *Just listened to about 7min of yelling . . . Zach to Amanda lol.*
>
> *Bahah. Did he find out shes hooking up with Aaron?! I'm waiting for this to be a legit thing. And when it is . . . popcorn!*
>
> *Not yet . . . haven't come across that one.*
>
> *Dang it.*
>
> *Hahah ill bring beer.*[52]

A bit of watercooler talk, some gossip. A year and a half later, it ended up in a court case.

But for the sleuths, open communication and collaboration was their superpower. They could talk; they could joke; they could hivemind. They weren't going to have to turn over their chats in discovery. They could keep trash-talking.

Publicly, DOJ relied on the sleuths more and more and even started using graphics designed by sleuths in court filings because they did a much better job than anything DOJ had at laying out what the Capitol looked like that day. It was 2023 before the Justice Department started syncing up videos in court exhibits, giving viewers a fuller representation of what was happening at any given moment, adding audio into CCTV footage. The kind of stuff sleuths had been doing for two years.

"They *are* these investigations," one official told me. "I am so incredibly grateful to the sleuths for everything they have done. But what an egg on the face of United States law enforcement."

------◆------

A couple of months after Reeder's sentencing, a Jan. 6 defendant named Matthew Perna took a plea. Before his initial arrest, the Penn State graduate and CBD marketer reportedly told the FBI that he'd been pushed into the Capitol building and that he had not intended to enter, but said he had "tapped on a window of the Capitol building" using a metal pole, because he was frustrated. In a somewhat unusual move, he pleaded guilty to four counts, including the felony charge of obstruction of an official proceeding, without a plea deal in place.

His agreed-upon statement of offense was straightforward. It stated that Perna was inside the building for about twenty minutes and that he'd later posted an eight-minute video to his Facebook account. "It's not over, trust me," he told his audience. "The purpose of today was to expose Pence as a traitor." In the video, he, like many Jan. 6 participants, blamed antifa.[53] His friend, Stephen Ayers, disagreed.

"I personally don't remember seeing like antifa there or anything like that," Ayers, who sat beside Perna in the video, later said. "I was just kind of going along and just basically agreeing with, you know, his statements."

The Justice Department had secured a significant concession from Perna, getting him to plead guilty to a felony charge. As it turned out, there may have been a reason Perna was so eager to plea without a plea deal: There was *a lot* that statement of offense left out.

Perna, according to his family, was a kind man who taught English in Thailand, lived in South Korea, and then moved back to the US when his mother was diagnosed with leukemia. He also, extensive video evidence illustrates, got extremely worked up on Jan. 6 and tried, repeatedly, to assault officers, committing the type of conduct he'd soon blame on antifa.

Perna, while outside the east doors leading to the Capitol rotunda, can be seen chucking what appears to be a water bottle at police. Then, outside the north doors as officers are under attack, Perna throws a metal flagpole like a spear, seeming to hit a door.[54]

Perna was set to be sentenced, but he got word that there were complications, that prosecutors might be seeking more time. He died by suicide on February 25, 2022. In an obituary, his family said he died of a "broken heart" and that his "community (which he loved), his country, and the justice system killed his spirit and his zest for life."[55]

"He felt there was no other way out, you know. It is sad. Very sad, because he was a good person," Ayers later said. "I don't know if, you know, waiting got to him."

Perna would be mourned, and his case used as an example of purported government overreach, but the broader public didn't understand the basic facts of the case. The Justice Department wasn't about to come out and smear the name of a dead man, and sleuths largely avoided harping on the issue. Federal prosecutors would just file a court document—"Government's Suggestion of Death and Request for Abatement of Prosecution"—to notify the court of Perna's death.[56]

When Sergeant Gonell returned to the tunnel late on the night of Jan. 6, he snapped several photos of the scene. In the small space between the double set of golden doors of the tunnel, where MPD officer Daniel Hodges was crushed by a riot shield held by a rioter with long hair and transition lenses as Hodges screamed out in pain,[57] Gonell snapped a photo of a bunch of debris on the ground: water bottles, broken sticks, a cord, a mini bullhorn, and a crumpled-up red-and-white scarf.

In 2017, the town of Skellefteå mailed out 934 scarves as Christmas gifts to people who had moved away from the region.[58] One of them had ended up in the Capitol tunnel. #SwedishScarf had dropped it.

———————◆———————

One of the most productive source relationships of my journalistic career started off with a joke tweet. In the wake of Jan. 6, footage had been circulating online. It showed a woman in sunglasses and a pink hat yelling at rioters through a broken window leading into the conference room. She took on an authoritative voice, the unmistakable tone of a frazzled parent trying to herd multiple young ones. She explained the layout of the floor below.

"So, people should probably coordinate together if you're going to take this building," she said. It was as though she was chaperoning an eighth-grade trip to the Capitol insurrection.

"$20 says this leads to both criminal charges and a PTA resignation," I tweeted after watching the video.

A message from the Deep State Dogs Twitter account landed in my inbox. "Hi Ryan. FYI only and off record, please. We don't want her to go off-grid. She has been identified. Trust me," the account said on Jan. 25. "I don't want to wait too much longer. The Twitter sleuths need a reward so they can move on to other things. And I don't want any more other women getting doxxed."

In the early days of the Jan. 6 manhunt, there was plenty of confusion. Some social media users were convinced that the woman with the bullhorn, #BullhornLady, as she was dubbed, had to be connected to a Republican member of Congress because she appeared to have inside knowledge of the Capitol. Some speculated she was the mother of Rep. Lauren Boebert, who was elected after running a Colorado restaurant called Shooters Grill, where staff members were encouraged to openly carry firearms. Boebert took office three days before the riot and tweeted about House Speaker Nancy Pelosi being "removed from the chambers" after rioters breached the building.

The person behind the Deep State Dogs wanted to make sure this budding community of online sleuths was staying focused on the task at hand, not getting distracted. Pretty soon I was on the phone with Forrest Rogers.

Rogers was one of the few sleuths willing to go on the record. He was living in Germany at the time of the attack, and he spent hours going over footage. "Alexa, calm my dog," he'd say to his Bose Sound System, cuing up a playlist filled with soothing music to drown out the disturbing audio of many of the scenes. "I co-opted my pooch's music to lower the tension," Rogers said, recalling how he'd watch scenes of the breathtaking violence on the upper West Terrace to Tchaikovsky's "Dance of the Sugar Plum Fairy."

He was in the business sector at the time, but Rogers had a background in journalism. In his youth, he'd been trained by investigative journalists from the Southern Poverty Law Center who'd started a newspaper in the mountains of Georgia when Rogers was just sixteen. That's where he got his first taste of investigative work. Two of the men he'd worked with there would go on to establish Klanwatch at the SPLC. That experience continued to be his lodestar even decades later.

"The FBI is overwhelmed. It's like ten thousand people storming Walmart, and there is only one cash register open," Rogers told me. He credited experienced online investigators like John Scott-Railton, a senior researcher at the Citizen Lab at the University of Toronto, with influencing the direction of the budding community that would come to be known as the Sedition

Hunters as best he could, encouraging good habits and discouraging bad ones, like tossing out unconfirmed names on social media.

"Railton was the perfect beekeeper who was a master at guiding the swarm into forming a new 'Sedition Hunter' colony," Rogers said.

Rogers and his team had identified #BullhornLady as Rachel Powell, a mother of eight from Pennsylvania who'd used a battering ram to smash a window and a bullhorn to give orders to the crowd around her. He was worried about stepping on the FBI investigation and wanted to give the bureau time, and I was unable to convince him to let me break the news. Ronan Farrow of the *New Yorker* was evidently more persuasive, and Rogers ended up giving him the scoop.[59] Farrow contacted Powell, who admitted to much of her behavior that was caught on video. (Later, a judge wondered why Powell would have talked to Farrow about her conduct, "no matter how charming that reporter might be.")

I put my feelings of journalistic jealousy aside, figuring there were plenty more IDs to come and I could position myself to get the next one. Rogers delivered.

"Look at Taser Prick," he said.

———————◆———————

Danny Rodriquez, the California man who'd traveled to DC with his friend Ed Badalian, ready for "revolution," had attended Trump's speech at the Ellipse on Jan. 6. Trump was about halfway through an hour-long speech, and he'd lost the thread a bit, as he often did.

"You know, on Twitter, it's very hard to come onto my account. It's very hard to get out a message. They don't let the message get out nearly like they should. But I've had many people say, 'I can't get on your Twitter.' I don't care about Twitter. Twitter's bad news. They're all bad news," Trump said.

"But you know what, if you want to, if you want to get out a message and if you want to go through Big Tech, social media, they are really, if you're a conservative, if you're a Republican, if you have a big voice, I guess they call it shadow banned, right? Shadow banned. They shadow ban you, and it should be illegal," Trump continued.

Rodriguez was standing and listening when he noticed the camera filming him. He smiled and held up four fingers, mouthing, "Four more years." Then he tapped a friend, *watch this, watch this.* He turned to the camera. He said Joe Biden's name and then made a cutting motion over his throat three times.

"I've been telling these Republicans, get rid of Section 230," Trump continued, referring to a law that helped shield Twitter and other companies from being sued over their users' posts. "And for some reason, Mitch [McConnell] and the group, they don't want to put it in there and they don't realize that that's going to be the end of the Republican Party as we know it, but it's never going to be the end of us. Never. Let them get out. Let, let the weak ones get out. This is a time for strength."

After the speech, Rodriguez made his way over to the Capitol. He ran into Gina Bisignano, whom he knew from the Trump rallies in California. They joined the mob on the inauguration platform.

Bisignano was one of the first to the tunnel. She and Rodriguez and had been split up, but she was filming on her phone when they ran into each other amid the chaos in the tunnel.

"We're not going to let our country go!" she said, as men grabbed police shields and passed them back. "They're not going to steal our votes!"

"DJ! DJ!" she said.

"Hell yeah, we're fucking doing it!" Rodriguez said.

Soon, Capitol Police officer Moore was dragged down the stairs, face first. "I just felt that that particular day, I wanted to leave it out there on the West Front," said Moore.[60] Fanone was abducted and had a stun gun driven into his neck.

The YouTube video, posted by a videographer for a far-right website, was a gold mine. It was shot in high definition, and it captured some of the most violent scenes in front of the tunnel. In Germany, Rogers sat at home with his dog and watched it, often frame by frame.

At full speed, you could easily miss it. But Rogers spotted the black device—an electroshock weapon—in the MAGA-hatted man's hand and then saw him drive it into Officer Mike Fanone's neck. He posted an edited, slowed-down version of the video on Twitter.

"Let's find this coward!" he posted under @1600PennPooch. "This is Officer Michael Fanone being tased in the neck. Fanone suffered a heart attack as the result of this assault. I think Taser-Prick is an appropriate name."

Before long, Rogers and the other members of Deep State Dogs had figured out everything the man was wearing, from the branding on the glasses to the camouflage InfoWars shirt he had on underneath his jacket.

Twitter did its thing. A Twitter user—an antifascist activist from California named @waterspider_, who knew Rodriguez from pro-Trump rallies in the LA area—soon saw the videos. She'd already identified Rodriguez as the man seen in the mob pushing against police in the tunnel and smashing out a window inside the Capitol, attempting to let other rioters inside. Now he'd been identified as the man who nearly killed a police officer.

The FBI was alerted in late January, but weeks went by without any action. After Rogers's tip, I worked with a colleague to confirm Rodriguez's identity, and the story ran in late February.

"It's the unknown that's sometimes frustrating," Rogers told me. "The FBI can't say to me, 'Okay, Forrest, we've got it under control, your team can move on to the next one.' But this is what kind of slows down the process for us."[61]

Rogers knew the FBI had a "mammoth" task ahead. They just wanted to be helpful. "We are here in a support function to provide them the information that is necessary to bring these people to justice," Rogers said.

———◆———

Sometime in March, after his interview at the FBI and after our story identifying Rodriquez ran, somebody asked Fanone if they could borrow his handcuffs, which he mailed out to California via FedEx. Around 6:00 a.m. on the last day of March,[62] FBI special agents arrived at the two-bedroom bungalow in Fontana, California, where Rodriquez lived with his mother. There was a Trump sticker on the front window. The FBI came through the back sliding glass door. He was arrested, was handed off to awaiting FBI special agents, and soon found himself in an interview room. Rodriguez gave a full confession to the FBI, describing himself as "not smart," "so stupid," "an asshole," and "a piece of shit."

"Are we all that stupid that we thought we were going to go do this and save the country and it was all going to be fine after? We really thought that. That's so stupid, huh?" Rodriguez told the FBI, discussing the Capitol attack.

"I understand what it—it's very stupid and ignorant, and I see that it's a big joke, that we thought that we were going to save this country, we were doing the right thing and stuff. I get it," he said.

Not long after Rodriguez's arrest, Fanone was at an event on Capitol Hill and found a manila envelope on his seat with his name on it. He opened it up, finding that his handcuffs had been returned to him, along with a Polaroid of the cuffs on Rodriquez's wrists.[63]

Rodriguez's lawyers would later argue that when he drove a stun gun into Fanone's neck, he "did so in reasonable reliance upon a grant of authority from a government official," having listened to Trump tell his supporters that the election was stolen.

"As the leader of this country at that time, former President Trump told individuals that the election had been stolen and that, if they did not fight, they would lose their country. He relayed this message to individuals while telling them all that they were going to march to the Capitol."[64]

I spoke with Fanone a few months after Rodriguez's arrest. Forrest Rogers was in town, and we grabbed dinner at an Italian place near a stretch of bars and nightclubs, where Fanone told us stories about his career in policing, how he used to buy crack undercover around the corner, how the neighborhood had changed. Fanone later told me how he had some sympathy for Rodriquez. He wanted him held accountable, sure, but that didn't mean "he's a moron and a misfit, and he was like many people looking for comradery, he was looking for something to belong to," Fanone said.

Fanone was also blown away by the work of the online investigators who'd identified his assailant, and who he knew were providing tremendous help to the investigation. "I grew up going to Baltimore Orioles games with my dad, and every time someone hit a home run and a fan caught it . . . they would say, 'Give that fan a contract,'" Fanone said. The FBI was working on something along those lines.

The cases continued to churn. Sometimes the feds just got lucky. The FBI got a tip about Michael Roche on Jan. 8 from a tipster who said they saw a video on Facebook that showed Roche talking about entering the Capitol. "We all started praying and shouting in the name of Jesus Christ, and inviting Christ back into our state capitol," Roche said, referring to the US Capitol. Images on Facebook showed him posing with the QAnon Shaman, and video showed him on the dais inside the Senate chamber.

Later, a group of deputy US Marshals and members of the Tennessee Bureau of Investigation arrived at an apartment in Murfreesboro, attempting to locate a missing child. Michael

Roche and a woman were inside, and the woman told law enforcement that Roche had been all over the news and that they were expecting law enforcement to come for him. After deputies placed Roche in handcuffs as a safety precaution, Roche said he thought the US Marshals were there to arrest him for storming the Capitol. A warrant check came back negative, and the deputies uncuffed Roche. A deputy marshal gave the FBI's Nashville office a heads-up, and Roche was arrested in April.[65]

Yet even with high-profile arrests happening nearly every day, some Jan. 6 rioters seemed to move on with their lives in brazen ways. Sam Lazar, known to sleuths as #FacePaintBlowHard because he was dressed in military gear on Jan. 6 and had his face painted in camouflage colors despite operating in an urban environment, attended an event with Rudy Giuliani on May 15. He snapped a photo with Pennsylvania gubernatorial candidate Doug Mastriano, who himself was on the restricted grounds of the US Capitol on Jan. 6.[66]

Lazar was arrested on July 27,[67] the day that Fanone, Gonell, Capitol Police officer Harry Dunn and MPD officer Daniel Hodges testified at the first public hearing of the Jan. 6 committee. Lazar, who was seen on video telling rioters to take officers' guns and using pepper spray against officers on the police line, was ordered jailed until trial.[68]

The identifications and arrests kept chugging along, and soon the sentencings started. By the six-month mark, the feds had made five hundred arrests, but there were plenty more important tips locked up in the bureaucracy.[69]

"If someone tried to design the whole system in which all of these Sedition Hunters are operating, you couldn't create it," said Chris Sigurdson, a Canadian man who submitted numerous tips to the FBI and ended up being cited in an FBI affidavit. "It emerged organically, not just out of this event, but from Charlottesville and that whole experience of trying to identify people involved with that."

By this point, many of the sleuths had established relationships with individual FBI special agents. Some had even signed a contract, if you will, signing up as confidential human sources to the FBI. The bureau came to rely on the sleuths to build their cases, not only to identify suspects they didn't know about but to identify the actions of rioters they'd already charged and to trace their steps throughout the Capitol that day. Sleuths who were working with the FBI felt a duty to provide reliable information and took pride in making sure all the information they provided was reliable.

"If we screw this up," one key sleuth later told me, "they will never trust a group like us again."

⸻

On Valentine's Day 2023, a member of the US Marshals walked over and looked underneath the defendant's chair, checking for weapons. Danny Rodriguez, wearing a green prison jumpsuit, walked in from the holding cell behind the judge's bench and greeted his lawyers. He looked about the same. All that was missing was the MAGA hat.

"Okay, so you have a lot of signing to do," his lawyer told him. He started going through the paperwork.

*The defendant applied the electroshock weapon to the back of Officer Fanone's neck. Officer Fanone subsequently lost consciousness and was later admitted to Washington Hospital Center for treatment for his injuries.*[70]

Rodriguez was sworn in. At the end of that interview with the FBI after his arrest, when Rodriguez described himself as so stupid, he told the FBI special agent placing handcuffs around his wrists that he'd take his case to trial, that he'd see how a jury felt. Now, two years later, Rodriguez did the smart thing. He pleaded guilty.

By this point, as the government approached the halfway mark in the statute of limitations, it was abundantly clear: the feds were never going to arrest everyone who committed a crime that day.

The FBI's number of open domestic terrorism cases exploded by 357 percent between 2013 and 2021, from 1,981 to 9,049.[71] The FBI was nearing the 1,000 mark on cases by early 2023. The total number of people who could be charged—meaning they entered the Capitol, committed property destruction, or engaged in violence outside—was north of 3,000.[72] So much was charged so quickly early on, when the court system was strained by COVID-19 and when cases were delayed over discovery issues, that they'd spent year two of the investigation playing catch-up.

Trials and convictions were going well. There was only one case with a full acquittal: Trump-appointed judge Trevor McFadden found a federal defense contractor with a top secret security clearance not guilty on all the misdemeanor counts he faced.[73] McFadden made no secret that he was mad about how DOJ was handling the cases of Jan. 6 rioters compared to the handling of cases that didn't involve the storming of the US Capitol during a critical moment in the process of transferring power. Even though he thought a defense contractor more likely than not knew that he wasn't supposed to enter the US Capitol when there was a blaring alarm and the window to the door was smashed, McFadden said there was enough reasonable doubt that he could find his way to a not guilty verdict. He sent his message. Not guilty across the board, McFadden declared.

Some Jan. 6 participants who haven't been arrested are waiting for the clock to run out when the statute of limitations expires on Jan. 6, 2026. So how many more arrests are coming?

In a letter to the chief judge for DC's federal court, US Attorney Matthew Graves said that while it was "incredibly difficult" to predict how many more cases were coming given the "nature and the complexity of the investigation," 700 to 1,200 cases could be in the pipeline.

"We expect the pace of bringing new cases will increase, in an orderly fashion, over the course of the next few months," Graves wrote.[74]

Judge Beryl Howell, who was on her way out of the role of chief judge by March 2023, said in a statement to reporter Zoe Tillman that the court "continues to manage its caseload and trial calendar efficiently, notwithstanding the delays occasioned by the pandemic.

"So far, the court has been able to manage the increased criminal caseload well," Howell said. "Should a 'surge' of filings occur at a later date, the Court would assess what additional steps, if any, it should take."

Prosecutors knew they weren't going to be able to bring all of the cases, but there hadn't been widespread acknowledgment of that fact or any formal decision about which cases should get priority before the clock ran out.

"We're not going to get there," one official said. "At some point, someone is going to have to say that openly, and there will need to be a decision—either explicit or implicit—who do we decide to arrest?"

What would be of the most value to the nation? There were important decisions ahead.

"It's like the *Titanic*," said one law enforcement official. "It's obvious not everyone is getting off this boat, but how do you decide who's going to get off the boat?"

# CHAPTER 10

# "Information War"

It was the first Friday of August 2022. Erin Smith was sitting inside the attorney general's conference room on the fifth floor of Main Justice, the Justice Department headquarters building built in 1935. *This is it*, she thought. *This is my chance.* It had been nearly nineteen months since her husband, Jeff Smith, died by suicide nine days after he was assaulted during the attack on the Capitol. Jeff Smith had reported to the medical facility for police officers on Jan. 6, but the facility was overwhelmed with other officers who had also been injured. He was sent home with painkillers. Jeff didn't want to go into much detail about Jan. 6, but he told Erin it was the most chaotic situation he'd been in and felt it was something he wasn't trained for.[1]

Jeff had graduated from college in 2007 and tried to become a police officer, but there weren't many departments hiring around Chicago. When he heard there were openings at the Metropolitan Police Department in DC, he packed a duffel bag, threw it into his car, and went. Erin and Jeff had met on a dating app after he moved to the DC area. He was a funny guy, Erin said, always joking and dancing around. He could also be brutally honest, and that could rub some people the wrong way. Jeff loved his Ford

Mustang, and he always made sure they had an American flag flying outside the house. "You knew what you were getting with Jeff," Erin said a few months after his death. "He was just an all-around great person."

On Jan. 6, when Jeffrey heard the call of shots fired over the radio, he didn't know who'd been shot, whether it was an officer or a rioter.[2] He didn't tell this to Erin, but the first thing he'd encountered upon walking into the unfamiliar Capitol—a building he'd last entered when he was ten years old—was officials surrounding a body. It was Ashli Babbit, the QAnon follower and military veteran who'd been shot minutes earlier after jumping through a broken window leading into the House Speaker's Lobby, where members of Congress were still evacuating. Jeffrey and his crew made their way upstairs, stepping over the marble eastern staircase featuring some dark, old stains, where blood had seeped in.

On that staircase, nearly 131 years earlier, thirty-eight-year-old former representative William Preston Taulbee was fatally shot by a reporter from his home state of Kentucky, Charles E. Kincaid. A few years earlier, the correspondent for the *Louisville Courier-Journal* and the *Louisville Times* had scooped the news that Taulbee had had an extramarital affair with a young woman from the US Patent Office, writing in the language of the day that they were found "in a compromising way" and "were rather warmer than they were proper."[3] One headline: "Kentucky's Silver-Tongued Taulbee Caught in Flagrante, or Thereabouts, with Brown-Haired Miss Dodge."[4] Taulbee was in his midthirties, and Miss Dodge was a teen at the time, described in an account as "a little beauty, bright as a sunbeam and saucy as a bowl of jelly." Taulbee apparently got her a job in the Patent Office, and they'd taken up in the model room at the Interior Department.[5] After the scandal, Taulbee became a lobbyist, and Kincaid kept reporting. They'd run into each other in the halls, and the larger Taulbee would bully Kincaid, pulling on his ear or nose, warning him he should be armed. On Feb. 28, 1890, Taulbee threw Kincaid, who weighed less than one hundred pounds, around by the collar. Kincaid went home and got his pistol. A shot rang out, and policemen came running. Violence was very common in Congress in the 1800s, with canings and fistfights and stabbings.[6] But shootings in the halls of Congress were still surprising.

"For the first time in the memory of man a gunshot was heard in the National Capitol today, and the marble steps of the staircase leading from the House floor to the restaurant below were stained with human blood," read one account.[7] A jury acquitted Kincaid of murder after he argued self-defense.

When Jeffrey reached the top of the "Bloody Stairs" on Jan. 6, moments after passing a dying Ashli Babbitt, he came to the second floor, where several Capitol Police officers had unholstered their guns. Around the corner, someone had set off a fire extinguisher, filling the hallways with an uneasy haze.

It felt like a movie, Jeffrey told Erin, adding that he'd been punched as well as hit in the face shield with a flying metal object. He never got into the details; they generally didn't discuss Jeffrey's work. When he arrived home at 2:00 a.m., Erin was just glad he was back. The next morning, he was complaining about his head and his neck, and Erin saw some bruising around his eye. He wasn't the same Jeff Smith as he had been before. Jeff was short-tempered; he was pacing at night; he wasn't sleeping well. No more cracking jokes, no more dancing around the house. He lost his appetite; he didn't feel like hanging out; he didn't want to take the dog for a walk.

A week after he arrived home from his ordeal at the Capitol, Jeffrey reported back to the clinic on Jan. 14[8] and was ordered to return to work. Erin tried to make the morning of Jan. 15 as smooth as possible, preparing his breakfast, making his lunch, walking him out to his car. She kissed him goodbye, told him she loved him. That was the last time she ever saw him.

An officer showed up at Erin's door. At first, the police told Erin that her husband had died in a car crash on the George

Washington Parkway. Erin had to call to deliver the news to his parents. Then the officer told them it was a death by suicide, and Erin had to call again. He was thirty-five, and he'd been at the Metropolitan Police Department for twelve years.

Not long after Jeffrey's death, Erin was standing at a pharmacy counter when she discovered that the benefits she received through her husband's work had been cut off. "You can't really process anything," Erin said of the time after Jeffrey's death. "I didn't know what to think. I didn't know what to say. I didn't know what to do. I just tried to get through each minute, while realizing that no one's coming home."

It wasn't easy to find out what had happened to Jeffrey. Erin and her attorney, David P. Weber, had to fight. Weber's wife, also an attorney, had read a story I'd written about the Sedition Hunters six months after the attack, and Weber reached out for help.[9] He told me that they'd been stonewalled by the Metropolitan Police Department and really had no clue what had happened to Jeffrey that day other than what he'd written on an injury form and what he'd told Erin.

Forrest Rogers and the Deep State Dogs were soon on the case. Forrest compared the broader manhunt to Netflix on steroids, in the middle of the pandemic. There were tens of thousands of episodes to binge-watch, a whole cast of characters, and a mix of tragedy, comedy, and farce. This mission was serious, but the jokes helped. With the help of a couple of photos Jeffrey had sent Erin of himself at the Capitol on Jan. 6, they created a graphic to aid their search, noting his helmet number (4626), his radio (left side), his sunglasses (right above his body cam), and the equipment strap on his left leg. They also noted the COVID-19 mask he was wearing when he didn't have his gas mask on.

"Team Jeff (4626)," as they called themselves, got to work. "I finally found 4626 in a video!!!" one member wrote. Responses rolled in. 🔥🔥🔥🔥

"We found Jeff!" Forrest soon told me. "It was not easy." Soon they identified one of the men who was near Officer Smith when he appeared to go down in the crowd as police pushed the mob out of the building.

A few weeks after the Capitol attack, a woman named Elizabeth visited her chiropractor on Capitol Hill, at a yellow townhouse on East Capitol Street about five hundred yards from the Capitol grounds. From the street outside, you could see the Capitol dome behind a giant temporary fence, which had been erected to protect the Capitol in the aftermath of the attack. An outer perimeter fence would stay up until March, while another black fence surrounding the Capitol would remain in place for more than six months.

"I said I had been stressed out and upset and scared by the attack on the Capitol," Elizabeth said. She was sure her chiropractor, who had treated plenty of Hill staffers and knew the impact the "unforgiving" floors had, would have been mad about the Capitol attack too.[10] The man with his hands on Elizabeth's body saw it differently.

"While he was adjusting me, he said, 'I thought it was just a few broken windows,'" Elizabeth recalled. "I felt upset that he seemed clueless about how terrible it was. I felt upset that he didn't understand how important it was, what had happened," she said.

"Of course," Elizabeth later said, "he had been there."[11]

David Walls-Kaufman, the DC chiropractor, had been identified with the help of facial recognition technology. Soon the sleuths had confirmation it was him. "BOOM! WE GOT THE FUCKER IN THE SAME JACKET!" Rogers wrote me, linking to the YouTube video. The sleuths soon surfaced some of Walls-Kaufman's work, including a 2018 piece for the conservative American Thinker titled "12 Advances of Civilization by Flawed White American Males." Walls-Kaufman even gave an interview to a reporter for the NBC affiliate in South Florida, who found

Walls-Kaufman sitting at a café on Capitol Hill. He claimed he was not a Trump supporter but admitted he was on the Capitol steps.

"It was just so confusing, because a lot of people were, like, turning around and saying, you know, *These people are not with us*," Walls-Kaufman said, boosting the notion that the pro-Trump mob he joined to storm the Capitol had been infiltrated.

In fact, as video after video showed, Walls-Kaufman seemed to be everywhere on Jan. 6. He'd stormed the building through the rotunda doors on the east side, pulling up the hoodie of his red jacket and rushing through an opening. He pried open the door to House Speaker Nancy Pelosi's conference room to allow better flow into the room, where someone was stealing a laptop on the table and someone in the other room was smashing a mirror. He was right there when Babbitt was shot at the Speaker's Lobby. Then he was on the inauguration stage on the west side of the Capitol, standing on the bleachers overlooking the lower west tunnel, the site of some of the most violent attacks of the day. He remained on Capitol grounds long after nightfall.

Weber filed a lawsuit within forty-eight hours of learning Walls-Kaufman's identity. Before long, sleuths had also identified another man who was near Officer Smith, who was holding a cane designed to be used as a weapon and had tussled with officers. He was wearing a "Make Space Great Again" hat. A few folks tossed out nickname suggestions. Erin liked #AstroNOT.[12]

A facial recognition search pulled up images from the website of the Franklin County Republican Party in Washington State. Taylor Taranto, the group's webmaster, was seen posing with a cardboard figure at an event.

"When I saw him grinning with the Trump cutout, that's when I figured we probably had a good lead," the sleuth who found him joked. Taranto's biography said he had spent six years in the US Navy and enjoyed "making memes and homeschooling my children" when he wasn't "helping the President take down the deep state."

Finding confirmation wasn't too difficult there. Taranto had posted a video on his page "so someone will report me to the feds and we can get this party rolling!" It featured him inside the Capitol on Jan. 6. "*So . . .* we're in the Capitol building, the legislative building, we just stormed it," Taranto says in the recording.

I reached Clint Didier, chairman of the Franklin County Republican Executive Committee, when he was on his tractor. "They had buses full of these 'antifa' people posing as being Donald Trump supporters," Didier told me. Taranto had told him so, saying he'd seen "a bunch of buses coming in posing as Trump supporters who orchestrated this whole damn thing."

Didier, a former football player on the team now called the Washington Commanders, later said he wasn't sure if it was Taranto or someone else who'd filled him in on the ghost of antifa, but he called Taranto "a fine man" who "has some issues with PTSD" because of his time in the military.

Sitting in his loft off the Chesapeake Bay, Weber stayed up until 4:30 a.m. the night of the Taranto find, watching it unfold. The next day, he worked up an amended complaint that added Taranto as a defendant.

They still hadn't figured out who hit Officer Smith with a flying pole later that night, but the lawsuit allowed Weber to pry Officer Smith's body camera footage free from the bureaucracy. That's how they found out what had happened during the first fight inside and, later, after nightfall.

The body cam footage shows that Jeffrey was shaken by the encounter with Walls-Kaufman. He struggled to adjust his helmet; someone asked him if he was okay. Soon, near the main entrance to the House floor, he stopped again, and another officer assisted him. Jeffrey eventually caught up with his squad near the rotunda, where they were trying to force rioters out. Hours later, video shows Jeffrey outside, in the back of a group of officers, with an object cutting through the night sky. Someone tried to block it. It slammed Jeffrey in the face.

"Are you okay?" a startled colleague asked after the metal object nailed Jeffrey. "What the fuck was that?" Jeffrey asked, retreating toward the Capitol, where the inauguration platform was in tatters. He needed help.

Eventually, in March 2022, the DC Police and Firefighters' Retirement and Relief Board ruled that injuries Jeffrey Smith sustained on Jan. 6 had been "the direct cause" of his death.[13]

As of this writing, the rioter who threw the pole still hasn't been identified. There's not as much video to work from after the curfew was activated, and the government hasn't released much footage from the security cameras covering the area at night. The out-of-state officers who'd come in to help and who had taken over the front lines to relieve MPD officers who'd been fighting for hours, including the Virginia State Police, were not equipped with body-worn cameras. Most of the footage from MPD body-worn cameras, including Jeffrey's, only shows officers' backsides.

It would be nearly a year after his identification that Walls-Kaufman was arrested, in June 2022, about eight months after FBI special agents conducted surveillance on him in October 2021 to confirm his identity.[14] Despite the extent of his conduct, Walls-Kaufman was only charged with the standard four misdemeanors that most low-level defendants who entered the Capitol faced.

Taranto returned to DC in the summer of 2022, not long after Walls-Kaufman's arrest. He gave a tour of the Capitol grounds in a video titled "Storming the Crapitol 2.0" and posted it on YouTube. "Well, here I am! Obviously no one is here to arrest me," Taranto said. He approached a Capitol Police officer standing on the east side of the Capitol. "Hello, how you doing?" he asked the officer before shaking his hand.

"Still haven't been arrested," he said as he continued his walk. He said that antifa was there on Jan. 6, that an officer struck him. "I am walking freely across the grounds all by myself, without a walking cane even," Taranto said, narrating the video.[15]

Erin's visit to the Justice Department came in August 2022, a couple of months after Walls-Kaufman's arrest. She and other family members of officers who died by suicide were sitting in the attorney general's conference room. The official portrait of Elliott Richardson, Nixon's attorney general who resigned rather than fire the Watergate prosecutor, had been brought back from that hidden spot in the library and was back on the wall, along with three others. There was Robert Jackson, the FDR appointee, future Supreme Court justice, and "Patron Saint of the Rule of Law,"[16] whose 1940 speech[17] in the Justice Department's Great Hall about the role of the federal prosecutor was essentially DOJ canon. There was Robert F. Kennedy,[18] a Kennedy nepotism appointee who became attorney general at the age of just thirty-five, advanced the civil rights movement, and cracked down on the mob while holding Christmas parties for underprivileged children in the courtyard of the Justice Department,[19] and *maybe*—as legend has it—let his brother John F. Kennedy use the loft hidden away in the attorney general's suite for "liaisons" with Marilyn Monroe.[20] There was Edward Levi, who'd taken command of the Justice Department at what was previously considered a tumultuous period, post-Watergate, putting the department back on the right course and winning bipartisan praise, in the days before senators used oversight committee hearings to generate YouTube clips.[21] Garland had mentioned Levi in a virtual speech to Justice Department employees the day he was sworn in, when he was also briefed on the Capitol attack investigation by FBI director Chris Wray and other top officials.[22]

Erin and other family members of law enforcement officials had been brought in to discuss the Public Safety Officer Support Act of 2022, which was on the verge of passing Congress, and the Justice Department's administration of the program.[23]

Erin had fought for that bill for a long time, and it helped fundamentally change how suicide is dealt with in law enforcement, where there was often an extra stigma around suicide and mental health.[24] The meeting was important, no doubt, but Erin had a more pressing issue. They were having trouble getting through to prosecutors.

"I knew that this opportunity wouldn't come again," Erin said. "My feeling is these people are people just like you and me, right? They sit down for dinner every night; they watch TV. They just have a higher profile job." She wasn't nervous, she said; she just wanted to feel out the room, make sure she was picking the right moment.

"I think I've gotten to a point in this whole journey where I have nothing to lose," Erin said.

When the moment came, she told Garland that she needed help. That Jeffrey was attacked, that they had a video analysis done, that they were having trouble getting the video to the right person.

"He basically turned to Lisa Monaco and was like *You need to handle this*," Erin recalled. Monaco, the no. 2 official at the Justice Department, who oversees its day-to-day operations, jotted down her number for Erin on a Post-it note, and Erin passed along her lawyer's contact information.

Mission accomplished, they thought. They had the attention of the top leaders at the Justice Department.

"They reached out to me that very night," recalled Weber, the Salisbury University faculty member working as Erin's lawyer.[25] "Very late at night, they were burning the midnight oil."

Then they got word: The Justice Department would accept their file. All they had to do was purchase a thumb drive, transfer the documents to that thumb drive, get an envelope, go to the post office, stand in line, pay for the postage, send it off, and hope that it would actually reach someone on the other end and that they would actually examine the physical media. They'd wait and wait and wait for an answer.

The document they were trying to send was less than a gig. Too big for an email attachment but easily sent via Google or your pick of file-sharing services. They heard from the deputy chief of the Breach and Assault Unit that had been set up within the US Attorney's Office for the District of Columbia to handle these cases, who put them in touch with an FBI special agent. Weber suggested Dropbox. That was a no-go, said the bureau.

"Please place the files on either a CD, Flash drive or hard drive that have not been utilized for any other data, and mail it to the following," the FBI special agent wrote, giving the address for the FBI Washington Field Office on Fourth Street NW.

This was 2022, when most new computers didn't have CD burners. A hard drive seemed like overkill for a document that could be downloaded in a minute with a decent internet connection. Weber decided to put the less-than-one-gig file on a flash drive.

Weber was on Chincoteague Island in Virginia, and there was no office supply store. So he ordered a pack of USB flash drives on Amazon. Once they arrived, he used a converter to attach the USB drives to his computer (many computers don't have regular USB connections anymore) and loaded the file. He put it into a Priority Mail envelope, walked over to the only US Post Office on the island, and mailed it out. Weber, a former federal investigator himself, had been dealing with another federal agency that was able to accept digital evidence, and he said that the experience reflected his perspective of the bureau, that they were always a few steps behind.

"This is an organization that, even in more recent times, has not been at the forefront of technology," Weber said. "They initially didn't even have emails, then only some agents had emails, then the emails that they did have were configured in a very strange way."

Sadly, this wasn't an uncommon scenario. The bureau has picked up thumb drives from other sources. Some tipsters would constantly watch the size of their files, knowing that the bureau couldn't use file-sharing systems and that there was an upper limit on inbox attachments.

The expert report that Weber had prepared for the government ultimately didn't change the trajectory of the case. Walls-Kaufman pleaded guilty to a low-level misdemeanor count of parading, demonstrating, or picketing in January 2023. In his agreed-upon statement of offense, Walls-Kaufman admitted that he had "scuffled" with officers as he was forced out of the Capitol.[26] A federal judge had scheduled Walls-Kaufman's sentencing for May 2023 but then received letters from Erin and Jeffrey Smith's parents that gave her some questions about the Justice Department's handling of the case. "The case seems a bit different than the case that I thought I was presiding over when I took the plea," US District Judge Jia M. Cobb said in court. "There's a big difference between a scuffle . . . and an allegation of someone using a baton to strike an officer."[27] She ultimately reset Walls-Kaufman's sentencing for June 2023. In the meantime, the building on East Capitol that housed his chiropractor practice was listed for sale.

Taranto, meanwhile, was back to posting conspiracy theories and was still waiting for his arrest.

"STILL WAITING FOR THIS SHOW TO GET ON THE ROAD . . . WHERE'S MERRICK?" he wrote on the Facebook account he shares with his wife in February 2023. In May 2023, Taranto returned to DC again, this time hanging out at "Freedom Corner," where supporters of Jan. 6 defendants regularly gathered in support of the Capitol defendants being held at the DC jail. The facility is located right across the street from the Congressional Cemetery, where J. Edgar Hoover is buried along with representatives, senators, and former DC mayor Marion Barry.

———◆———

What nobody outside the Justice Department knew when Erin Smith made her pitch to Merrick Garland was that DOJ and the FBI were at the very end of intense deliberations over an ongoing investigation into former president Donald Trump.[28] Less than seventy-two hours after the Friday meeting, on Monday morning, FBI special agents showed up to Mar-a-Lago in search of classified documents they believed the former president was withholding. FBI officials had wanted to slow the probe, and even to close it altogether because the Trump team had claimed a diligent search had been performed. The Aug. 8 search ultimately turned up over one hundred classified records, including eighteen that were marked top secret.

Days later, one of the thousands who'd shown up to the Capitol on Jan. 6 went to an FBI field office and tried to breach a screening center with a nail gun while holding an AR-15-style weapon.[29] Ricky Walter Shiffer, who'd called for "combat" and urged other Trump supporters to kill FBI personnel "on sight," was soon killed after a standoff with police.[30]

Steve Friend had clearly thought about this a lot. The FBI special agent, a former accountant and a recent transplant to Florida, did not like the bureau's handling of Jan. 6 cases. He didn't like how the bureau was organizing them; he didn't like how many people were being charged; he didn't like that Jan. 6 defendants would face trial before a panel of DC jurors. In August 2022—the same month as the Mar-a-Lago raid, as those threats against the FBI—Friend had an opportunity to do something about it, and he jumped.

Friend said he'd learned of upcoming arrests involving Jan. 6 suspects that had been scheduled for Aug. 24, 2022, and told his supervisor it was inappropriate to use an FBI SWAT team to arrest someone charged with misdemeanors. He was also concerned about "biased jury pools" in DC and "overzealous charging" by the Justice Department, and he told his chain of command that he "believed some innocent individuals had been unjustly prosecuted, convicted, and sentenced."[31]

"I told them that I would not participate in any of these operations," he recalled in a Sept. 21, 2022, declaration. That very day, an op-ed praising Friend as an "American hero" appeared in the *New York Post*. "FBI hero paying the price for exposing unjust 'persecution' of conservative Americans," wrote columnist Miranda Devine.[32]

Within a few weeks, Friend had joined Trump's social media network, Truth Social, using the same handle style as the former president. He was @Real_SteveFriend.

I read the *New York Post* piece and grew curious. The story featured no mention of the actual arrest at hand, which was pretty relevant to the claim that the FBI had overreached. All it mentioned was that the arrest had to do with misdemeanors. With a few of the facts in the article, I pieced together the case at issue. It involved five members of the B-Squad, a militia that had come prepared for battle in DC and that had been organized by a man who'd later run for Congress, Jeremy Liggett.[33] Three members had been arrested in Florida, one of whom—at least initially—was only charged with two misdemeanor counts. If you look at photos of that defendant, who was dressed in military gear and was seen assaulting a non–law enforcement official near the Capitol tunnel, it's easy to see why the FBI might use SWAT to execute an arrest. Friend said he wanted to be a "conscientious objector" to Jan. 6 cases, and he'd stumbled into a case involving a poster boy who was less than ideal if the goal was to illustrate FBI overreach.

The facts didn't prevent Friend from being celebrated in the conservative press by those who showed a stunning lack of curiosity about the actual details of the case that Friend had proclaimed was his final straw. The *New York Post* opinion piece ran with a faux candid photo his wife shot of Friend all geared up, standing in a field, staring at the middle distance. A similar shot would appear on the cover of a book that would be published at lightning speed, less than nine months after Devine's piece ran and Friend signed his declaration. Devine wrote the foreword. Before long, Friend had been hired by the Center for Renewing America, as a "Fellow on Domestic Intelligence and Security

Services." One of his fellow employees was Jeffrey Clark, the almost attorney general who had his home raided in the course of an inspector general investigation into efforts to use Justice Department power to overturn the 2020 election, the event that culminated with the showdown at the White House less than seventy-two hours before the Capitol attack.

Friend would soon become a star witness for Republicans when they took over the House and set up the Select Subcommittee on the Weaponization of the Federal Government, and Democrats would put together a report decimating Friend and his allegations.[34] His social media accounts, in which he regularly called for the FBI to be dismantled, certainly helped.

"I've become a single issue voter," Friend wrote in March 2023. "The threshold for any presidential candidate to garner my vote consideration is a plan to end the FBI. I hope there is a stable of candidates to assess."

The same month that Mar-a-Lago was raided and that Friend rebelled against Jan. 6 cases, Trump met with some advocates for Capitol attack defendants at his club in Bedminster, New Jersey, in his office near the pool. Folks were talking, someone told him. He needed to step up to the plate. Trump said he was upset over what was happening to Jan. 6 defendants, but it was clear he wasn't up to speed on the details. He asked if he could help bail people out, and someone had to explain to the former president of the United States that there was no cash bail in the federal system.

Like a typical Trump meeting, it didn't stay completely on topic. But it ended with him saying he was going to contribute. Finally! A donor with cash to spare. Maybe Trump's donation would best a $100,000 donation from Dinesh D'Souza, the right-wing pundit who was given a full pardon by Trump after he pleaded guilty to violating federal campaign election law and went on to make a conspiracy film about the 2020 election.[35]

It took a while for the check to arrive, but it finally did, at the end of 2022. The self-proclaimed billionaire had donated just

$10,000. Split that among one thousand defendants, and it works out to about $10, which is barely enough for one mediocre lunch at the Prettyman United States Courthouse, where hundreds of Trump supporters have been sentenced. (A Trump campaign official did not respond to requests for comment on the donation.)

Trump had his own expensive legal problems to worry about, but he was willing to boost the cause of Jan. 6 defendants. Trump gave speaking spots at a September 3, 2022, rally to Matthew Perna's aunt, Geri Perna, who would later tweet that "President Trump is the ONLY person who will give my nephew Matthew Perna a Posthumous Pardon" and that Trump "told me himself."[36] Cynthia Hughes, who founded the group Patriot Freedom Project after her nephew was arrested on Jan. 6 charges, also spoke at the rally alongside Perna. A few weeks later, Trump-appointed US district judge Trevor McFadden sentenced Timothy Hale-Cusanelli, an army reservist who became infamous for posing as Hitler in photos the government found on his phone, to four years in federal prison.[37] Hale-Cusanelli was convicted on five charges, including a felony count of obstruction of an official proceeding. His testimony on the stand during his trial, in which he claimed that he didn't know that Congress met at the Capitol, was a "risible lie," McFadden said, and an "obvious attempt" to avoid accountability. McFadden also did not believe that Hale-Cusanelli's Hitler photos were a joke, saying "the evidence shows" that the defendant had "sexist, racist, and antisemitic" views. As of 2023, he was serving his time at a low-level federal prison facility at Fort Dix. Given the time he served in pretrial detention, as well as "good time" credit if he avoids disciplinary infractions, Hale-Cusanelli's anticipated release date was April 13, 2024, in time for the New Jersey Republican primary.

———◆———

On Jan. 6, 2023, as the fight over the House speakership was under way, a live audio chat unfolded involving a remarkable assemblage of figures. It was hosted by Brandon Straka, the founder of the conservative #WalkAway movement, who had pleaded guilty as part of a plea deal, admitting that he was guilty of a misdemeanor count of disorderly or disruptive conduct on the Capitol grounds and had intended to "impede, disrupt, or disturb the orderly conduct of a session of Congress."[38]

In 2016, a few weeks after he voted for Hillary Clinton, Straka performed[39] a one-man show for his friends and family six days before his fortieth birthday. The show was revealing and self-deprecating. He told the audience how he wanted to become a "big, rich, famous, powerful actor" but had washed out in the entertainment space and was working as a hairstylist. He spoke about his prior addiction to cocaine and alcohol. "Whether I've known you for twenty years or twenty days, I probably owe you money," Straka joked. And he sung an adaptation of the song "It Sucks to Be Me" from the musical *Avenue Q*:

> *So I'm not famous, I really don't care.*
> *I'm so much happier blow-drying hair.*
> *And having roommates in my forties feels . . . just so right.*

A bit more than a year later, at forty-one, Brandon Straka found a path to the fame he'd sought. "I'm a GAY REPUBLICAN," he tweeted in March 2018. "I NEED FOLLOWERS!!! SHOW ME SOME LOVE, PLEASE!!!"[40]

In the viral video that landed him on Fox News, he put the date of his political conversion in 2017, sometime after his one-man show and his vote for Clinton. His rollout came in time for the 2018 midterms, and by the 2020 election cycle, he was a full-blown MAGA influencer.

He spoke at the Stop the Steal rally on Jan. 5. Then he showed up to the Capitol.

As he later admitted in his plea, he yelled "Go, go, go" to other rioters to encourage them to storm the building and said "Take it, take it" as members of the mob stole an officer's police shield, as he captured in a video he filmed.[41] Like many Jan. 6 defendants, he'd filmed the government's most damning evidence against him himself.

After the attack, Straka didn't understand why so many conservatives were running away from what had happened at the Capitol.

"I'm completely confused. For 6–8 weeks everybody on the right has been saying '1776!' & that if congress moves forward it will mean a revolution! So congress moves forward. Patriots storm the Capitol—now everybody is virtual signaling their embarrassment that this happened," he tweeted.

"It was not Antifa at the Capitol. It was freedom loving Patriots who were DESPERATE to fight for the final hope of our Republic because literally nobody cares about them. Everyone else can denounce them. I will not," he tweeted.

Straka's confusion was understandably shared by plenty of Jan. 6 defendants. For those who *truly believed* in their heart of hearts that the election was stolen—that those dastardly Democrats had managed to pull off the perfect crime, that America was being swept out from under them, and the White House would soon be occupied by a presidential usurper—mass civil disobedience, at bare minimum, was a completely rational response, and it was easy to justify more violent measures. What were they supposed to do, just pack up, go home, take the loss, get "em next time? The 2020 election was the "crime of the century," Trump told his followers. The "real insurrection," he'd say, happened on Election Day 2020. Why, his supporters wondered, were Republican figures so quick to condemn those who were willing to put some skin in the game to save America? "You have to break eggs to make [an] omelette," Jan. 6 defendant Shane Jenkins, who used a tomahawk axe to smash in a Capitol window and assaulted officers with a wooden desk drawer, a broken wooden pole, and sticks,[42] wrote in a Facebook post on Jan. 7. "Don't think 'omg those poor eggs.' It's deadly serious when people said 'give me liberty or give me death' people bled and died serving this country and I'll be damned if I stand silently buy while we are bought by China."[43]

At his sentencing in January 2022, Straka did what many criminal defendants do at their sentencings. He apologized. He told the judge how he'd grown up in a small town in rural Nebraska and that his dream in life "was to make a living as a performer," mentioning his singing and acting and his "small amount of work in film and television." He said he was "deeply sorry and shameful" for being present at an event that sent members of Congress running in fear.

"I am sincerely sorry to all of the people of America, even the ones who absolutely hate my guts and hated me long before January 6," he said[44] at his sentencing hearing.[45] "I'm sorry that I was present in any way at an event that led anybody to feel afraid, that caused shame and embarrassment on our country, and that served absolutely no purpose other than to further tear away at the already heart-breaking divide in this country."

A judge gave him probation, with a few months of home detention. Now he was hosting a Twitter Space and becoming the de facto spokesman for Jan. 6 defendants, despite spending very little time behind bars.

"There's a very small handful of congresspeople who've expressed any interest in helping us, and most of them are the people who are holding out right now," Straka said. He said he'd had conversations "regularly" with Rep. Matt Gaetz, who at that moment was holding control of the House speakership in his hand.

I stared at the list of speakers and attendees. I noticed that a couple of sleuths were on there too, as they often were, listening, taking notes, recording. Sometimes they'd go in stealth mode, so no one knew they were listening.

Kyle Seraphin, a suspended FBI special agent who was making a name for himself in conservative outlets, hopped on the call, though Straka didn't seem to know him.

Since his initial round of interviews with conservative media personalities, Seraphin had become one himself. He launched a podcast called *The Kyle Seraphin Show*, where he interviewed people who shared his views while displaying an American flag, a rifle, and a bulletproof vest in the background. Seraphin's dad, Charlie, had a long career in radio and now hosted his own program on a small station in Arizona,[46] so Seraphin didn't skimp on the production quality.

"Prepare to hear the truth from a real whistleblower and American patriot," an announcer boomed at the beginning of his shows. "Here's civil liberties enthusiast, Second Amendment defender, and indefinitely suspended FBI agent Kyle Seraphin."

Seraphin was initially more of a Truth Social guy, but his audience grew much more quickly on Twitter. He became prolific on the platform and hopped into Twitter Spaces all the time.

"I just want to say my heartfelt, just, sorry for all of the things that you guys have gone through. I saw it on the back end working for the FBI," Seraphin told the group. "I've got friends who retired over all of your situations. . . . It's not universal within the agency there; there's a big divide that's happening."

Seraphin told them the FBI "needs to be broken into pieces" and described the Jan. 6 defendants as victims. He also explained what he would have done had he been asked to conduct interviews with Jan. 6 defendants, describing a hypothetical situation in which he arrived at a Jan. 6 suspect's house.

"I'd be shaking my head about whether he should talk to me or not," Seraphin said. "That would go probably nowhere, because, you know, I'm not going to pursue a trespassing case."

Rewind that a moment. What Seraphin is saying here—and not for the first time, either—is remarkable. An FBI special agent— he was still an employee, though unpaid at the time—was saying that he would purposefully undermine a criminal investigation. It wasn't the only time: a month later, in another Twitter Space, he encouraged a Jan. 6 participant who hadn't been arrested to pipe up if the FBI came calling.

"Obviously, they're going to have this recording, because it's out there in the world, brother, but there's no reason to talk to anybody, and if they don't have a court order, they don't need to come in and they can talk through a closed door," Seraphin said. "I offered to take all these interviews in Las Cruces, because I'm more than happy to walk up to people and just shake my head no while I'm saying, 'Would you like to talk to me?'" It's a good move to "just shut your mouth" when approached by the bureau, Seraphin said.

Seraphin had joined the bureau at thirty-five and was one of the oldest people in his FBI academy class. Here he was just a few years later, calling FBI headquarters "the death star" and advocating for the bureau's elimination.[47] He said he was "kind of a tumbleweed" who went to work for a furniture company in San Francisco after college and then ended up doing airtime sales for a radio company. He went back home to Texas for a bit, then went to Kansas City and ran a restaurant, and then went to San Diego as a "couch-surfing bum" before getting hired by a movie studio in Los Angeles. He then enlisted in the air force for four years, got out at the age of thirty-one, came back to Texas and got married, worked as a paramedic, and got hired on by the FBI. He graduated from Quantico in November 2016, the month Trump was elected.

Seraphin was assigned to the Washington Field Office and put on a counterintelligence unit. It was not his cup of tea, and he was bored, finding himself staring at a blank screen for hours a day. He transferred to a surveillance unit in June 2018 and was immediately assigned to a white supremacy case out of Alaska.

Seraphin enjoyed playing the role of troll. While he was at the bureau, he said he changed his gender to female just to mess with officials. Seraphin said he was one of the highest-ranking "female" FBI special agents on the fitness test in 2021.

"My FBI documents stated very clearly that I identified as female when I was at work," Seraphin said. "When I was in the office, there was no doubt that starting at a very early date with the FBI, I identified as a female." It just seemed like a reasonable thing to do, he said.

"I told my boss that I always felt like kind of a bitch when I was working for the FBI, and I updated my gender to match," Seraphin said. "I was rated as almost one of the top female physical performers in 2021 for a thirty-nine-year-old female."

Seraphin said he "was in the top five or so for pushups, for sit-ups, for the sprint, and you know I'm not that fast compared to dudes, but as a woman with a beard and, like, grown-man legs, I identify as a female whenever I have to sprint, and I crush it."

He stayed at WFO until May 2021, when he took a voluntary transfer to Las Cruces, a satellite of the Albuquerque Field Office. A few months later, he objected to the Biden administration's COVID-19 vaccine mandate and filed for a religious exception. He ended up taking personal leave in November 2021 so that he didn't have to show up and take COVID-19 tests. He thought he was pretty much done with the FBI at that point.

He wasn't. He eventually returned to the office wearing a "Let's Go Brandon" T-shirt his dad had given him, clothing that referenced a joke that started when an NBC Sports reporter at the Talladega Superspeedway for a NASCAR event mistook a crowd chant—"Fuck Joe Biden"—for a chant in support of Xfinity Series race winner Brandon Brown, whom she was interviewing.[48] A meme was born.

"I support all Brandons, I'm a big fan of Brandons," Seraphin later said, sarcastically. "I'm not 100 percent sure if there's a political nature to supporting someone named Brandon, almost named my son Brandon, big fan."[49]

Seraphin said that the senior agent in the Las Cruces FBI office bought a number of "Let's Go Brandon" T-shirts and hoodies and then wore them into the office. "Now, she's a Hispanic female, so it's possible that she doesn't have to answer to the rules," Seraphin said. "But as a Native American female, I feel like I should also not be beholden to those kind of rules."

He also mentioned that he posted a cartoon of Donald Trump riding on a tank and carrying a golden bazooka around the FBI office, as well as an image of Ronald Reagan riding on a dinosaur and firing an Uzi.

On Feb. 9, 2022, a law enforcement officer with the Las Cruces Police Department responded to a report of shots fired near a school. He found Seraphin in the midst of target practice. "Everybody is freaking out over there at the school, man. You can hear it over there like nobody's business," the officer said. "Probably not the best place to be shooting." Seraphin pushed back; he said that he was out there to work and train and that he had some time off so he was using it. He said he was with the FBI. He said he didn't have the time to drive out to the shooting range. Seraphin pulled out his FBI badge, but the officer wanted his license.[50] They departed after a few minutes. "On the same team," Seraphin said as the officer got back into his car.

An investigation was launched, and Seraphin thought it was just a cover, a way to go after him. He recorded his interviews with FBI and internal affairs officials. When one official indicated that they understood why the officer had responded to the sound of gunfire near a high school and was concerned, Seraphin doubled down, suggesting that the officer's question was a violation of federal civil rights law. "As far as I can tell, we could've opened a civil rights investigation under color of law, because he was purporting to have authorities he didn't have," Seraphin said.[51]

A few months later, Seraphin went public. He said he wasn't a big social media guy until that point, but he adapted quickly.[52]

One of the first interviews that Seraphin did was with Dan Bongino, the former Secret Service agent who quit in 2011 and ran three unsuccessful campaigns for Congress before turning himself into a media star.[53] Seraphin heaped praise on Bongino, clearly seeing his path from law enforcement to media mogul as an inspiration, and quickly fell into the role of digital shock jock himself. He upped his production game, replacing the background seen in his first video—an ill-lit family living room, with a grandfather clock—with a more modern look: first a shrine of weapons displayed on the wall with an American flag, then a wooden Betsy Ross flag display bathed in blue light.

Despite his clear right-wing views, Seraphin insisted that it was the rest of the world that had moved and that he wasn't an extremist. "I like to think I'm a centrist," Seraphin explained in an interview with his dad. "I just think that the world has shifted so far to the left now that I'm somewhat on the right of center—you know, we have the White House lit up with rainbow-colored lighting—and that's kind of wild for something that five years ago or eight years ago would have been totally absurd."

Seraphin regularly began weighing in on Jan. 6 cases and thought the bureau was blowing the event out of proportion. "The FBI and the DOJ got together and decided this was going to be the biggest case that's been worked," Seraphin said in one of his podcasts. "There's such an obvious and disgusting double standard being taken. It just it nauseates me."

Eventually, in 2023, Seraphin got the notification that his top secret security clearance had been officially revoked. In a letter, the bureau said Seraphin "demonstrated a repeated pattern of unwillingness to comply with numerous FBI rules and regulations," which included gun safety policies as well as his use of offensive language and improper disclosures.[54]

"During 2022, you demonstrated inappropriate behavior through your routine use of derogatory, racist, sexist, and/or homophobic language and comments which co-workers found offensive; and made unauthorized releases of sensitive government information," another FBI letter said. "In addition, you repeatedly appeared in public media and provided information used in news articles, video posts, and media reports. The information you provided disclosed the existence of FBI investigations, FBI methods and capabilities, and the identities of active FBI personnel without authorization."

Seraphin, who had no intention of returning to the FBI, took it in stride, joking that the bureau accusing him of using offensive language and improperly disclosing sensitive government information put him in the same category as a certain ex-president. "😂So, I'm Trump?😂," he wrote on Twitter.[55]

Meanwhile, the network of COVID-19 vaccine opponents that Seraphin had built was paying off. He said that he had a Signal group and contacts for hundreds of people and that he was being fed information from within the bureau that he was making public.

"You'd be delighted to see the kind of panic you induced at WFO this week," one FBI employee wrote Seraphin in a message that appeared to have been sent over Signal, after Seraphin posted a number of weekly emails that had been sent by a former WFO official. "So much management 'oh shit oh fuck' over the Captain's logs."[56]

Derrick Evans, who yelled "Derrick Evans is in the Capitol!" as he stormed the Capitol, was in the space too, and preparing to announce his plan to challenge a House incumbent in the Republican primary in his home state of West Virginia. A spokesman he'd hired had made an aggressive pitch to networks, trying to get a major show to run a feature on his announcement. But nobody seemed to take the bait. He eventually decided to announce his run on the anniversary of Jan. 6. He told me, in an hour-long interview that was embargoed until he made his announcement, that he wanted to be a combination of Reps. Matt Gaetz, Marjorie Taylor Greene, and Thomas Massie and Sen. Rand Paul.

"Southern West Virginia, this is Ultra-MAGA country, and the people of West Virginia deserve a firebrand Ultra MAGA representative, and that's what I'm going to give 'em," he told me. Evans still thought the election was stolen. He told me how much he'd learned about the criminal justice system while behind bars and how he befriended other Jan. 6 defendants behind bars.

Evans certainly had the politician spin routine down. Online, he was constantly playing down his behavior on Jan. 6 and making himself out to be a victim. "I served three months in federal prison for taking a guided tour through the Capitol," he tweeted. He regularly suggested that the police had simply let them in.

Evans's own video showed something much different, and the statement of offense that he signed when he admitted that he committed a crime did too.[57]

Evans was at the front of the pack when rioters breached the police barricades, and he noted how outnumbered officers were and how little chance they stood of holding the line. Then he joined the mob on the stairs and celebrated and encouraged the mob to break through the rotunda doors.

"The cops are running! The cops are running! Here we go! Here we go! Open the doors!" Evans yelled.

Brandon Straka was wrapping things up. But before he did, he'd scrolled down to check on his stats. He wanted to see how many people were tuned in and which users had joined. I'd been spotted.

"I see the NBC reporter Ryan Reilly here," Straka said. "Ryan Reilly has literally dedicated his entire life to obsessively writing about Jan. 6."

He went on like this. Derrick Evans dropped a few 100 percent emojis. Seraphin went with a heart.

Straka had gone through what the Walk Away website describes as a "quite remarkable" transition "from actor, singer, and hairstylist to an overnight political activist." He still had quite the flair for the theatrical. A while back, he sang the national anthem at an event at the Trump International Hotel while wearing a tux with white gloves and holding a cane.[58]

A few months earlier, Straka staged what is safe to say is one of the only pieces of performance art ever staged from a booth at the Conservative Political Action Conference (CPAC). There was a fake brick wall, and Straka was surrounded by bars on all sides.

He sat in the faux cell, barefoot, wearing a MAGA hat, and trying to look distraught and pretend he wasn't encircled by people with cell phones. "Where is everyone?" read the words on the chalkboard behind him.

Rep. Marjorie Taylor Greene entered the cell and squatted in front of him. Some people had thrown cash into his cell. CPAC attendees recited the Lord's Prayer.[59]

Now, in front of a group of several Jan. 6 defendants and a budding shock jock, the man who'd achieved the fame he'd sought for decades of his life with a viral video was lecturing me.

"All about the clicks, I guess," Straka said.

Strip away everything else, and there was a fair point at the core of Straka's complaints: had it not been for his public profile, Straka probably would not have been charged.

Straka was arrested early, just a few weeks after Jan. 6, when the federal investigation was primarily focused on the most viral riot participants. The video he recorded—quickly ripped off the internet by sleuths before he took it down—sealed up a case against him. It was only Straka's video—the one where, as he admitted in court, he "chimed in with the crowd, saying 'take it, take it'" as others yelled about taking the officer's shield—that gave the best evidence against him. Without "take it, take it," there's little to separate Straka from the thousands who were on the restricted grounds of the Capitol on Jan. 6 but never entered the building.

If someone happened to identify Straka today, there's very little chance he'd be charged. The sleuths may not even bother attempting to identify someone who they knew didn't enter the building. By the two-year mark, the parameters of chargeable conduct were pretty well defined: You either had to enter the building, or you had to commit violence or engage in property damage outside. Anything else wasn't worth wasting time over.

---•◆•---

Joseph McBride, the Jan. 6 lawyer who told me he didn't "give a shit about being wrong," was on another Twitter rant. It was mid-March 2023, and McBride wanted to draw attention to a document that the DC Metropolitan Police Department had prepared ahead of Jan. 6. He tweeted out a tightly cropped photo that showed a hand pepper spraying a line of officers at the Capitol on Jan. 6. There was a wristband around the person's wrist, not too dissimilar from the ones that undercover MPD officers were supposed to wear that day. The implication of McBride's tweets: an undercover MPD officer had pepper sprayed fellow members of law enforcement.

But I knew that photo. It was of an early arrestee who had been identified by online sleuths. He was known as Suitmacer, and he'd long been arrested. In fact, he had a plea agreement hearing scheduled on Monday, after the weekend.[60]

It was tough to tell when McBride believed what he was claiming and when he was just trying to muddy the waters and create disinformation about Jan. 6. I gave him a ring and asked if he knew that the person with the wristband had already been arrested before he dashed off his tweets. "No comment. No comment, Ryan Reilly," McBride said. I could practically hear his smirk through the phone.

"We have a job to do. There's an information war, right?" McBride said. "My job has always been to stick up for people that the world hates and to do so with the greatest amount of vigor and force."

I kept pressing. How did that photo come to his attention? Why so tightly cropped? Did he really not know that the person in the photo, who was from New York City just like McBride, had long been identified and charged?

"I would never knowingly put any false information out into the public; that's not what I do. And that's the truth," McBride said. "Now, off the record, we're off the record, right?" I agreed to go off the record and can't use it here.

A few weeks later, one of the most conservative judges on the bench, Trump-appointed US District Court Judge Trevor McFadden, ordered McBride to explain why he shouldn't be referred to a disciplinary board for unnecessarily delaying the trial of a detained Jan. 6 defendant. McFadden's order pointed to McBride's claims about his personal health that necessitated a delay in his defendant's trial and then brought up what McBride had been up to on social media, including visiting Mar-a-Lago for Trump's 2024 announcement. McFadden admonished McBride, saying he should "scrupulously adhere to his duty of candor to this Court and others going forward," but ultimately decided disciplinary actions weren't necessary.[61]

---

More than twenty years ago, the FBI director said the bureau was "years behind where it should be" on tech.[62] Recruiting tech wizards to the FBI was a major challenge, and it remains one today. The bureau's systems are perpetually out of date.

"Legacy information technology systems are a significant barrier to evidence building. Several components reported that potentially useful data is trapped in unusable forms such as word processing documents," the Justice Department's 2022–2026 Strategic Plan stated. "Other components reported that outdated data systems made responding to new information requests difficult and resource intensive. Upgraded technological solutions for data collection, data storage, and data analysis could have substantial value."[63]

The truth is that the FBI needs real scrutiny. There are fundamental, structural issues that need to be addressed, and it's unclear when that could happen. Simply tossing more money at the bureau isn't going to help fix the deeper issues. But the FBI isn't being targeted by smart investigations into the bureaucratic failures leading up to Jan. 6. Instead, the focus and energy is internet garbage: Republicans attacking the bureau and asking officials about the conspiracies their constituents believe, Democrats spending much of their time defending the bureau and fighting a never-ending battle to kill conspiracy theories that have online immortality. It's Brandolini's Law, a.k.a. the bullshit asymmetry principle: "The amount of energy needed to refute bullshit is an order of magnitude bigger than to produce it."[64]

The Jan. 6 committee was extremely effective at focusing the nation's attention on Trump's actions and at creating gripping televised hearings that set a new standard for congressional hearings. What got left out, though, was an in-depth examination of what exactly went wrong with law enforcement ahead of Jan. 6 and how those problems could be fixed.

While the Jan. 6 committee had a "blue team" dedicated to examining law enforcement failures, they never got their time in the sun, nor did other teams who dedicated countless hours to investigating the Capitol attack. There was no public hearing focused on law enforcement failures, and the only information critical of the FBI that seemed to trickle out came when the committee was going after the Secret Service.

Frustrated committee staffers said, only half-jokingly, that they hadn't signed up for Liz Cheney's 2024 presidential campaign. Some committee staffers, I was told, even had mugs mocked up that featured the late-aughts "Disaster Girl" meme— a smiling young girl posing in front of a house fire—only with

Cheney's face instead.[65] The message: Cheney was burning down their work and loving every minute of it. (That wasn't the only unofficial Jan. 6 committee gear: there were also sweatshirts that featured a Trump post that referred to the committee as the Unselect Committee of Partisan Hacks.)

Still, staffers were hopeful that the committee's final report would reflect the full scope of the committee's work. A few weeks before it was released, they learned it wouldn't. The final report, like the hearings, was shaping up to be "all Trump," and many staffers were upset.[66]

Cheney, having lost the Republican primary for her House seat to a Trump-backed opponent named Harriet Hageman in August, had plenty of time on her hands and was driving the committee's work. She thought of the other issues, like law enforcement failures, as much less important than the main goal and maintained what *New York Times Magazine* described as a "Captain Ahab–like focus on Donald Trump as a singular threat to American democracy."[67]

After the report came out, reporters were left to find the little nuggets buried in the transcripts that the committee released to the public. Some of the work of the "blue team" did make it through to the final report, but it was relegated to the appendix, and the punches were pulled. The report noted that law enforcement had "multiple streams of intelligence predicting violence directed at the Capitol prior to January 6th" and that, while some of it was fragmentary, "it should have been sufficient to warrant far more vigorous preparations for the security of the joint session." There were broad, generic recommendations, including a call for federal government agencies to "move forward on whole-of-government strategies" to combat violent extremism and "review their intelligence sharing protocols to ensure that threat intelligence is properly prioritized and shared with other responsible intelligence and security agencies on a timely basis" that could've been ripped right from a dusty copy of the Sept. 11 commission report.

There was little that was concrete or actionable, no legislative solutions or recommendations for executive action proposed. Even the appendix on intelligence failures couldn't end with a clear message. Instead, it wrapped up with a platitude about how "the best defense" against a president inciting an attack on his own government "will come not from law enforcement, but from an informed and active citizenry," as though eighth-grade social studies teachers were as much responsible for securing the Capitol from future political mob violence as were federal law enforcement.[68]

The internet is both to blame for Jan. 6 and responsible for helping solve it. As cheesy as it sounded, the report isn't wrong: Americans who care about democracy do need to find ways to fight back against dangerous bullshit, even if that work can be soul-sucking. Perhaps, with increased internet literacy from digital natives, a smaller percentage of the electorate will be gullible enough to fall for the ham-fisted conspiracy theories that duped many during the 2020 election cycle, especially if they aren't being backed by a sitting president. But digital tools are getting more complex, the electorate more siloed, and the path ahead is uncertain.

There are always going to be people who want to believe the bullshit and people who will tell them those lies are true, either because they really believe the lies or because they have no principles. Neither the FBI nor DOJ nor the judiciary branch will fix all the nation's woes. A multimillion-dollar settlement with a network that knowingly spread lies about the election may rein in one channel and dethrone a cable news king, but that wouldn't stop a conspiracy-minded billionaire from buying a social media platform, lifting the floodgates on misinformation, and offering that former cable news king a home and a new megaphone with fewer constraints.

As of this writing, the most thorough investigation into the failures of federal law enforcement agencies in the lead-up to Jan. 6 came from the US Government Accountability Office, which is an agency that falls under the legislative branch but is a step removed from day-to-day politics. The GAO found that the bureau "did not process certain referrals from social media platforms according to policies and procedures and, as a result, it failed to share critical information with all relevant partners" and called for changes at the bureau.

"If the FBI does not process tips or information according to policy and procedures, information can get lost or may not be developed into threat products that the FBI can share with partners," the report stated. "According to FBI officials, the FBI is in the process of assessing lessons learned and making improvements following the events of January 6. In particular, the FBI is conducting several reviews focusing on challenges the FBI faced in communication, information sharing, and analysis."

The FBI acknowledged the GAO's findings that "despite collecting and sharing significant pieces of threat reporting, the FBI did not process all relevant information related to potential violence on January 6," and said the bureau "continues to be introspective" about the lessons of the day.

"Our goal is always to disrupt and stay ahead of the threat, and we are constantly trying to learn and evaluate what we could have done better or differently, this is especially true of the attack on the Capitol," wrote Larissa L. Knapp, the executive assistant director of the FBI's National Security Branch. "While the FBI and our law enforcement partners were aware of and certainly planned for a response to potential violence in the [National Capital Region] on January 6, the FBI was not aware of actionable intelligence indicating that a large mob would storm the Capitol building." Nothing in the GAO report, she said, changed that.[69]

———— ♦ ————

A few months into the third year of the Jan. 6 probe, some of the earliest cases were coming to fruition. There were the marquee trials against members of the Oath Keepers and Proud Boys, with the Justice Department securing fourteen seditious conspiracy convictions. Many of sleuths' earliest finds were getting to the sentencing stage. Logan Barnhart, the model/construction worker who dragged an officer down the stairs, had pleaded guilty.[70] He was eventually sentenced to three years in federal prison.[71] Barnhart spoke to me after his sentencing, declining to say whether he still believed the election was stolen, but saying he'd do his time and try to rebuild his life.

"If there's anything I've learned from all this, it's that there's a lot more to life than politics, there's a lot more to life than being on social media," Barnhart told me outside the courthouse. "I have my family, and I just want to be with them."

A sentencing hearing for Matthew Beddingfield, a North Carolina man who was out on bail on a first-degree attempted murder charge when he traveled to DC with his father and then stormed the Capitol,[72] was on the books. He'd pleaded guilty to a felony count of assaulting, resisting, or impeding officers, admitting that he used his flagpole to jab at and strike one police officer, swung the flagpole at another, and then threw a piece of the flagpole at the line of police officers before storming the building.[73] He had a sentencing range of thirty-seven months to forty-six months in federal prison.[74]

The gulf between cases was huge. In the same story that identified Beddingfield in March 2021, I referenced two other men who'd been identified by online sleuths.

One man, whose photo was on the FBI's website, was a perfect match for a profile image on Classmates.com. He'd deleted his personal Facebook page, but there were plenty of photos available of him online. Online sleuths referenced the man as #Tunnel-Commander, because he seemed to be guiding the mob in the western tunnel that day. In this case, an old FBI confidential human source who had provided reliable information to the FBI since March 2016 and was motivated by financial compensation identified the suspect using "publicly available methods," according to an August 2021 search warrant.[75] David Mehaffie was arrested in August 2021,[76] had a bench trial before Judge McFadden, and was sentenced to fourteen months in federal prison.[77]

The other man, seen pepper spraying officers, was a perfect match for a man who'd given an interview at a 2020 Trump rally as well as appeared at a Tea Party rally a decade earlier. He was a huge Trump fan. When I made calls to his family and friends, I could not nail down whether he was actually in DC on Jan. 6. One person simply told me it would be "up to him to tell you." I could never get him on the phone. More than two years later, he still had not been arrested.

---◆---

In March 2023, yet another tip from a sleuth sent me dashing to the courthouse. Samuel Lazar, the rioter who'd screamed "Take their guns!" through a bullhorn as rioters attacked cops and then attended a 2021 event with Rudy Giuliani while his photograph was featured on the FBI's Capitol Violence page, was appearing for a sentencing hearing. That was strange; there hadn't been anything on the docket, and there hadn't even been a plea hearing yet.

It made some sense though: there hadn't been activity on Lazar's docket since the summer of 2022. By the time I arrived, the hearing had wrapped up, but I'd spotted his family there, including his sister who'd been given the nickname Gal Dice Clay for her demeanor, which reminded sleuths of the stand-up comedian. I overheard a group of Lazar supporters who had made the hearing discussing Lazar's sentence and a transfer to a halfway house. The outcome suggested that Lazar had become very useful to the government.

The hearing wasn't *sealed* per se, I learned. Theoretically, if you just so happened to be wandering the halls of the courthouse, you might be able to go in. Lazar had been held at the Lewisburg Federal Prison before his court appearance, but it was unclear where he ended up. When I checked the inmate locator on the Bureau of Prisons' website, it said that inmate 56948-509 was no longer in custody and that his release date was unknown. Nothing about Lazar's case hit the docket.

What Lazar ended up getting, and what cases he may have aided, was a lingering mystery. His sister, accompanied by Lazar's young fiancée, did not want to answer questions when I approached her outside the courthouse. But Lazar's sister had posted letters online that he'd written from prison. There were some benefits to being off the internet, he wrote in one.

"Inside here its surprisingly refreshing to be able to speak to people face to face rather then through text or emojis, it feels human. Maybe we all should put down our devices and talk to each other face to face more, often we may find we have a lot more in common then the media tells us," he wrote.[78] He was looking forward to the future.

"2023 will be a year full of joy, prosperity and surprises!" he'd written in a letter his sister posted. #FacePaintBlowhard then added a smiley face.

———————— ✦ ————————

For months, Micki Witthoeft, Babbitt's mother, had been a regular presence in federal court, sitting in on Jan. 6 trials while wearing a T-shirt or wristband memorializing her daughter. On the two-year anniversary of Jan. 6, the conspiracy website Gateway Pundit named Witthoeft "JANUARY 6TH HERO OF THE YEAR," declaring Babbitt a "slain warrior" and her mother "an unlikely civil rights leader." Witthoeft had contact with Trump, who called into a rally outside the DC jail, though his remarks were interrupted by a counterprotester in a chicken mask who ran up behind the group, holding a cardboard speech bubble reading "fart noises." Randy Ireland, a former Proud Boy who had transformed himself into a Jan. 6 prisoner advocate and who got himself kicked out of court one day for calling Mike Fanone a "piece of shit" during a sentencing hearing,[79] tried to block the man-chicken, with minimal success.

Witthoeft had been living in the DC region for months, with an anonymous benefactor putting her and other Jan. 6 supporters up at an Airbnb in Virginia. Ireland was one of her housemates. She had morphed into a full-time activist, and on the two-year anniversary of her daughter's death, she deployed one of the most critical tools in a Capitol protester's toolbox: a high-profile arrest that draws media attention.

It certainly worked on me. I got a notification that Witthoeft had been arrested and scrambled over to the Capitol. When I arrived on the scene about twenty minutes later, I found protesters gathered at the location of the first breach of the Capitol perimeter on the west side of the Capitol, right near Peace Circle. The Jan. 6 supporters were being trolled by antifascists. One of their signs read "Sedition: When you're white they let you do it." A catchy 2009 tune by British singer Lily Allen, which had become a staple of DC counterprotests, blasted out over a speaker. "Fuck you (Fuck you), fuck you very, very much."

She probably wouldn't have described it this way, but Witthoeft had engaged in an act of civil disobedience. Video of Witthoeft's arrest at the Capitol showed that it went down as most do on Capitol Hill: voluntarily. She had refused multiple orders to get out of the street as they obstructed traffic.

Witthoeft handed the roses she was carrying to Nicole Reffitt, put her McDonald's coffee cup on the ground, and turned around. Two officers glanced at each other, and one of them took her into custody. "Capitol Police suck ass!" she said as she was patted down and put in the back of a police car.

A small demonstration by Jan. 6 supporters barely would've made a blip in the media, especially with the chaos unfolding in the House over McCarthy's race for Speaker. But Witthoeft's arrest got a lot of attention. The *Washington Post*, the *Hill*, and the *New York Daily News* all wrote it up. Witthoeft appeared on Newsmax.[80]

In a segment on her arrest on *Fox News Primetime*, Tucker Carlson said Babbitt was "murdered" without justification, and aired an edited video that cuts out the part where Witthoeft turned around and placed her hands behind her back. "BABBITT'S MOTHER ARRESTED WHILE HOSTING MEMORIAL," read the banner.

Aaron Babbitt, Ashli's husband, joined the program, floating the conspiracy that Capitol Police had conspired to arrest his mother-in-law on the two-year anniversary of his wife's death.

"I'm not going to say it's a coincidence that she was finally arrested on Jan. 6, but that doesn't really smell right," Aaron Babbitt said. "Arrest of Ashli Babbitt's Mom on Jan. 6 Anniversary 'Doesn't Smell Right': Widower," read the headline on the Fox News website.

On social media, Witthoeft didn't volunteer that she had made the choice to be arrested. Instead, she shared the Fox News segment that suggested there was something suspicious and sinister happening. The notion that the Capitol Police had conspired to arrest Ashli Babbitt's mother on the two-year anniversary of her death was now canon in MAGA world, a conspiracy theory that wouldn't die.

A couple of months later, I learned that Witthoeft was set to meet with House Speaker Kevin McCarthy.[81] The meeting came after McCarthy commented that the Capitol Police officer who killed Babbitt "did his job" on Jan. 6. Trump didn't like that and commented on Truth Social that he disagreed with what McCarthy said about the officer who "shot and killed Great Patriot Ashli Babbitt." McCarthy didn't want to detail the meeting, saying only that Witthoeft had requested it, as though any American could just sign up to get a meeting with the Speaker of the House.

"I felt like it was a good meeting," Witthoeft said afterward. "I thought Speaker McCarthy was delightful."

Reffitt, the wife of the first Jan. 6 defendant convicted by a jury, came over to say hi with her dog Oliver. We'd been in several courtrooms together, and I'd interviewed her on the phone a few days earlier after she'd called me a "little seditious hunter" before begrudgingly referring to a "decent" article I wrote about how the Jan. 6 committee had buried the evidence of law enforcement failures. But this was really the first time we'd spoken face-to-face.

Her husband, Guy, now fifty, had been transferred to low-security federal prison in El Paso, Texas, with a release date set for 2027. A few weeks before the Jan. 6 anniversary, a federal grand jury in Texas had indicted him on a gun charge: the FBI said they turned up an unregistered silencer when they searched his home after the attack.

Nicole was living at an Airbnb in Virginia along with the ex–Proud Boy and Witthoeft, Babbitt's mom. When Witthoeft chose to be arrested, she handed her bag to Reffitt.

"There's very different personalities going on here," she told me from the house when we spoke a few days earlier. "But you kind of make it work."

Nicole was pretty media savvy at this point. She was also, frankly, likable. On some topics, you could reason with her. She was willing to concede her naivete about the criminal justice system before her husband got caught up in it. She loved her kids, even the one who'd turned her husband in to the FBI.

She could be funny too. When counterprotesters chanted that the pro-Trump demonstrators were "a bunch of freaking losers," Nicole playfully bounced along to the beat as Oliver led her around on his leash.

Nicole had now spent five months living in the DC area, and she'd sat in on more court hearings than many reporters on the beat. She'd stuck around after her husband was sentenced in

August to more than seven years in federal prison, which was still far short of the fifteen years federal prosecutors sought under a domestic terrorism enhancement.

"I just couldn't sit still at home; I just was not settled," Nicole said. "I thought I would be settled after Guy was sentenced; I would have a direction about what to do." It never came. So she decided to become an advocate for other families, showing up to court to be a friendly face or a shoulder to cry on for families when their loved ones were sentenced to prison.

"I just felt like they needed a voice," she said. "I've got thick skin, so I thought who better than me to try to help them?"

Nicole thought that many Jan. 6 defendants were being overcharged.

"I see a lot of good things in the judiciary system, also. It's not like they're just out there blindly doing this; I think the bigger problem is with the DOJ and with the FBI," Nicole said.

Nicole even spent Christmas outside the DC jail. But now she was at a pivot point.

For most of their lives, Guy had been the breadwinner, but that came crashing down after COVID-19. They wiped out their savings. Nicole had been a manager at a Kohl's but now was paying the bills with money from a fundraising site. Later in the month, she planned to head back to Texas and downsize her home.

She wasn't sure what was next for her.

"I don't know what to do, to be honest, Ryan," she said. "I mean, I'm being very honest here. It's a lot of anxiety. You just don't prepare for something like this."

Overall, though, there wasn't much news here, and the small groups were dissipating. I filmed a few videos and then cut out to head back to the bureau. I was about to hop back on my bike when I heard what sounded like a group of protesters. I looked up and saw an American flag leading the pack.

*Oh gosh*, I thought. *More of them?*

My eyes immediately went to the man leading the march, who was sporting an animal on his head. He was tough to miss, dressed like a frontiersman, and I quickly clocked him as a Jan. 6 participant someone dubbed Little Crockett. An even better nickname is Scut Farkus, after the fur-hatted antagonist in *A Christmas Story* who bullies hero Ralphie until he snaps, tackling Farkus into the snow and punching him again and again. Picture a slightly older Farkus without braces, and you've got Little Crockett.

Although Little Crockett was at the Capitol on Jan. 6, he's unlikely to face charges. He was on the restricted Capitol grounds along with thousands of others, but there's no evidence that he entered the building or destroyed property. He only came onto my radar because he attended that September 2021 rally in support of Jan. 6 defendants, where, with his hands shaking uncontrollably, he performed an ode to Babbitt before assembled reporters:

> *She stood against violence, no weapon she held;*
> *She stood for our liberty up on that hill;*
> *They needed a body to cover their crimes;*
> *An innocent woman would serve them just fine.*[82]

Now, here he was marching at the head of a crowd. But something was a bit off. The crowd members weren't wearing MAGA swag; it didn't even look like there was any red in the whole crew. As they neared, I realized what I was looking at.

They were eighth graders, at the Capitol on a field trip. Here was Little Crockett, a Jan. 6 participant, leading them around the grounds of the Capitol, Pied Piper–style, trailed by their chaperones.

The group arrived near Peace Circle to await their circling bus. The man lingered, awkwardly, for a bit as the eighth graders stood around.

"Thank you!" someone said, in what seemed like a hint that their interaction was over. "Bye!" chimed in another. "We love you!" yelled one.

"Have a great day!" he told them, waving, as he walked west, toward Pennsylvania Avenue. "God bless you all!"

Neither the eighth graders nor their chaperones seemed to know what exactly was up. Unless you took a close look at the "Second Annual Hero's Day Ceremony J6th 2023" sign affixed to his back, you might not realize what this was about. Capitol tours didn't mention the Jan. 6 attack,[83] and the small demonstration was easy to miss. The kids just saw a guy in a funny hat. For the students, this was basically the equivalent of the naked cowboy in Times Square.

After he walked away, I talked to the chaperones, who were wondering who exactly that guy was. They were surprised to learn he was a Jan. 6 supporter.

"We don't support that," one of them said, just in case their shock upon learning his background didn't register with me.

I'd filmed the end of his interaction with the students. My instinct, of course, was to post it. It was a weird moment, a mix of disturbing and funny. But then I thought about how images of those eighth graders could've followed them around on facial recognition websites for years. I posted video of Little Crockett, but I left them out.

# Epilogue

## April 4, 2023
## Washington, DC

A few hours before the forty-fifth president of the United States arrived at a courthouse in Manhattan for his first appearance in the criminal case of *People of the State of New York v. Donald J. Trump*, a member of the "Patriots 45 MAGA Gang" was in a federal courtroom in DC receiving life-altering news. A federal judge had convicted him of two felonies.[1]

Now, standing outside the Prettyman federal courthouse with a GPS monitor freshly affixed to his ankle, twenty-eight-year-old InfoWars fan Ed Badalian was explaining to me why he believed Trump supporters could have arrested Nancy Pelosi and Joe Biden on Jan. 6.

"Any person has the right to arrest anyone if they see them committing a crime or if they have knowledge of them committing a crime," he claimed. "A citizen's arrest isn't a kidnapping, in that sense; it's bringing them to justice."

As Badalian saw it, citizens could have seized Pelosi and Biden on suspicion of "election interference" and delivered them to law enforcement.[2] The speaker of the House and the president-elect should have been detained, Badalian said, while investigators looked for evidence of whether they helped suppress stories about Hunter Biden's laptop during the 2020 campaign.

US District Judge Amy Berman Jackson, who would assume senior status—semiretirement, with a lightened caseload—in a few weeks, wasn't buying it. "No, Mr. Badalian, the Constitution does not give you the right" to "effect an utterly unlawful citizen's arrest of Nancy Pelosi," Jackson told him while delivering her verdict in the bench trial. After his codefendant Danny Rodriguez reached a plea deal, Badalian had elected not to go before a jury, choosing to try his luck before a judge. It didn't go so well.

"The defendant appears to be a very self-satisfied young man, impressed with his own intelligence and strategic acumen," Jackson said. "He seemed to think that charming me, impressing me with how smart he was, was the way to go. For example, the defendant told me, *Oh, I couldn't call and turn in Danny because I didn't know Danny had assaulted officers; that was hearsay.* Actually, Mr. Badalian, it wasn't." She said Badalian appeared to be telling a "reverse engineered" story on the stand, and the "hubris" he displayed on the stand matched up with "the hubris he showed all along; that he had the right to come to the nation's Capitol and arrest people."

Badalian, she said, "knew exactly what Jan. 6 was all about" and "went to Washington with clear intent, the single-minded purpose to try to keep the transfer of power from taking place." Badalian's "intentions were clear, and he expressed them repeatedly," she said.

"It may have turned out that he's not so brave, or he's not such a great military tactician after all, or that he turned out to be more talk than action. He tended to cheer on others, rather than doing any dirty work himself," Jackson said. "Or it may be that he was 100 percent ready to take action with bare hands, but the 'traitors and tyrants' never presented themselves." Whatever the reason for his failures, that wasn't an excuse, she said.

It was the type of verbal filleting that reporters had come to expect from Judge Jackson, who had overseen some of the biggest cases of the Trump era and the Mueller investigation, is considered by more than one reporter who covers the DC federal courthouse to be the sharpest judge on the bench, and is someone with little tolerance for bullshit.

In 2019, Jackson sentenced Manafort to several years in federal prison, quipping that his corruption supported an "opulent" lifestyle with "more suits than one man can wear" but more seriously criticized the "disregard for facts" in his court filings that indicated "ongoing contempt" and a "belief that he had the right to manipulate" the court proceedings.[3] Trump pardoned Manafort two weeks before the Jan. 6 attack.[4]

She oversaw the Roger Stone case too and sentenced him to forty months in federal prison in early 2020. "The truth still exists," Jackson said during Stone's sentencing. "The truth still matters."[5] Trump commuted that sentence in July 2020 and then gave Stone the full pardon on Dec. 23, 2020, the same day as Manafort.[6]

Jackson, a Baltimore native who received both her undergraduate and law degrees from Harvard, joined the US Attorney's Office in 1980[7] and went on to receive Justice Department awards for her work on murder and sexual assault cases.[8] President Barack Obama nominated her to the bench during his first term. But she wasn't isolated in some partisan bubble. Her husband was in the George W. Bush administration. When her son appeared on *Jeopardy* in 2015—winning thirteen consecutive games and more than $400,000 during a run that made him one of the biggest winners in the show's history—he described his mom as "white, liberal and Jewish" and his dad as "Black, Christian and conservative."[9]

She didn't suffer fools gladly, but she'd often pair a verbal thrashing with a sentence on the more lenient end, or with an accommodation. In one of her earlier high-profile cases, in 2013, she sentenced former Rep. Jesse Jackson Jr. to two and a half years in federal prison and his wife, Sandra Stevens Jackson, to a year behind bars, saying they'd used campaign accounts as a "personal piggy bank" and said she couldn't find her way to giving the couple probation because she'd have trouble explaining it to those who donated to the campaign. "The ethical standard has got to be simply higher than unindicted," Judge Jackson said.[10] But she also allowed the couple to stagger their sentences to make the transition easier on their children.

One of Judge Amy Berman Jackson's clarion calls about the ongoing threat after Jan. 6 came during a sentencing hearing for Kyle Young, a violent rioter known to sleuths as #AscendDad because he was accompanied by his sixteen-year-old son. In a moment that was caught by Forrest Rogers after the government released three hours of CCTV footage from inside the tunnel, Young handed off an electric shock weapon to Danny Rodriguez before he drove the stun gun into Fanone's neck.[11] That moment was left out of his plea agreement but included in his sentencing memo. Young also assaulted Fanone himself, restraining the officer as Rodriguez drove the stun gun into his neck and members of the crowd yelled, "Kill him!" and "Get his gun!"

Fanone brought prepared remarks to the hearing, but he also had some thoughts as he sat inside the packed courtroom.

"Your honor, this is not my first rodeo. I have been here in this courthouse many, many times before," Fanone said. He'd been assaulted before in the course of his duties, Fanone said, but what happened on Jan. 6 was different. "Those individuals were trying to escape. And I get it. Jail sucks. But this case is unique. The assault on me by Mr. Young cost me my career. It cost me my faith in law enforcement and many of the institutions I dedicated two decades of my life to serving."

Young, he said, was a "career criminal" who was busy committing felonies while Fanone served DC and the country with distinction. But the career he loved was over now, thanks to people like Young. He asked Jackson to send Young away for ten years. Then Fanone read the last words he'd jotted down on the piece of paper.

"I hope you suffer," he said. That's when Randy Ireland, the former Proud Boy, not so quietly called Fanone a "piece of shit," sparking a brief tense moment as Fanone stopped in his tracks and looked at the man a few rows back. Others filled the gap between the two men, and a US Marshal ordered Ireland to leave the courtroom.

Soon it was the defendant's turn to speak. Young apologized to Fanone and the other officers, told Jackson that he was "very ashamed," and said he was particularly broken up over the impact on his family. "My kids have to go to school and hear about their dad, what he did," Young said. "I just wish I could take it back. And I know I can't."

Jackson took a brief recess. Soon she reassumed the bench, calling Young "a one-man wrecking ball" who made decision after decision after decision to engage in violence that day. Then she turned to the broader picture.

"The essence of a democracy is that each citizen gets to express his views, exercise his choice with the right to vote," Jackson said. "What you were trying to do was undo the indisputable result of other people's votes and to do it by force. And you were trying to stop the singular thing that makes America America, the peaceful transfer of power. That's what 'stop the steal' meant. And that's exactly the opposite of what the Constitution means."

Young's lawyers, like many defense teams, had tried to portray Jan. 6 as something unique, something that wouldn't be replicated anytime soon. Jackson wasn't so sure.

"I wish I could feel comforted by that optimism," Jackson said. "But it's not as if the divisions in our country have eased in any way," adding that the "heated, inflammatory rhetoric" that brought Young to DC had not subsided, nor had the "steady pumping of misinformation on the air and online" abated.

"The threat did not evaporate or dissipate just because the election got certified. The lie that the election was stolen or illegitimate is still being propagated. Indeed, it's being amplified, not only on extremist social media sites, but on mainstream news outlets. And worse, it's become heresy for a member of the former's president's party to say otherwise," Jackson said.

Politicians, Jackson said, were cynically manipulating and stoking anger for their own ends, and "actively shunning the few who think standing up for principle is more important than power and have stepped forward to educate the public and to speak the truth." Government officials were under threat for investigating Jan. 6 and Trump, she noted.

"Some prominent figures in the Republican Party and the former president himself are cagily predicting or even outright calling for violence in the streets if one of the multiple ongoing investigations doesn't go his way," she added.

"The judiciary, if no one else, has to make it clear, it has to be crystal clear, that it is not patriotism, it is not standing up for America to stand up for one man—who knows full well that he lost—instead of the Constitution he was trying to subvert," Jackson said. "It is not justified to take to the streets, to descend on the nation's capital, or to attack law enforcement officers doing their sworn duty at the behest of that one man."

Jan. 6, and the "effort to keep that spirit alive," Jackson said, were "the utter antithesis of what America stands for. It is the pure embodiment of tyranny and authoritarianism."[12]

Jackson sentenced Young to more than seven years in federal prison.[13] Jackson would continue to ring the alarm about the outstanding threat the next month, when she sentenced another defendant who assaulted Fanone, Albuquerque Head, to more than seven years in prison.

"People need to understand that they can't do this," she said. "They can't try to force their will on the American people once the American people have already spoken at the ballot box. That's the opposite of democracy, it's tyranny. And the threat to democracy, the dark shadow of tyranny, unfortunately, has not gone away.

"There are people who are still disseminating the lie that the election was stolen. They're doing it today," she continued. "And the people who are stoking that anger for their own selfish purposes, they need to think about the havoc they've wreaked, the lives they've ruined, the harm to their supporters' families, even, and the threat to this country's foundation."[14]

A defense attorney once noted that Jackson's demeanor was the same "whether it is in a sparsely populated courtroom or one packed with national media."[15] As the Jan. 6 trials continued, as the intense media interest in the cases slowly faded and the 2024 election cycle ramped up, it was a theory I'd get to test firsthand.

The court gallery was nearly empty when Jackson delivered her verdict in Badalian's case. She allowed Badalian to go home until his sentencing but ordered him to wear a GPS ankle monitor, which Badalian said, on his way out of the courtroom, amounted to "cruel and unusual punishment."

Outside the courthouse, after posing for photos with his ankle monitor, Badalian assured me that he wouldn't be showing up to his sentencing hearing in a few months claiming that he'd seen the light, that he no longer believed the election was stolen.

"I think that would be bizarre," Badalian said. "That would seem like a lie."

The man who Badalian said gave him his marching orders didn't leave court with an ankle monitor on April 4. He got to skip the handcuffs and the mug shot too. And once his arraignment was over, the former president would hop right back on his private plane and head back to Palm Beach International Airport and his Mar-a-Lago compound.

Trump's next court date was set for Dec. 4, 2023, months after the first GOP primary debate and just weeks before the first Republican primary voters cast their ballots.[16] First, though, the judge in Trump's hush money case issued an unusual warning.

"Defense counsel, speak to your client and anybody else you need to, and remind them to please refrain, please refrain from making statements that are likely to incite violence or civil unrest," New York Supreme Court Justice Juan M. Merchan told Trump's lawyer. "Please refrain from making comments or engaging in conduct that has the potential to incite violence, create civil unrest, or jeopardize the safety or well-being of any individuals. Also, please do not engage in words or conduct which jeopardizes the rule of law, particularly as it applies to these proceedings in this courtroom."

Trump called Merchan a "Trump-hating judge" that night, and District Attorney Alvin Bragg a "criminal."[17] On his return to Mar-a-Lago the night of April 4, Trump DJed for aides and guests, using an iPad to play "Justice for All," the song that featured Trump alongside the J6 "prison choir."[18]

That night, I spoke with Gina Bisignano, the Beverly Hills cosmetologist who'd called for "strong, angry patriots" over a bullhorn as police came under attack in the tunnel, telling the crowd that they weren't going to let their "Trumpy bear" be pushed out of office as mascara ran down her face.

Bisignano was another early Jan. 6 arrestee, and she initially had it rough. She'd been locked up pretrial and was accidentally held at a jail for two weeks even though a judge had ordered that she be brought to DC.[19] By late February 2021, about a month after her arrest, Judge Carl J. Nichols ordered Bisignano released, with conditions. "No access to social media. No communications with anyone who was at the event on January 6, 2021," he wrote.[20]

Those restrictions would prove challenging for Bisignano, whom no one would call camera shy. Jan. 6 had not been Bisignano's first brush with infamy. A few weeks before the Capitol attack, she was featured yelling slurs outside the home of LA County's health director during a protest over coronavirus restrictions. "Karen Goes on Homophobic Rant During L.A. Protest," read the TMZ headline, featuring Bisignano in a red Trump sweatshirt.[21] She'd dipped her toe into reality television too, filming a trailer for a show called *Real Patriots of Beverly Hills*, costarring a man named Eric Christie, who'd later be charged in connection with Jan. 6, as well as a Bryna Makowka, who'd dated Badalian.[22]

I'd written and tweeted about developments in Bisignano's case, and we ended up talking regularly. She was a big sharer and could reveal a lot, often more than she probably should have. Once, while watching a documentary on the Jan. 6 attack, she snapped a photo of her television as footage played of Rodriguez driving a stun gun into Fanone's neck. I could see her reflection on the television screen and that she was only wearing a bra.

Bisignano, during a sealed hearing in August 2021, pleaded guilty to several counts, including felony obstruction of an official proceeding and civil disorder.[23] A few months later, after Bisignano testified before a federal grand jury, three people she knew from the Beverly Hills pro-Trump scene were indicted: Rodriguez, Badalian, and the man known as #SwedishScarf, whose name was redacted from the indictment and on the court docket.[24]

Cooperating with the government didn't mean Bisignano had seen the light. Far from it. She still thought the election was stolen, and she still spouted bizarre conspiracy theories. I'd occasionally try to lure her back from the ledge and away from the tall tales that landed her in this predicament, once suggesting, half-seriously, that we should visit Comet Ping Pong—home to the #Pizzagate conspiracy theory—so she could see, in fact, that there was no basement full of children, as a gunman who stormed the neighborhood family pizza shop believed when he visited in 2016.[25]

When I spoke with Bisignano the night of Trump's arraignment, she called Trump "a fucking icon" and "an alpha male" who made her feel safe. She wasn't so sure Trump even slept with Stormy Daniels.

"He probably maybe squeezed her tit or something," Bisignano said. "And he's like 'shut her the fuck up.'"

Bisignano also had some thoughts on Judge Jackson. A few weeks earlier, during Badalian's bench trial, Bisignano testified for the prosecution under her cooperation agreement. She had no issue identifying Badalian in videos from Jan. 6, but her testimony was less certain on the issue of whether she felt pressured by Badalian to get rid of evidence from the Jan. 6 attack. That charge mostly related to a visit that Rodriguez, Badalian, and #SwedishScarf made to Bisignano's house after they arrived back in California. #SwedishScarf—or Jeff, as Bisignano knew him—had come into her home, unplugged her Amazon Alexa devices, and "indicated through miming," as the indictment stated, that Bisignano should not speak out loud.

"I want to help you delete everything and to transfer the files to a secure hard drive," Jeff had written on one of Bisignano's notepads.

Bisignano, while on the stand, had trouble recalling where Badalian was during some of the conversations she remembered with Rodriguez and Jeff. Still, Bisignano initially thought that Jackson liked her, that her testimony wasn't too painful. She'd misjudged.

"Gina was a hot mess," Jackson said as she delivered her verdict in Badalian's case. "She was possibly one of the worst witnesses I've ever sat next to in this courtroom. She came in beaming. She struck a pose as if this was about to be her breakout performance, a star turn. At times she seemed to be smiling and flirting with the agent, possibly hoping to please him." (I was sitting behind the agent, alone in the courtroom gallery, and was under the impression Bisignano was smiling at me, and Bisignano said that was the case.)

Bisignano, Jackson said, "appeared torn between her obligation to cooperate with the government" and her desire to "maintain her standing with her peers," meaning the "patriot" community in California. That was certainly true: Bisignano was very attuned to what I'd write about her testimony. She wanted to tell the truth, she told me, but she also liked Badalian and didn't want to get him into any more trouble.

"All of these competing impulses, which you could practically see flickering across her face, produced testimony which was internally inconsistent and inconsistent with her prior sworn testimony on the matter," Jackson continued. "Her claimed lack of memory was probably the most incredible aspect of her testimony since it came and went depending on who was asking the questions."

Given her testimony, Jackson found Badalian not guilty of the charge of tampering with documents or proceedings but guilty on several other counts. She set his sentencing hearing for July. The sentencing of Rodriguez—the man who drove a stun gun into Fanone's neck and kicked off my reporting on the work of the Sedition Hunters—was set for June. Under his plea deal, his sentencing range was between six and a half years and more than ten years in federal prison.

But there was still the matter of #SwedishScarf, or Jeff, the third unnamed codefendant in the case against Rodriguez and Badalian, whose background proved very intriguing. Facial recognition revealed that #SwedishScarf appeared as a background actor in a number of music videos, including a video for LeAnn Rimes, the 2003 John Mayer song "Bigger Than My Body," and "Bad Day," the 2005 earworm by Canadian singer Daniel Powter. He also appeared in the 2011 film *The Artist*. When Justin Bieber made a cameo on the show *CSI*, #SwedishScarf popped up in the background.

Bisignano revealed to me that she had a romantic interest in Jeff and that they'd kissed when he came back over to her place for a second time after Jan. 6, this time without Rodriguez and Badalian. Jeff arrived on a skateboard, Bisignano said, and used the callbox downstairs to get buzzed in. Jeff told her that he didn't have a phone, that he was off the grid. "I thought it was very inconvenient to date somebody without a telephone," she said.

Badalian told me that an FBI agent had mentioned (mistakenly, he thought) #SwedishScarf's real name during the trial. His ears had perked up, he said, because he heard it as something like "Paul Pelosi," as in the name of the eighty-two-year-old husband of Nancy Pelosi, the woman he'd wanted arrested on Jan. 6. Paul Pelosi's name had been in the news: a few months before Badalian's trial, a 2020 election conspiracy theorist named David DePage broke into Pelosi's home, took him hostage, and then attacked him with a hammer when the police showed up.[26]

The sleuths got to work, trying any variation of the name they could find. Polacek, Pelishek, Palosik, Paloczyk, Palosyk, Paloczek, Peloshek, Paloshek, Palocsik, Poloczek, Pulisić, Pulisich, Paulisic, Pulišić, Paulisick, Pulizic, Paulissich, Poulisis, Pulicich, Palasik, Palochik.

Nothing. Soon, there was a breakthrough. A Swedish filmmaker I'd spoken with was working on a documentary about #SwedishScarf, and he'd updated the credits on the film's IMDB page. There was #SwedishScarf's name—Paul Belosic—listed in the role of "self" in the forthcoming movie "The Swedish Scarf." Belosic's full category of roles as a background actor wasn't listed on his IMDB page, but it noted he'd worked as a production assistant in a 2003 TV movie called *Maximum Thrust* and played the role of a "clean-cut" airman in a 2004 short called "Radius," which only seemed to be available on DVD.

Soon I was on the phone with Bianca Nasca, a love and sex coach from Hollywood. Back in 2021, she'd identified Belosic in a tweet that many sleuths had missed. She filled me in on how she'd known Belosic since middle school, how they'd roomed together as young adults, and how they'd done some extra work in Hollywood together too. They'd filmed some beer commercial—she couldn't recall whether it was Budweiser or Miller Lite—and a Rob Zombie music video too, she thought.

She said Belosic was adaptable and known for his ability to get access to big events. "He basically could get in anywhere: Oscar parties, the *Vanity Fair* party, you name it. Paul didn't need an invite, he didn't need a wristband, he just got in," Nasca said.

Nasca said she was a "backseat Democrat" until Trump, when she felt like she needed to start paying attention. She'd known Belosic as a libertarian, as someone who liked the idea of being away from it all. He talked about building an Earthship, an "off-the-grid-ready" home designed to resist extreme temperatures. She tried to convince him not to vote for Trump in 2020, to no avail. When she realized he was going to vote for Trump that year, "I couldn't even hear him out," Nasca said. She cut off the conversation, and then they didn't talk again for a bit, until the start of 2021.

"I think he wished me a happy New Year, like, five days before the insurrection," Nasca said. "That was it. I haven't heard from him ever since then."

Nasca had spoken with the FBI repeatedly, spoke with another reporter based in LA, and had filmed scenes for Mattias Löw, the Swedish filmmaker working on the Swedish Scarf documentary. FBI special agents, Löw told me, had even flown out from Los Angeles to meet him at the airport when he'd flown into New York, questioning him about any contacts he'd had with Belosic. For now, he was still on the run.

Something caught my eye as I looked into Belosic's background: he'd been at a protest concert the band Rage Against the Machine held across the street from the 2000 Democratic National Convention. After the concert, police said, masked individuals began throwing items at police, who cut off the power and then responded with force.

"They came in with tear gas, they came in with rubber bullets, shooting at us," Belosic told one reporter. "There was nowhere to go here. Everything was blocked off. It was chaos."[27]

He told the *Village Voice* he'd been struck in the back with a police baton after he and several others were cornered by law enforcement under a freeway overpass. "We had a right to be here," said Belosic, then twenty-six. "Instead, it was chaos, with children being shot with rubber bullets."[28]

If you pulled it all together, it was grist for a conspiracy theory: a Hollywood actor, previously part of a group of black-clad protesters that battled police at a concert for a left-wing band, had—twenty years later—incited the mob on Jan. 6, leading Trump supporters into battle. That he was indicted by a federal grand jury would do little to dissuade those who believed the FBI to be master manipulators. *Why hasn't he been arrested?* they'd ask. *Why is the FBI hiding his name?*

By 2023, Forrest Rogers was working at the *Neue Zürcher Zeitung*, the German-language newspaper of record in Switzerland. Living in Zurich, he'd returned to the roots of his youth, working as an open-source reporter, covering the Ukraine war, helping produce a documentary on Jan. 6, and working as a Swiss Army knife across the newsroom on breaking news and long-term investigations. Rogers had previously lived in the Los Angeles area and was fascinated to learn more about the background of #SwedishScarf as the investigation unfolded.

"He's a struggling actor, so I wouldn't put him in the category of being in the Brad Pitts of the world," Rogers said. "Maybe he'd be flattered that, after all those years of being an uncredited extra in movies, music videos and commercials, that there were indeed people who wanted to know his name."

Rogers wished that the FBI would just rip the Band-Aid off and publicly name Belosic, saying it was "somewhat baffling" that they wouldn't just name someone who'd already been charged. There didn't seem to be much reason to keep it secret: Belosic clearly knew he was wanted, and throwing some publicity at the case might help the bureau determine his whereabouts. Plus, publicly stating that the FBI was on the case would help take the wind out of the sails of at least one lingering conspiracy.

That could be helpful, because Jan. 6 conspiracy theorists now had their own men inside the bureau, willing to boost the narrative of Jan. 6 defendants. Just ahead of his book release, former FBI special agent Steve Friend testified alongside two other FBI employees before the House Select Subcommittee on the Weaponization of the Federal Government, formed just after McCarthy became speaker. One of the other men, Marcus Allen, was an FBI analyst whose security clearance was suspended when the bureau's security division learned he had "espoused conspiratorial views both orally and in writing and promoted unreliable information which indicates support for the events of January 6th" that "raise sufficient concerns" about his "allegiance to the United States" and his judgment.[29] Another special agent, Brett Gloss, was suspended because he'd been present on the restricted grounds of the Capitol on Jan. 6, the FBI said, and then misled the bureau about his actions. He wasn't the only person with FBI ties there: the bureau had arrested Jared Wise, a former FBI supervisory special agent, saying that the former counterterrorism official had urged other members of the mob to kill officers that day.[30]

As the narrative of Jan. 6 rioters gained traction within the House GOP, it was also getting some support from overseas. Russia had released a list of five hundred Americans now banned from Russia, and it singled out several individuals involved in the Jan. 6 investigation. Alongside former president Barack Obama, Stephen Colbert, Joe Scarborough, and Rachel Maddow, the list mentioned what the Russian foreign ministry described as "those in government and law enforcement agencies who are directly involved in the persecution of dissidents in the wake of the so-called Storming the Capitol." The list included special counsel Jack Smith; US attorney for the District of Columbia Matthew Graves; the newly appointed head of the FBI's Washington Field Office, David Sundberg; FBI deputy director Paul Abbate; former acting attorney general Monty Wilkinson, who headed up the executive office for United States attorneys; and Michael Byrd, the officer who shot Ashli Babbitt.[31] Russia's announcement came the same week that Rep. Marjorie Taylor Greene introduced articles of impeachment against Graves as well as FBI director Chris Wray.[32] The right-wing website Gateway Pundit, which had become something of an in-house rag for Jan. 6 defendants and was willing to promote fundraisers for any Jan. 6 defendant regardless of their level of violence, celebrated the news.[33] The *New York Times*, noting the inclusion of Trump enemies New York attorney general Letitia James and Georgia secretary of state Brad Raffensperger, wrote it was "particularly striking" that President Vladimir Putin was "adopting perceived enemies of former President Donald J. Trump as his own."[34]

———— • ————

Stewart Rhodes wasn't about to go quietly.

The Oath Keepers founder, like his co-defendants, had been allowed to wear a suit during his trial, and authorities tried—as best they could—not to give any indications to the jury that Rhodes and most of his co-defendants had been in government custody. During his testimony he tried to woo jurors: poking fun at his weight; making jokes about his sexual relationship with Kellye SoRelle; telling them his comments after Jan. 6 about how they should have hung Nancy Pelosi from a lamppost came after he'd "had a couple of drinks at dinner." His full name, he told them, was Elmer Stewart Rhodes III, but he went by Stewart "for obvious reasons." He seized opportunities to remind jurors that he was a quarter Mexican and that the Oath Keepers had Black members, and criticized the murder of George Floyd, saying that other officers on the scene should have lived up to their oaths by pushing Derek Chauvin off of Floyd as Chauvin drove his knee into Floyd's neck.[35]

The charm offensive hadn't worked at trial, and now Rhodes had dropped the act. When he appeared for his sentencing on May 25, 2023, he was wearing prison orange. One of his attorneys had told a few of us reporters the day before Rhodes's sentencing that the statement Rhodes would deliver would be everything we wanted it to be, meaning it was everything that they wished he wouldn't say. Rhodes had given up on the notion that what he said to the judge at his sentencing would make much of a difference in the outcome. The government wanted to send the fifty-eight-year-old away for twenty-five years, effectively a life sentence. Rhodes, sporting the eye patch he wore because he'd shot his own eye out in 1993, wanted to go out with a bang.

"I'm a political prisoner," the former army paratrooper, Yale Law graduate, and Ron Paul congressional staffer said. "Like President Trump, my only crime is opposing those who are destroying our country."

Rhodes said the government had taken some of his recent remarks out of context. When he spoke about "regime change," he said, he simply meant that he hoped that "Trump wins in 2024." Rhodes had struggled to get into Trump's inner circle before the Capitol attack, failed to reach him on Jan. 6, and the message he wanted to pass along to Trump after Jan. 6—calling on Trump to "save the Republic," by arresting members of Congress, and warning Trump that Trump and his family would "die in prison" if Joe Biden took office—didn't get through.[36] Now Rhodes was at risk of dying in prison himself. His fate would rest on the outcome of the 2024 election, and he wanted to cast his lot with all the Jan. 6 political prisoners, looping in his seditious conspiracy case with the defendants who got probation on a misdemeanor picketing or parading charge.

"I consider every J6er a political prisoner, because all of them are being grossly overcharged," Rhodes said.

The judge wasn't buying it.

"You, sir, present an ongoing threat and a peril to this country and to the republic and to the very fabric of this democracy," US District Judge Amit Mehta told Rhodes. "You are not a political prisoner."

Mehta called Rhodes "charismatic and compelling" and said that was what made him especially dangerous, predicting that Rhodes would be ready to take up arms against the government the moment he was released from federal prison.

Mehta gave Rhodes a record-setting sentence of eighteen years in federal prison.[37] Federal prosecutors had asked for terrorism sentencing enhancements before, but Rhodes's case marked that a judge gave an upward departure from the sentencing guidelines to a Jan. 6 defendant because their crime met the legal definition of terrorism.

Rhodes was defiant until the end, comparing himself to a Soviet dissident and anti-communist named Alexander Solzhenitsyn, who spent years in the Gulag. Rhodes promised he would "expose the criminality of this regime" from federal prison. Just like Patrick Stein—one of the three Trump-supporting militiamen who was sent away for thirty years for plotting to slaughter a community of Muslim refugees—he showed no remorse.

Tony Mattivi, the federal prosecutor from Kansas who helped secure the convictions of those three militiamen, had retired from the Justice Department not long before Jan. 6. Mattivi had hopes of being the next Kansas attorney general, but he lost the Republican primary to Kris Kobach, who had stronger ties to Trump and, like the former president, was seasoned at making claims about mass voter fraud that didn't pan out. That was what he did, in his previous stint as Kansas secretary of state, when he implemented a strict voter ID law.[38]

I caught up with Mattivi around the time of the two-year anniversary of the Jan. 6 attack, and we talked about Dan Day, the FBI informant in the militia case, who had died in June 2022 after years of health issues. Day's death, at fifty-four, came just a few months after the premiere of an ABC documentary about the case called *The Informant: Fear and Faith in the Heartland*. One of Day's last posts on Facebook was an endorsement of Mattivi's campaign. "He is the most qualified!" Day wrote.

Voters in the Kansas Republican primary went a different way. Kobach won, with 42 percent of the vote. Mattivi finished a distant third, with 19 percent. I asked Mattivi how he dealt with a Republican base that believed the lies that they were being fed about voter fraud from the former president.

"I dealt with it by not talking about it, because in a Republican primary in Kansas, that's a minefield," Mattivi said. "There are people that I like and respect, but do not understand how they can continue to believe that the election is stolen. They believe it to their core, and I just don't understand."

Mattivi supported many of the policies of the Trump administration. He wanted strong immigration laws and strong national security laws, and he was happy with Trump's Supreme Court picks. But he had a fundamental disconnect with how people could believe the election had been stolen or, worse, support what happened on Jan. 6.

"How can anybody who claims to be an American patriot think they are doing the right thing by interrupting the electoral process?" Mattivi said. "That's banana republic stuff."

A few weeks after we talked, Kobach nominated Mattivi to serve as director of the Kansas Bureau of Investigation (KBI). He was confirmed in February 2023. Among his biggest priorities, Mattivi said in an early interview, was making sure the KBI had the "infrastructure in place to keep up with the increasing demands of technology."[39]

———— ◆ ————

I was busy reporting on the verdict in the Proud Boys seditious conspiracy trial when the text from Bisignano arrived. She was due in court soon, and she was asking for the address of the courthouse. When her ride pulled up outside, she came over and said hello, then went inside the courthouse for her hearing.

For the third time in just a few months, federal prosecutors had secured guilty verdicts for members of a far-right organization in connection with the Capitol attack. Enrique Tarrio was convicted of seditious conspiracy alongside Ethan Nordean, Joe Biggs, and Zachary Rehl. Dominic Pezzola, who was seen smashing a window with a stolen Capitol Police shield, was acquitted on the seditious conspiracy count but found guilty of obstruction of an official proceeding, conspiracy to prevent members of Congress and federal law enforcement officers from discharging their duties, civil disorder, and destruction of government property.[40]

The sleuths played a role in the Proud Boys trial too, surfacing footage that appeared to show Rehl—the head of the Philadelphia chapter of the Proud Boys and the son and grandson of Philadelphia cops—aiming a can of pepper spray at police on the west side of the Capitol on Jan. 6. The sleuths found the footage over the course of a long weekend and sent it to prosecutors, who disclosed it to the defense just before Rehl's cross-examination. When asked if he'd pepper sprayed officers, Rehl offered no denial, instead stating that he did not "recall" doing so.[41] Not long after the verdicts, a federal grand jury indicted Shane Lamond, the police officer who'd tipped off Tarrio about law enforcement's plan to arrest the Proud Boys leader before Jan. 6, in connection with the burning of the Black Lives Matter flag.[42]

The outcome of the Oath Keepers and Proud Boys seditious conspiracy trials was a far cry from 1988, when an all-white jury acquitted thirteen white supremacists at a seditious conspiracy trial in which defendants were accused of plotting to kill an FBI special agent and a federal judge. One of those acquitted, former Ku Klux Klan grand dragon Ray Beam Jr., "celebrated the acquittal by going to a Confederate memorial opposite the court and claiming victory against what he called the 'Zionist Occupation Government,'" one account recalled.[43] After the trial, a

thirty-year-old juror named Carolyn S. Slater started a relationship with David McGuire, a twenty-five-year-old defendant. "He
told me he fell in love with me the day they picked the jury," she
later recalled, saying they shared a laugh during the trial when
they both yawned at the same time.[44] They later married and set
up another juror, twenty-four, with a fifty-year-old white supremacist named David Lane, who authored the "fourteen words," a
white nationalist catchphrase that states, "We must secure the
existence of our people and a future for white children."[45] The
other juror became pen pals with Lane, a member of the white
supremacist terrorist group the Order who was incarcerated following his conviction in connection with the murder of a liberal
Jewish talk show host in Denver named Alan Berg.[46] Years later,
some scholars would tie the "14 words" slogan to the Proud Boys
creed, in which new members of the organization declare themselves to be "a Western chauvinist" who refuses "to apologize for
the creation of the modern world." One extremism expert said
that the Proud Boys slogan "about Western chauvinists creating
the modern world cannot be disentangled from the idea that
Western is a code word for 'white people.'"[47]

Bisignano wasn't supposed to be in touch with other Jan. 6
participants or using social media, but she did both regularly.
She'd even set up her friend with Jake Lang, a Jan. 6 offender
whose actions at the Capitol were not in doubt. "This is me," he
wrote in one Instagram post, using a hand pointing emoji (👈)
to point to himself using a police shield as a weapon against the
police line.[48] He wasn't backing away from his actions that day.

"I consider myself doing exactly what the founding fathers
would have done, and I don't play the watered-down version of
Jan. 6," Lang said in one interview in 2023. "I think that was
well within our rights and constitutional duty to overthrow the
chains of tyranny." He considered Jan. 6 the first battle in the
second American Revolution. "I'm a real 1776 patriot," he said,
reading a quote from Thomas Jefferson: "The tree of liberty must

be refreshed from time to time with the blood of patriots and tyrants. It is its natural manure."[49]

Judge Carl Nichols, a Trump appointee, had summoned Bisignano back to DC because of what happened when she traveled there for her testimony in the Badalian trial. Bisignano had taken a little detour, violating her pretrial conditions "by attending the 'January 6 Block Party' outside the D.C. jail," prosecutors alleged.

Nichols ended up letting Bisignano stay on pretrial release but gave her another warning about staying away from social media. He also allowed her to withdraw her guilty plea on one of the charges she faced, a count of obstruction of an official proceeding, and set a bench trial for Oct. 30, 2023. Nichols was a skeptic of using the obstruction of an official proceeding charge in Jan. 6 cases, and litigation over the issue was playing out in a federal appeals court. Ultimately, the federal appellate-level decision could have massive implications not only for the hundreds of rioters who'd been charged under the statute but also for Trump himself. Obstruction of an official proceeding was one of the few charges that some legal experts agreed could apply to the former president's actions on Jan. 6.

Bisignano had reached a breaking point with her attorney. In court, he'd told Judge Nichols that Bisignano was "of limited intelligence," and he soon sent her a letter stating that her testimony in the Badalian trial had been "absolutely disastrous." In the letter, he told Bisignano that he'd "graciously" capped his fee at $75,000 but could no longer honor the cap given the work he expected was ahead of them, including that upcoming trial. A few weeks after the hearing, Bisignano would fire the lawyer, and he'd sent along a final invoice totaling more than $93,000, with an outstanding balance of over $23,000. She'd still have to find a new lawyer for her trial in October.

I caught up with Bisignano after the hearing, finding her at a soda machine downstairs. I wondered whether the message about her internet activity had gotten through to her or if she was going to find it difficult to stay away from the lure of social media. She said she could really do it this time, speaking about social media the same way that she spoke of a second Trump term: worth the wait. Heck, maybe if she dragged her appeal out long enough, Trump would be back in office, and he'd pardon all the Jan. 6 defendants. Staying away from social media for a bit, she said, would make her return even more grand.

Her comeback, she told me, would be bigger and better than ever.

———————◆———————

The vibes were off at the Prettyman federal courthouse on April 27, 2023. On the fourth floor, jurors in the Proud Boys seditious conspiracy trial were beginning their first full day of deliberations. One floor below, a former vice president was appearing before a federal grand jury as part of Special Counsel Jack Smith's investigation into Trump.

When Dan Scavino, the ex-president's caddie-turned-social-media-guru, appeared before the grand jury the next week, he'd have to walk in the front door, past the cameras camped out in hopes of a shot of witnesses. Not Mike Pence. On the morning of April 27, a pair of black SUVs approached and rolled into the parking garage below the court, the barricades lowering as the vehicles approached. There, in the same courthouse where countless Jan. 6 defendants who wanted him dead went on trial, he testified before a federal grand jury for hours.

News of Pence's appearance wouldn't break until the end of the day. In the meantime, the FBI Washington Field Office tweeted out two striking images of the 537th person added to the FBI's Capitol Violence website.

No. 537 on the FBI list was a woman wearing a white coat and black gloves. She was carrying a black Dolce & Gabbana purse.

In one image, with her eyebrow arched, she looked dead at the camera like she was Jim from *The Office*. In another, she was standing near the Capitol building, appearing to direct rioters with a stick.

Resting atop her head: a pink beret.

The FBI Washington Field Office's tweet, sent just after noon on Thursday, April 27, went viral.[50] The images did not show what the woman did at the Capitol, so many viewers assumed she didn't do anything serious. Some Donald Trump supporters pounced, calling this another instance of FBI overreach, a reason to defund the bureau.

The jokes flooded in too. One Twitter user dubbed the woman "Insurrection Eva Braun," another compared her to Carmen Sandiego, someone called her "fascist Matilda," and several users made jokes about her being a character from a Wes Anderson movie. "Emily in-carceration," read one of the joke tweets, referencing the show *Emily in Paris*. There were comments on her looks. Millions of men probably thought they could "fix" her, someone joked. (There were, indeed, several "I can fix her" tweets and at least one "I can change her" post.)

Recent tweets from the FBI Washington Field Office had gathered between ten thousand and twenty thousand views. The tweet about the woman in the pink beret received more than 7.2 million.

Among those millions of views was the friend of a California clothing designer I'll call Michael. That Saturday, he was standing in the checkout line at a JOANN Fabric and Craft store with his buddy, waiting to purchase a replacement part for his sewing machine, when his buddy saw something funny on his phone.

"He's always on Twitter, and he said something like, 'Yo, check out this chick,'" Michael told me in an interview the next day.

"I stopped dead in my tracks," he said. "I'm like, 'That's Jenny.'"

To the Sedition Hunters, who'd now spent more than eight hundred days on their mission, Jenny was known as #PinkBeret, an elusive rioter they weren't able to find through any of their normal methods.

#PinkBeret was everywhere that day, the sleuths had found. There she was, at the initial breach of the police line, by the Peace Monument. There she was, on the front lines of the attack, cheering as rioters tore apart a black fence so they could chuck the pieces at the police line. There she was, holding the door open for other rioters at one breach point, entering the building, then entering the building again from a second breach point. There she was, inside as men in military gear chase police officers under a garage door. There she was, smoking a victory cigar on the east side of the Capitol. There she was, appearing to walk away with a large black bag from the pile of media equipment that rioters were hell-bent on destroying. "Traitors get the fucking rope," someone yelled as rioters smash equipment and #PinkBeret looks on in high heels.

Sleuths had attacked it from all the angles, but no luck. One sleuth searched for pink berets so much that he began to get targeted ads for the caps, including a pink one adorned with small white puffs.

That all changed over the weekend. Michael said that he dashed off a tip to the FBI after he saw the bureau's tweet and was waiting for a special agent to reach out. Later, on Saturday evening, he took to his own Twitter account, quote-tweeting the FBI's post.

"I use[d] to date this girl in 2019 LOL," he tweeted, attaching an old pic of Jenny, wearing a red ski hat.

The sleuths saw Michael's tweet and got to work. They ran a facial recognition check, got a match, found more photos, and found plenty of material to confirm the ID, including a post in which she sold a (slightly damaged) Dolce & Gabbana purse that appeared to match the one she wore to the Capitol.

After Michael's tweet began to pick up steam, he decided to delete it, saying that things were getting "crazy." But he agreed to speak with me about his "fling" with Jenny using a pseudonym.

Michael, who's based in Los Angeles, and Jenny, who's from Sacramento, met online, and they were "hitting it off really well." In early 2019, when they were in their early twenties, Jenny flew down to LA. "We weren't, like, trying to get married or anything," Michael said. "We were hooking up for a few months."

Toward the end of those months, Michael said, Jenny posted on his Discord that she was reading Adolf Hitler's 1925 manifesto. They got into a discussion about it that revealed more of Jenny's far-right politics, Michael said.

"I was just instantly turned off, like, 'Yo, I don't think this is gonna work out,'" he said. "You're, like, reading *Mein Kampf*, you think immigrants don't deserve X, Y, Z." (One of the social media accounts linked to Jenny also made reference to Hitler.)

After their relationship fizzled, Jenny stuck around in the Los Angeles area. Michael thought Jenny had stayed in LA because her prior marriage, at eighteen, had ended poorly, and she had a strained relationship with her family. She didn't have much to return to in Sacramento, he thought.

He kept in touch with Jenny, occasionally exchanging messages even though their interests diverged. "She's super into politics, and I didn't know anything besides the fact that Trump lost," Michael said. But he knew she was in Washington on Jan. 6 and even asked her if she was on the no-fly list in a message he wrote a few days after the attack, on Jan. 10, 2021.

"Nope, cause I didn't go into the [Capitol]," she wrote, falsely.

"But you still crossed state lines to riot breaking the 1968 anti riot statute :/," Michael replied.

"I was there to support the president. Not to partake in that riot. I support the police," Jenny replied.

In this case, federal authorities acted swiftly. Jennifer Inzunza Vargas Geller was charged eleven days after appearing on the FBI's Twitter feed.[51] Some sleuths were frustrated that federal prosecutors and the FBI could move so quickly to charge Vargas Geller, and yet several violent Jan. 6 participants had been identified for more than two years without action. Even still, it gave them yet another boost, a little extra momentum to keep the investigation moving.

"You can outrun the cops," one YouTube user commented, speaking about the #PinkBeret case, "but you can never outrun the internet."[52]

———— ◆ ————

On June 21, 2023, a couple of weeks after Donald Trump was indicted on thirty-seven counts by a federal grand jury in Miami in connection with his handling of classified documents, the man who drove a stun gun into Fanone's neck on Trump's behalf returned to federal court in DC. Federal prosecutors were seeking fourteen years in federal prison for Danny Rodriguez, which would be the longest sentence given to a defendant who struck a plea deal rather than try his luck at trial.

Fanone was well-practiced at the routine by now. After prior sentencing hearings, Fanone had gifted me the written remarks he'd prepared: five pages at one hearing, four pages at another. By Rodriguez's sentencing, he had it boiled down to a single sheet. In Judge Amy Berman Jackson's newly assigned courtroom, Fanone spoke about how officers were "overrun and assaulted by President Donald J. Trump's army of insurrectionists;" he remarked on the "pathetic" life story Rodriguez recounted in his FBI interview; and he spoke about the enormous impact that Jan. 6 had on his life. Then he called for Trump's indictment for Jan. 6. "If it is true that no one in this country is above the law, then those responsible should be brought to justice, not because of public sentiment, but because the law demands it," Fanone said. "In the fight to preserve our Republic, there can be no spectators."

Assistant US attorney Kim Paschall, who was winding down her Jan. 6 cases, was up next. This was terrorism, she said, and

Fanone experienced terror. But it wasn't just Fanone who was a victim that day, Paschall said. "We, all of us, the citizens of this District and this country, are the victims of this criminal conspiracy. We all now must live in a post-January 6 world. We live in a world that Mr. Rodriguez and his fellow rioters created," Paschall said. "The country needs to see a sentence from this court knowing that the judicial system does not stand back, the judicial system does not stand by while our democracy and those who defend it are attacked."

Eventually, it was Rodriguez's time to speak. He'd been locked up for over two years already, but said he'd only written his remarks that morning. His comments were all over the map. He talked about how he always thought life was unfair; how he'd be "wasting some good years of my life in a cell;" how he'd probably go back to living with his mom and "driving a forklift with my GED" if Judge Jackson let him out; and how he didn't like the 1988 song "Fuck Tha Police" by N.W.A.

As Rodriguez went on and on, a frustrated Fanone walked out. I followed him into the hallway, and made a *Billy Madison* reference, joking that everyone in the courtroom was now dumber for having listened to Rodriguez's rambling, incoherent speech. He smiled, and said he was just about to text me that GIF. Soon we went back inside for the sentence. Judge Jackson said Rodriguez was of "average intelligence," and that no mental impairments helped explain his behavior. She couldn't get over the fact that Rodriguez's assault on Fanone took place under a Blue Lives Matter flag. "I don't think ironic is a good enough word to describe that," she said. She emphasized that the threat to democracy and the "shadow of tyranny" had not gone away.

"January 6 may have been two-and-a-half years ago, but the claims that the election were stolen, they're still being repeated by people and media outlets that are well aware they aren't true," she said, noting that other losing candidates had seized on "the fraud canard" as a tactic. "And it seems to work because—with only a few exceptions—others in the public eye are not willing to risk their own power or their popularity by calling it out," she said. "The statute talks about respect for the law, and it offends the rule of law just as much to look away, to fail to speak the truth. So we have to do it here."

Jackson then sentenced Rodriguez to twelve-and-a-half years in federal prison.[53] As federal marshals led Rodriguez out of the courtroom to spend the next decade of his life behind bars, he yelled out two words, as if to underline Judge Jackson's warning about the ongoing threat to the nation.

"Trump won!"

# Update, July 2023

Most of the federal laws violated on Jan. 6 have a five-year statute of limitations. If investigators were "in the first quarter" of the investigation in early 2021, as former FBI special agent in charge D'Antuono said at the time, then July 2023 was halftime. By the time thousands of Americans gathered on the west front of the US Capitol on July 4 to celebrate the 247th anniversary of America's independence (with performances by Boyz II Men; the Sesame Street Muppets; Babyface; the Broadway cast of *A Beautiful Noise, The Neil Diamond Musical*; the band Chicago; and Alfonso Ribeiro of *The Fresh Prince of Bel-Air* and *Dancing with the Stars* fame), the FBI had made 1,069 arrests in connection with the Capitol attack.

Over the course of those first thirty months, about 594 defendants pleaded guilty to at least one federal charge, and other ninety-eight were found guilty at contested trials. Of the 561 defendants who were sentenced, 335 were given periods of incarceration. At that point, online sleuths said, nearly 1,000 additional Jan. 6 participants who committed chargeable conduct at the Capitol had been identified, but not yet arrested. That number included more than 100 rioters who were featured on the FBI's Capitol Violence webpage. "The Department of Justice's resolve to hold accountable those who committed crimes on January 6,

2021, has not, and will not, wane," the Justice Department said in a statement on July 6.[1]

Peter G. Moloney, the funeral home co-owner referenced in the preface, who sprayed officers with wasp spray, was arrested in Long Island June 7.[2] Jay Johnston—a Hollywood actor who appeared in *Mr. Show*, *Arrested Development* and *Bob's Burgers* as well as in *Anchorman: The Legend of Ron Burgundy*—was also arrested that day.[3] Johnston had already hired Stanley Woodward, a lawyer who'd represented Kelly Meggs of the Oath Keepers along with a number of Trump associates. The day of Johnston's arrest, Woodward was accompanying Taylor Budowich—a former Trump spokesman and founder of the MAGA Inc. super PAC—to his testimony before a federal grand jury in Miami.

The next day, June 8, the grand jury indicted Trump and another one of Woodward's clients, Walt Nauta, in connection with the handling of classified documents found at Mar-a-Lago. Trump faced thirty-seven felony counts. The court docket wouldn't be unsealed until the next day, but Trump broke the news of his own indictment on his social media platform, Truth Social.

On June 13—as former President Donald Trump made his first court appearance in federal court in Miami—a federal judge in DC sentenced DC chiropractor David Walls-Kaufman to sixty days of incarceration.[4] Taylor Taranto, Walls-Kaufman's co-defendant in the civil suit filed by Taylor Taranto, who had still not been charged, showed up to court to observe the proceeding. I watched as he was pulled aside by US Marshals after he used his cell phone in the courtroom, and then listened in as he spoke with officers outside and identified himself by name.

Two weeks later, after Trump posted a screenshot that included former President Barack Obama's address on his Truth Social media platform, Taranto drove his van to the neighborhood and began casing the location. He was arrested, and officers found two guns and 400 rounds of ammunition inside his van. At first, he was charged with the same four misdemeanors as Walls-Kaufman; prosecutors later added in weapons charges in connection with the incident at Obama's home. US Magistrate Judge Zia Faruqui later ordered Taranto detained until trial, saying the country had failed Taranto and noting that those who helped inspire his actions were nowhere to be found. "Where are the people telling you do things?" Faruqui asked Taranto, rhetorically. "Where are they? They're not here."[5]

Kellye SoRelle, the Texas attorney affiliated with the Oath Keepers who mistook a Detroit news station's wagon full of media equipment for a box of ballots, was declared "incompetent to stand trial" on June 20. She would be "committed to the custody of the Attorney General for hospitalization for treatment in a suitable medical facility for the purposes of competency restoration" as soon as space became available, Judge Amit Mehta ordered.[6]

On June 22, Trump spoke at a fundraiser held at Bedminster in support of Jan. 6 defendants. "I'm going to make a contribution," Trump told the crowd. He did not specify the amount. Earlier in the day, former Trump deputy director of election day operations G. Michael Brown appeared before the federal grand jury in Washington that was hearing evidence about efforts to obstruct the transfer of power.[7]

Tyler Bensch, the militia member associated with members of the "B-Squad" who former FBI special agent Steve Friend didn't want to transport, was not sent to the metaphorical "gallows" as Friend worried, but was instead sentenced to sixty days of home detention by Judge McFadden. At the sentencing on July 7, the judge said he was giving Bensch a "break" in large part due to him being nineteen at the time he deployed pepper spray at the Capitol. Bensch's lawyer said that the young man still hoped to have a career in law enforcement, and that it wasn't off the table since he was only convicted of misdemeanors.[8]

Matthew Beddingfield, who stormed the Capitol while he was out on first-degree attempted murder charge, was sentenced to over three years in federal prison on July 11.[9]

Kevin Lyons, the Chicago man who stole the picture of Nancy Pelosi and John Lewis (and later posed for a photo with the stolen item in the back of his Uber), was sentenced to more than four years in federal prison on July 14. Judge Beryl Howell said it was "bizarrely ironic" that Lyons called officers protecting the Capitol "fucking Nazi bastards," when Jan. 6 was "about the mast fascistic" thing she'd seen.[10]

The same day, federal authorities charged Spencer Geller—#PinkBeret's now-husband—with several crimes. They'd both been in DC on Jan. 6, but split up when they got to the Capitol. Geller was among the first to breach the barricades at Peace Circle, helping push down the bike racks. The couple was now living overseas in Asia, according to sleuths, and had a child.[11]

On July 18, Trump took to Truth Social to announce that he had received a target letter from Special Counsel Jack Smith's office, indicating that he'd almost certainly face charges in connection with Jan. 6 and his efforts to stop the transfer of power. That day, federal authorities unsealed charges against a right-wing extremist who had carried a tiki torch in Charlottesville in 2017, and then went on to storm the Capitol on Jan. 6. Meanwhile, Rachel Powell—aka "Bullhorn Lady"—was found guilty following a bench trial. Several of her children accompanied her to the hearing, including one son who was wearing a "Make America Great Again" hat that Trump had gifted to him at the fundraiser in Bedminster. Powell's sentencing was set for Oct. 17, 2023.

# Acknowledgments

J.K.: This book would not exist without you. Pitching publishers while wrangling our own (emotional) terrorists/young children in a tiny apartment during a global pandemic was not my most well-developed plan, but you rolled with it. Writing most of the manuscript in a few months was less than ideal, but you made it work. You offered this tremendous support while balancing a job that is, by any objective measure, far more challenging than my own. It will be tough to repay the debt I owe you (especially given the scarcity of Taylor Swift tickets), but I promise to try. Thanks for building a life with me. ♥ [*Paul Rudd "Hey, Look at Us" meme.*]

Z & P: Our bike rides, dance parties, s'more sessions, breakfast taco runs, movie nights, ice cream outings, pillow battles, and Lego building marathons helped me get through this book. Thanks for your laughs, your hugs, your drawings, and your surprising insights and wisdom. I've loved watching you rise to meet new challenges and develop into creative, curious, and empathetic humans. Sorry for calling you emotional terrorists in that last paragraph; for all the times I'm going to work in a book plug when I talk to you about the dangers of social media; and for *[gestures broadly at everything]* the general state of the world. I'm so proud to be your dad.

Dad: Where do I start? Thanks for always showing up, for putting your kids first, and for never letting us forget how much you love us. Thanks for modeling the importance of hard work and perseverance as well as the power of a bit of well-calibrated humor in even the toughest situations. Kudos on your social media monitoring skills (you might be able to market that). C.: I knew I was onto something with this topic because you were actually into a story I was covering. Thanks for always reminding me that I'm not cool. I'm extremely proud of you, your work, and the life you're building. Grammie and Poppie: I can't thank you enough for your invaluable help you've offered to our family from the beginning, but especially in these past three years. Thanks to you, jokes about in-laws are completely lost on me. M & J, N & P, and David: Thank you, thank you, thank you. One thing that happens when you write a book about finding people online is that you get a bit wary about publicly mapping out your entire personal social network, so I'll resist the urge to name-drop more of our incredible friends here, but I know how lucky we are.

Thank you to the amazing team at NBC News Digital— including Sarah Mimms, Ginger Gibson, Amanda Terkel, Liz Johnstone, Tom Namako, and Catherine Kim—for recognizing the importance of the Jan. 6 beat, and for giving me the time and space I needed to write this book. It's been a privilege to work with the NBC News justice team, including Ken Dilanian, Julia Ainsley, Mike Kosnar, Daniel Barnes, Jonathan Dienst, Laura Jarrett, Tom Winter, Andrew Blankstein, and the legendary Pete Williams. Thanks also to Gary Grumbach, Julia Jester, Fiona Glisson, Victoria Ebner, Liz Brown-Kaiser, and Olympia Sonnier, as well as to the leadership in the DC bureau, including Ken Strickland, Chloe Arensberg, and Carrie Budoff Brown. I've learned a ton from reading, talking with, and competing against members of the DOJ beat past and present, including but not limited to Matt Apuzzo, Mike Balsamo, Katie Benner, Carrie Johnson, Sadie Gurman, Josh Gerstein, Adam Goldman, Mike

Levine, Eric Tucker, Alexander Mallin, Aruna Viswanatha, Evan Perez, Byron Tau, Devlin Barrett Rob Legare, Lindsay Whitehurst, and Perry Stein. Thanks also to the Prettyman courthouse press corps, including but not limited to: Roger Parloff, Brandi Buchman, Jordan Fischer, Spencer Hsu, Michael Kunzelman, Zoe Tillman, Ryan Barber, Tom Jackman, Rachel Weiner, Kyle Cheney Hannah Rabinowitz, Casey Gannon, Holmes Lybrand, and honorary member Alan Feuer.

My former Ferguson cellmate Wesley Lowery offered sage advice on this entire process, and he also put me in touch with Anthony Mattero, our mutual literary agent at CAA, who has now cornered the market on books written by journalists arrested inside a McDonald's in a St. Louis suburb in 2014. As it turns out, Anthony and I first met a couple of decades back: he was an editor on the high school newspaper when I was a freshman. Thanks, Anthony, for having faith that my writing had marginally improved since then. (Speaking of school: I owe my love of history, politics, writing, and journalism to teachers like Miss Ambrose, Mr. O'Connell, Mr. Connors, Mr. Rupertus, and Mr. Whelan.) Thanks to Ben Adams and the team at Public Affairs for believing in this project and helping a first-time author learn the ropes. Ben, appreciate you helping me realize that, while a book coming out in 2023 could not possibly be a completely comprehensive account of an investigation that's only halfway over, it could be a pretty dang good snapshot. Thanks also to Bob Barnett of Williams & Connolly for helping me navigate the world of television.

Finally, I'm grateful to the sources who trusted me with their stories, even at personal risk. Thanks especially to Forrest Rogers for luring me into the wildest story of my career. A special shoutout to the Loud Page Turners, ReillyHive, the Get Along Gang, and The Crying Booth (formerly known as InPACERrection).

# Notes

## A Note on Sourcing

This book was informed by hundreds of interviews with dozens of sources over the course of more than two years. Most of the names of the "Sedition Hunters" I used throughout this book are pseudonyms. I'm deeply indebted to those who shared their stories with me, which informed the book in ways that may not be obvious to the reader.

## Preface

1   Ryan J. Reilly, "The FBI Hasn't Arrested Hundreds Who Joined the Capitol Mob on Jan. 6. Just Ask This MAGA Comedian," NBC News, July 16, 2022, https://www .nbcnews.com/politics/justice-department/fbi-hasnt-arrested-hundreds-joined -capitol-mob-jan-6-just-ask-maga-com-rcna30509.

2   Ryan J. Reilly, "Arrested after a Fatal Stabbing, a Utah Man Told Police He Brought a Gun to Capitol Riot," NBC News, Feb. 19, 2022, https://www.nbcnews.com/poli tics/politics-news/arrested-fatal-stabbing-utah-man-told-police-brought-gun-capitol -riot-rcna16351.

3   Ryan J. Reilly, "The Feds Say They're in for the Long Haul in the Jan. 6 Investiga- tion. There Is a Time Limit," NBC News, Jan. 6, 2023, https://www.nbcnews.com /politics/justice-department/feds-say-long-haul-jan-6-investigation-rcna61728.

4   C-SPAN (@cspan), "Attorney General Merrick Garland on January 6th investiga- tion: 'We do not do our investigations in public. This is the most wide-ranging inves- tigation and the most important investigation that the Justice Department has ever entered into . . . We have to get this right,'" Twitter, July 20, 2022, https://twitter .com/cspan/status/1549826196719935488 (ellipsis in the original).

5   US v. Lints, Statement of Facts, June 24, 2022, https://www.justice.gov/usao-dc /press-release/file/1516841/download.

6   US v. Smith, Statement of Facts, May 3, 2022, https://www.justice.gov/usao-dc/case
    -multi-defendant/file/1499036/download.

7   US v. O'Brien, Statement of Facts, Jan. 4, 2023, https://storage.courtlistener.com
    /recap/gov.uscourts.dcd.250674/gov.uscourts.dcd.250674.1.1.pdf.

## Chapter 1

1   Jessica Pressler, "The Plot to Bomb Garden City, Kansas," *New York*, December 2017,
    https://nymag.com/intelligencer/2017/12/a-militias-plot-to-bomb-somali-refugees
    -in-garden-city-ks.html.

2   Ryan J. Reilly, "Trump Supporters Have Been Primed for His Bogus Voter Fraud
    Claims for Years," *HuffPost*, Oct. 18, 2016, https://www.huffpost.com/entry/trump
    -voter-fraud_n_58062ef6e4b0b994d4c16848.

3   Matt Viser and Tracy Jan, "Warnings of Conspiracy Stoke Anger among Trump
    Faithful," *Boston Globe*, Oct. 15, 2016, https://www.bostonglobe.com/news/politics
    /2016/10/15/donald-trump-warnings-conspiracy-rig-election-are-stoking-anger
    -among-his-followers/LcCY6e0QOcfH8VdeK9UdsM/story.html.

4   Nick Corasaniti and Maggie Haberman, "Donald Trump Suggests 'Second Amend-
    ment People' Could Act Against Hillary Clinton," *New York Times*, Aug. 9, 2016, https://
    www.nytimes.com/2016/08/10/us/politics/donald-trump-hillary-clinton.html.

5   US v. Allen, "Defense Curtis Wayne Allen's Sentencing Memorandum," Oct. 29,
    2018, https://www.documentcloud.org/documents/5024394-Curtis-Allen.html.

6   US v. Allen, "Appellant's Opening Brief and Required Attachments," US Court of
    Appeals for the 10th Circuit, Nov. 11, 2019, https://storage.courtlistener.com/recap
    /gov.uscourts.ca10.79369/gov.uscourts.ca10.79369.10110257781.0.pdf.

7   Author interview, 2018.

8   US v. Stein et al., "Consolidated Brief for the United States as Appellee," US Court of
    Appeals for the 10th Circuit, May 8, 2020, https://storage.courtlistener.com/recap
    /gov.uscourts.ca10.79369/gov.uscourts.ca10.79369.10110345319.0.pdf.

9   Patrick Stein, text messages, obtained from US Attorney's Office for the District of
    Kansas, 2018, https://www.documentcloud.org/documents/5023812-Patrick-Stein
    -text-messages?responsive=1&title=1.

10  Liberty Pledge, "The Plan: Action of the Liberty Restoration Committee Operating
    Under the Authority of We the People," 2016, https://web.archive.org/web/2016100
    4024942/http://www.thelibertypledge.com/the-plan.html.

11  US v. Stein, "Memorandum and Order," US District Court for the District of Kansas,
    Oct. 23, 2018, https://storage.courtlistener.com/recap/gov.uscourts.ksd.116005/gov
    .uscourts.ksd.116005.43.0.pdf.

12  US v. Stein, FBI Recording, Aug. 8, 2016, https://storage.courtlistener.com/recap
    /gov.uscourts.ksd.114049/gov.uscourts.ksd.114049.449.2.pdf.

13  Ryan J. Reilly (@ryanjreilly), "'The only good Muslim is a dead Muslim.'—Patrick
    Stein, an alleged domestic terrorist and big @realDonaldTrump fan http://huff
    .to/2e0qDTc," Twitter, Oct. 14, 2016, https://twitter.com/ryanjreilly/status/787060
    282073899009.

14  Imtiya Delawala, "What ABC News Footage Shows of 9/11 Celebrations," *ABC
    News*, Dec. 4, 2015, https://abcnews.go.com/Politics/abc-news-footage-shows-911
    -celebrations/story?id=35534125.

15 Christopher Mathias, "FBI Informant: I Was Prepared to Kill Militiaman to Stop Him from Killing Muslims," *HuffPost*, March 28, 2018, https://www.huffpost.com /entry/fbi-informant-kansas-militia-muslim-massacre_n_5abc40f1e4b03e2a5c7 91690.

16 Steve Fry, "Topeka Prosecutor Receives Award for Work on USS *Cole* Terrorist Attack Case," *Topeka Capital-Journal*, April 18, 2015, https://www.cjonline.com /story/news/politics/state/2015/04/19/topeka-prosecutor-receives-award-work-uss -cole-terrorist-attack-case/16631872007/.

17 Ryan J. Reilly, "Obama's Guantanamo Is Never Going to Close, so Everyone Might as Well Get Comfortable," *HuffPost*, Feb. 16, 2013, https://www.huffpost.com/entry /obama-guantanamo_n_2618503.

18 US v. Loewen, Plea Agreement, June 15, 2015, https://www.justice.gov/opa/file/452 876/download.

19 US v. Booker, Criminal Complaint, April 10, 2015, https://www.justice.gov/sites /default/files/opa/press-releases/attachments/2015/04/10/booker_complaint.pdf.

20 "Trial and Terror," *Intercept*, last updated Nov. 14, 2022, https://trial-and-terror .theintercept.com/.

21 Ryan J. Reilly, "FBI: When It Comes to @ISIS Terror, Retweets = Endorsements," *HuffPost*, Aug. 17. 2015, https://www.huffpost.com/entry/twitter-terrorism-fbi_n _55b7e25de4b0224d8834466e.

22 Seth G. Jones, Catrina Doxsee, and Nicholas Harrington, "The Escalating Terrorism Problem in the United States," Center for Strategic and International Studies, June 17, 2020, https://www.csis.org/analysis/escalating-terrorism-problem-united-states.

23 Ryan J. Reilly, "FBI Director James Comey Still Unsure If White Supremacist's Attack in Charleston Was Terrorism," *HuffPost*, July 9, 2015, https://www.huffpost .com/entry/james-comey-charleston-terrorism_n_7764614.

24 Ryan J. Reilly, "There's a Good Reason Feds Don't Call White Guys Terrorists, Says DOJ Domestic Terror Chief," *HuffPost*, Jan. 11, 2018, https://www.huffpost.com /entry/white-terrorists-domestic-extremists_n_5a550158e4b003133ecceb74.

25 US v. Stein, Consolidated Brief for the United States of America, May 8, 2020, https://www.justice.gov/crt/case-document/file/1335136/download.

26 US v. Stein et al., Exhibit 26, Dec. 3, 2016, https://storage.courtlistener.com/recap /gov.uscourts.ksd.114047/gov.uscourts.ksd.114047.473.26.pdf.

27 Stein, texts.

28 US v. Stein, FBI Recording, Oct. 14, 2016, https://storage.courtlistener.com/recap /gov.uscourts.ksd.114049/gov.uscourts.ksd.114049.449.4.pdf.

29 Ryan J. Reilly, "Trump's FBI Attacks Are Helping Accused Terrorists Defend Themselves in Court," *HuffPost*, March 28, 2018, https://www.huffpost.com/entry/dome stic-terrorism-kansas-militia-trump-muslim_n_5ab926b9e4b008c9e5f9fbd4.

30 Ryan J. Reilly, "'Locker Room Talk': Trump Fans Charged in Anti-Muslim Terror Plot Say It Was Just Bluster," *HuffPost*, March 23, 2018, https://www.huffpost.com /entry/domestic-terrorism-kansas-militia-plot_n_5ab43c40e4b0decad04865e0.

31 Ryan J. Reilly and Christopher Mathias, "Right-Wing Extremists Guilty in Terror Plot Against Muslim Refugees," *HuffPost*, April 18, 2018, https://www.huffpost.com /entry/trump-muslim-militia-terror-plot-kansas_n_5ad78882e4b029ebe0207801.

32 Ryan J. Reilly, "Exclusive: 'Everyday Guy' Describes How He Brought Down an American Terrorist Cell," *HuffPost*, April 20, 2018, https://www.huffpost.com /entry/right-wing-terrorism-dan-day-fbi-informant_n_5ad80fa7e4b03c426dab314c.

33  US v. Allen, Government's Sentencing Memo, Oct. 29, 2018, https://storage.courtlis tener.com/recap/gov.uscourts.ksd.114049/gov.uscourts.ksd.114049.449.0.pdf.

34  Ryan J. Reilly and Lee Moran, "Trump Associate Roger Stone Indicted in Mueller Investigation," *HuffPost*, Jan. 25, 2019, https://www.huffpost.com/entry/roger-stone -indicted-mueller-investigation_n_5b928d30e4b0cf7b003f3321.

35  Ryan J. Reilly, "Russia's Not to Blame for Trump Fans' Anti-Muslim Terror Plot, Judge Says at Sentencing," *HuffPost*, Jan. 25, 2019, https://www.huffpost.com /entry/kansas-terror-plot-trump-fbi_n_5c4b1d86e4b06ba6d3bc4088.

36  US v. Allen, Transcript of Proceedings, filed Feb. 12, 2019, https://storage.courtlis tener.com/recap/gov.uscourts.ksd.114048/gov.uscourts.ksd.114048.504.0.pdf.

37  "Bomb Plotter Sentenced for Possession of Child Pornography," Justice Department, Feb. 22, 2019, https://www.justice.gov/usao-ks/pr/bomb-plotter-sentenced-possession -child-pornography.

38  Dan Friedman, "The Acting Attorney General Helped an Alleged Scam Company Hawk Bizarre Products," *Mother Jones*, Nov. 14, 2018, https://www.motherjones .com/politics/2018/11/the-acting-attorney-general-helped-an-alleged-scam-com pany-hawk-bizarre-products/.

39  Federal Trade Commission v. World Patent Marketing, Complaint, March 6, 2017, https://www.ftc.gov/system/files/documents/cases/1723010complaint.pdf.

40  "Inventvillage.com announces the marketing launch of the MASCULINE TOI-LET," Press Release, InventVillage, Nov. 20, 2014, https://www.prlog.org/12395827 -inventvillagecom-announces-the-marketing-launch-of-the-masculine-toilet.html.

41  Ryan J. Reilly, "Matthew Whitaker's Time as Trump's Attorney General Was Unique. His Book Isn't," *HuffPost*, May 21, 2020, https://www.huffpost.com/entry /matt-whitaker-book-doj-trump-mueller_n_5ec5a522c5b6ab6fffae03f2.

42  Neal K. Katyal and George T. Conway III, "Trump's Appointment of the Acting Attorney General Is Unconstitutional," *New York Times*, Nov. 8, 2018, https://www .nytimes.com/2018/11/08/opinion/trump-attorney-general-sessions-unconsti tutional.html.

43  "Three Southwest Kansas Men Sentenced to Prison for Plotting to Bomb Somali Immigrants in Garden City," Justice Department, Jan. 25, 2019, https://www.jus tice.gov/opa/pr/three-southwest-kansas-men-sentenced-prison-plotting-bomb -somali-immigrants-garden-city.

44  US v. Wright, Defendant's Sentencing Memo, Oct. 30, 2018, https://storage.courtlis tener.com/recap/gov.uscourts.ksd.114048/gov.uscourts.ksd.114048.504.0.pdf.

# Chapter 2

1  GTL, home page, https://www.gtl.net/ (accessed 2023).

2  Anthony Accurso, "Prison Telecom Giant GTL Agrees to $67 Million Settlement in Class-Action over Inactive Account Seizure Policy," Prison Legal News, April 1, 2022, https://www.prisonlegalnews.org/news/2022/apr/1/prison-telecom-giant-gtl -agrees-67-million-settlement-class-action-over-inactive-account-seizure-policy/.

3  Derrick Bryson Taylor, "George Floyd Protests: A Timeline," *New York Times*, Nov. 5, 2021, https://www.nytimes.com/article/george-floyd-protests-timeline.html.

4  Kim Barker, Mike Baker, and Ali Watkins, "In City After City, Police Mishandled Black Lives Matter Protests," *New York Times*, March 20, 2021, https://www.ny times.com/2021/03/20/us/protests-policing-george-floyd.html.

5    FBI, "David Bowdich Named Deputy Director of the FBI," press release, April 13, 2018, https://www.fbi.gov/news/press-releases/david-bowdich-named-deputy-director-of-the-fbi.

6    David Bowdich, interview by Select Committee to Investigate the Jan. 6th Attack on the US Capitol, Dec. 16, 2021, https://www.govinfo.gov/content/pkg/GPO-J6-TRANSCRIPT-CTRL0000034627/pdf/GPO-J6-TRANSCRIPT-CTRL0000034627.pdf.

7    Mattathias Schwartz, "William Barr's State of Emergency," *New York Times Magazine*, June 1, 2020, https://www.nytimes.com/2020/06/01/magazine/william-barr-attorney-general.html.

8    David Johnston, "New Attorney General Shifts Department's Focus," *New York Times*, March 3, 1992.

9    *Verdict with Ted Cruz* Podcast, "Bill Barr Is the Honey Badger ft. Attorney General William Barr | Ep. 34," YouTube, June 25, 2020, https://www.youtube.com/watch?v=_HVqRE-6bkc.

10   Adam Goldman, Katie Benner, and Zolan Kanno-Youngs, "How Trump's Focus on Antifa Distracted Attention from the Far-Right Threat," *New York Times,* Jan. 30, 2021, https://www.nytimes.com/2021/01/30/us/politics/trump-right-wing-domestic-terrorism.html.

11   Paul Cronin, "The Time That Bill Barr Faced Down Protesters—Personally," Politico, June 7, 2020, https://www.politico.com/news/magazine/2020/06/07/barr-protesters-columbia-1968-304556.

12   Schwartz, "William Barr's State of Emergency."

13   William Barr, *One Damn Thing After Another: Memoirs of an Attorney General* (New York: William Morrow, 2022).

14   Lois Beckett, "Anti-fascists Linked to Zero Murders in the US in 25 years," *Guardian*, July 27, 2020, https://www.theguardian.com/world/2020/jul/27/us-rightwing-extremists-attacks-deaths-database-leftwing-antifa.

15   Jonathan Greenblatt, "Right-Wing Extremist Violence Is Our Biggest Threat. The Numbers Don't Lie," Anti-Defamation League, Jan. 24, 2019, https://www.adl.org/resources/news/right-wing-extremist-violence-our-biggest-threat-numbers-dont-lie.

16   "Anarchist/Left-Wing Violent Extremism in America," George Washington University Program on Extremism, November 2021, https://extremism.gwu.edu/sites/g/files/zaxdzs2191/f/Anarchist%20-%20Left-Wing%20Violent%20Extremism%20in%20America.pdf.

17   William T. Cullen, "The Grave Threats of White Supremacy and Far-Right Extremism," Feb. 22, 2019, https://www.nytimes.com/2019/02/22/opinion/christopher-hasson-extremism.html.

18   Harry Jaffe, "The Trump Appointee Who's Putting White Supremacists in Jail," *Washington Post Magazine*, Aug. 7, 2019, https://www.washingtonpost.com/news/magazine/wp/2019/08/07/feature/the-trump-appointee-whos-putting-white-supremacists-in-jail/.

19   "Three Members of California-Based White Supremacist Group Sentenced on Riots Charges Related to August 2017 'Unite the Right' Rally in Charlottesville," July 19, 2019, https://www.justice.gov/usao-wdva/pr/three-members-california-based-white-supremacist-group-sentenced-riots-charges-related.

20   US Attorney's Office, "U.S. Attorney Thomas T. Cullen Announces Departure," press release, Sept. 10, 2020, https://www.justice.gov/usao-wdva/pr/us-attorney-thomas-t-cullen-announces-departure.

21   "A Comparison of Political Violence by Left-Wing, Right-Wing, and Islamist Extremists in the United States and the World," Proceedings of the National Academy of Sciences of the United States of America, July 18, 2022, https://www.pnas .org/doi/full/10.1073/pnas.2122593119.

22   William Rosenau, *Tonight We Bombed the U.S. Capitol* (New York: Atria Books, 2019).

23   Bryan Burrough, *Days of Rage: America's Radical Underground, the FBI, and the Forgotten Age of Revolutionary Violence* (New York: Penguin Books, 2015).

24   *Report of the Select Committee on Assassinations of the U.S. House of Representatives* (Washington, DC: US Government Printing Office, 1979), https://www.archives .gov/research/jfk/select-committee-report/part-2e.html.

25   Ryan J. Reilly, "Lead Cop in Felony Trial for Trump Inaugural Protesters Sent Anti-activist Tweets," *HuffPost*, Dec. 12, 2017, https://www.huffpost.com/entry/gregg -pemberton-trump-inauguration-felony-trial_n_5a301402e4b046175433e4f3.

26   Ryan J. Reilly, "Inside the Trial That Could Determine the Future of Free Speech in America's Capital," *HuffPost*, Dec. 10, 2017, https://www.huffpost.com/entry/pro testing-dc-trump-inauguration-trial_n_5a1e1e84e4b0d724fed48d32.

27   Ryan J. Reilly, "All 6 Defendants Not Guilty in Key Felony Trial of Trump Inauguration Protesters," *HuffPost*, Dec. 21, 2017, https://www.huffpost.com/entry/trump -inauguration-protesters-not-guilty_n_5a37e6c2e4b040881becafe2.

28   Ryan J. Reilly, "Justice Department Still Can't Convict Any Trump Inauguration Protesters at Second Trial," *HuffPost*, June 7, 2018, https://www.huffpost.com /entry/j20-trial-trump-inauguration-protesters_n_5b19779be4b0bbb7a0da99c8.

29   Ryan J. Reilly, "DOJ Drops All Charges Against Remaining Trump Inauguration Defendants," *HuffPost*, July 6, 2018, https://www.huffpost.com/entry/j20-trump -inauguration-protesters_n_5b3fc53ee4b05127ccf1ee6a.

30   Aram Roston, "Before Jan. 6, FBI Collected Information from at Least 4 Proud Boys," Reuters, April 26, 2021, https://www.reuters.com/world/us/exclusive-before -jan-6-fbi-collected-information-least-4-proud-boys-2021-04-26/.

31   US v. Biggs, Defendant Biggs' Opposition to Motion to Revoke Pretrial Release, March 29, 2021, https://storage.courtlistener.com/recap/gov.uscourts.dcd.229062 /gov.uscourts.dcd.229062.42.0.pdf.

32   Julia Ainsley, "FBI's Wray Says Antifa More an Ideology Than a Group, Undercutting Statements by Other Trump Officials," NBC News, Sept. 17, 2020, https://www .nbcnews.com/politics/congress/fbi-s-wray-says-antifa-more-ideology-group-under cutting-statements-n1240317.

33   Colby Itkowitz and Devlin Barrett, "Trump Suggests He'd Consider Removing FBI Director over Unfavorable Testimony," *Washington Post*, Sept. 18, 2020, https:// www.washingtonpost.com/politics/trump-wray-fbi-testimony/2020/09/18/2c7d16 a2-f9f3-11ea-89e3-4b9efa36dc64_story.html.

34   Ryan J. Reilly, "Trump Justice Department Gears Up for War on Antifa 'Terrorists,'" *HuffPost*, June 1, 2020, https://www.huffpost.com/entry/trump-justice-de partment-william-barr-antifa-domestic-terrorism_n_5ed4ffe6c5b6013c6590479f.

35   Katie Benner, "Planeload of 'Thugs'? Barr Skirts Trump's Claim but Suggests Rioters Targeted D.C.," *New York Times*, Sept. 2, 2020, https://www.nytimes.com/2020 /09/02/us/politics/barr-trump-protest-violence.html.

36   Ed Campuzano, "Man Fatally Shot After Pro-Trump Caravan Was Patriot Prayer 'Friend and Supporter' Aaron Danielso," *Oregonian*, Sept. 1, 2020, https://www.ore

gonlive.com/portland/2020/08/man-fatally-shot-after-pro-trump-caravan-was
-patriot-prayer-friend-and-supporter.html.

37  Department of Justice, "Statement by Attorney General William P. Barr on the
Tracking Down of Fugitive Michael Forest Reinoehl," press release, Sept. 4, 2020,
https://www.justice.gov/opa/pr/statement-attorney-general-william-p-barr-tracking
-down-fugitive-michael-forest-reinoehl.

38  Evan Hill, Mike Baker, Derek Knowles, and Stella Cooper, "'Straight to Gunshots':
How a U.S. Task Force Killed an Antifa Activist," *New York Times*, Oct. 13, 2020,
https://www.nytimes.com/2020/10/13/us/michael-reinoehl-antifa-portland-shoot
ing.html.

39  "Review of U.S. Park Police Actions at Lafayette Park," US Department of the Interior Office of Inspector General, June 8, 2021, https://www.oversight.gov/sites
/default/files/oig-reports/DOI/SpecialReviewUSPPActionsAtLafayetteParkPub
lic.pdf.

40  Carol Leonning and Phillip Rucker, *I Alone Can Fix It* (New York: Penguin Press, 2021).

41  Jonathan Allen, "Trump and Tear Gas in Lafayette Square: A Memo from the Protest Front Lines," NBC News, June 2, 2020, https://www.nbcnews.com/politics
/white-house/memo-front-lines-different-america-n1222066.

42  Department of Justice, "Justice Department Announces Civil Settlement in Lafayette Square Cases," press release, April 22, 2022, https://www.justice.gov/opa/pr
/justice-department-announces-civil-settlement-lafayette-square-cases.

43  Ryan J. Reilly (@ryanjreilly), "From a source, here's a BOP SORT team posing with
AG William Barr. Photo appears to have been snapped in the DOJ courtyard," Twitter, June 11, 2020, https://twitter.com/ryanjreilly/status/1271156287556943875.

44  Ryan J. Reilly and Tara Golshan, "William Barr's Vast, Nameless Army Is Being
Brought to Bear on D.C. Protesters," *HuffPost*, June 3, 2020, https://www.huffpost
.com/entry/william-barr-justice-department-protests-trump_n_5ed8330ec5b6513f
dffca21a.

45  Ryan J. Reilly, "Federal Prison Riot Team That Patrolled D.C. Streets Injured Colleagues in Training Exercises," *HuffPost*, June 18, 2020, https://www.huffpost.com
/entry/federal-prison-sort-bop-dc_n_5eeb774ec5b6c8594c7f2679.

46  Department of Justice, "More Than 300 Facing Federal Charges for Crimes Committed During Nationwide Demonstrations," press release, Sept. 24, 2020, https://
www.justice.gov/opa/pr/over-300-people-facing-federal-charges-crimes-committed
-during-nationwide-demonstrations.

47  Ryan J. Reilly, "How Segregationists Rushed Through the 1968 Rioting Laws DOJ
Is Using in 2020," *HuffPost*, Sept. 24, 2020, https://www.huffpost.com/entry
/anti-rioting-act-civil-disorder-law-doj-barr-trump-consitutional_n_5f6a012cc5b6
55acbc701ca2.

48  Ryan J. Reilly, "Trump Justice Department Forms Task Force to Investigate Antifa,
Boogaloo," *HuffPost*, June 26, 2020, https://www.huffpost.com/entry/doj-task-force
-antifa-boogaloo_n_5ef680c0c5b6ca97090fa730.

49  Brandon Shields, "'I Can't Breathe' Protest Stays Peaceful at Mall, Courthouse,"
*Jackson Sun*, June 1, 2020, https://www.jacksonsun.com/story/news/2020/06/01
/george-floyd-protest-black-lives-matter-jackson-old-hickory-mall/5306415002/.

50  Brandy Zadrozny and Ben Collins, "Antifa Rumors Spread on Local Social Media
with No Evidence," NBC News, June 2, 2020, https://www.nbcnews.com/tech/tech
-news/antifa-rumors-spread-local-social-media-no-evidence-n1222486.

51 Brandy Zadrozny and Ben Collins, "In Klamath Falls, Oregon, Victory Declared over Antifa, Which Never Showed Up," NBC News, June 6, 2020, https://www.nbc news.com/tech/social-media/klamath-falls-oregon-victory-declared-over-antifa -which-never-showed-n1226681.

52 Isaac Stanley-Becker and Tony Romm, "Armed White Residents Lined Idaho Streets amid 'Antifa' Protest Fears. The Leftist Incursion Was an Online Myth," *Washington Post*, June 4, 2020, https://www.washingtonpost.com/national/protests -armed-white-vigilantes/2020/06/04/09e17610-a5bb-11ea-b619-3f9133bbb482 _story.html.

53 Tammy Butry, YouTube, Oct. 22, 2020, https://www.youtube.com/watch?v=c2DNz 2KWOc8.

54 US v. Bronsburg, Filing, Sept. 21, 2022, https://storage.courtlistener.com/recap/gov .uscourts.dcd.227847/gov.uscourts.dcd.227847.70.0.pdf.

55 Reilly, "Trump Justice Department Gears Up.'"

56 Affidavit for Search Warrant, State of Tennessee, Madison County, June 2, 2020, https://storage.courtlistener.com/recap/gov.uscourts.tnwd.89461/gov.uscourts .tnwd.89461.29.1.pdf.

57 Ryan J. Reilly, "DOJ's Antifa Push Spurs Trump Appointee to Charge a Band's Bassist over a Bag of Weed," *HuffPost*, Oct. 16, 2020, https://www.huffpost.com/entry /trump-barr-antifa-dunavant-tennessee-justin-coffman-bassist_n_5f89a212c5b67d a85d1d7515.

58 Aruna Viswanatha and Sadie Gurman, "Barr Tells Prosecutors to Consider Charging Violent Protesters with Sedition," *Wall Street Journal*, Sept. 17, 2020, https:// www.wsj.com/articles/barr-tells-prosecutors-to-consider-charging-violent-protesters -with-sedition-11600276683.

59 State of Tennessee v. Justin Wade Coffman, Indictment, Oct. 5, 2020, obtained by author, https://www.documentcloud.org/documents/7273530-Justin-Coffman-Indict ment.html.

60 Department of Justice, "Madison County Man Charged in Both Federal and State Courts for Unlawful Possession of Firearms and Possession of a Hoax Device During Civil Unrest," press release, Oct. 14, 2020, https://www.justice.gov/usao-wdtn/pr /madison-county-man-charged-both-federal-and-state-courts-unlawful-possession -firearms.

61 Ryan J. Reilly, "Judge Rejects Trump Appointee's Attempt to Keep Band Member Jailed over Photo Shoot," *HuffPost*, Oct. 30, 2020, https://www.huffpost.com/entry /justin-coffman-trump-justice-department-dunavant_n_5f972154c5b6b74d85f 37700.

62 Ryan J. Reilly (@ryanjreilly), "Justin Coffman, the bassist the Trump administration tried to keep locked up for months over a photoshoot for his band, just cast his vote. He was wearing a Black Lives Matter shirt," Nov. 3, 2020, Twitter, https://twit ter.com/ryanjreilly/status/1323699444915851265.

63 US v. Coffman, Order Granting Motion to Suppress Evidence, July 16, 2021, https:// storage.courtlistener.com/recap/gov.uscourts.tnwd.89461/gov.uscourts.tnwd.89461 .30.0.pdf.

64 Ryan Devereaux, "Brooklyn Man Was Arrested for Curfew Violation. The FBI Interrogated Him About His Political Beliefs," *Intercept*, June 4, 2020, https://thein tercept.com/2020/06/04/fbi-nypd-political-spying-antifa-protests/.

65  Ryan Devereaux, "He Tweeted That He Was the Leader of Antifa. Then the FBI Asked Him to Be an Informant," *Intercept*, June 9, 2020, https://theintercept.com /2020/06/09/antifa-fbi-tweet/.

66  Chris Brooks, "After Barr Ordered FBI to 'Identify Criminal Organizers,' Activists Were Intimidated at Home and at Work," *Intercept*, June 12, 2020, https://theinter cept.com/2020/06/12/fbi-jttf-protests-activists-cookeville-tennessee/.

67  Christopher Mathias, "Violent Proto-fascists Came to Portland. The Police Went After the Anti-fascists," *HuffPost*, Aug. 5, 2018, https://www.huffpost.com/entry /portland-patriot-prayer-proud-boys-police-antifascists_n_5b668b7de4b0de86f4a22faf.

68  "Distracted Boyfriend," Know Your Meme, Aug. 22, 2017, https://knowyourmeme .com/memes/distracted-boyfriend.

# Chapter 3

1  Team Trump (@TeamTrump), "Team Trump announces 'Lawyers for Trump,' a coalition to mobilize support for President Trump & secure four more years of strong leadership to uphold our nation's Constitution & preserve a government of, by, & for the people!" Twitter, July 23, 2020, https://twitter.com/TeamTrump/status/1286 269832502095872.

2  Ronna McDaniel (@GOPChairwoman), "Joe Biden's agenda is a threat to our country. Today, the Trump campaign launched #LawyersforTrump to help protect our votes, the rule of law, and re-elect @realDonaldTrump! https://lawyers.donald jtrump.com," Twitter, July 23, 2020, https://twitter.com/GOPChairwoman/status /1286352856770125825.

3  DNC v. RNC, US District Court for the District of New Jersey, Feb. 11, 1982, https:// www.brennancenter.org/sites/default/files/legacy/Democracy/dnc.v.rnc/1981%20 complaint.pdf.

4  Ian T. Shearn, "Voter Suppression in New Jersey: A Pivotal Episode That Still Reverberates," NJ Spotlight News, Nov. 3, 2020, https://www.njspotlightnews.org/2020 /11/voter-suppression-in-nj-kean-florio-election-rnc-flying-squads-roger-stone-john -roberts/.

5  Jeffrey Toobin, "The Dirty Trickster," *New Yorker*, May 23, 2008, https://www.new yorker.com/magazine/2008/06/02/the-dirty-trickster.

6  Manuel Roig-Franzia, "The Swamp Builders," *Washington Post*, Nov. 29, 2018, https:// www.washingtonpost.com/graphics/2018/politics/paul-manafort-roger-stone/.

7  Christina Wilkie, "Republicans Tried Trump's 'Ballot Security' Strategy 35 Years Ago. Here's What Happened," *HuffPost*, Nov. 1, 2016, https://www.huffpost.com /entry/roger-stone-donald-trump-ballot-security_n_5818f6a1e4b0f96eba966f03.

8  Jane Perlez, "Democrats Will Sue G.O.P. over Voting Patrol in Jersey," *New York Times*, Dec. 14, 1981, https://www.nytimes.com/1981/12/14/nyregion/democrats -will-sue-gop-over-voting-patrol-in-jersey.html.

9  Selwyn Raab, "G.O.P. Relieves Security Official in Jersey Voting," *New York Times*, Nov. 12, 1981, https://www.nytimes.com/1981/11/12/nyregion/gop-relieves-security -official-in-jersey-voting.html.

10  Matt Katz, "Armed Men Once Patrolled the Polls. Will They Reappear in November?" WNYC News, Sept. 1, 2016, https://www.wnyc.org/story/armed-men-once -patrolled-polls-will-they-reappear-november/.

11  Ian T. Shearn, "Roger Stone: The Ultimate Dirty Trickster, Formed by Watergate and Tempered in New Jersey," NJ Spotlight News, July 13, 2020, https://www .njspotlightnews.org/2020/07/roger-stone-the-ultimate-dirty-trickster-formed -by-watergate-and-tempered-in-new-jersey/.

12  Clayton Knowles, "Many Under 21 Register to Vote Here," New York Times, Jan. 5, 1971, https://www.nytimes.com/1971/01/05/archives/many-under-21-register-to-vote -here-many-city-youths-register-to.html.

13  Shearn, "Voter Suppression in New Jersey."

14  Adam Clymer, "G.O.P. to Expand to Other States 'Ballot Security It Used in Jersey," New York Times, Nov. 9, 1981, https://www.nytimes.com/1981/11/09/nyregion/gop -to-expand-to-other-states-ballot-security-it-used-in-jersey.html.

15  Selwyn Raab, "Queries Arise on Background of Ballot Task Force Official," New York Times, Nov. 11, 1981, https://www.nytimes.com/1981/11/11/nyregion/queries -arise-on-background-of-ballot-task-force-official.html.

16  Jane Perlez, "Kelly Reported on Reagan's Appointees," New York Times, Nov. 13, 1981, https://www.nytimes.com/1981/11/13/nyregion/kelly-reported-on-reagan-s -appointees.html.

17  Matthew C. Quinn, "John Kelly: The Key to Jersey Controversy," United Press International, Nov. 14, 1981, https://www.upi.com/Archives/1981/11/14/John-Kelly -The-key-to-Jersey-controversy/6053374562000/.

18  Shearn, "Voter Suppression in New Jersey."

19  David W. Dunlap, "Poll Team Chief in Jersey Leaves His G.O.P. Post," New York Times, Dec. 31, 1981.

20  The Prowler, "Ghosts from the Past," American Spectator, March 10, 2008, https:// spectator.org/43964_ghosts-past/.

21  Lisa Riordan Seville, "These Lawyers Remade the Supreme Court. Now They're Fighting to Limit Voting," NBC News, Nov. 1, 2020, https://www.nbcnews.com /politics/2020-election/these-attorneys-remade-supreme-court-now-they-re -fighting-limit-n1245469.

22  Photo gallery, "Pope Benedict XVI's Birthday," Washington Life Magazine, May 2008, https://issuu.com/washingtonlife/docs/may2008/38.

23  "Gift from Jack Kelly and Gail Weiss to Support Notre Dame International Security Center," Notre Dame News, October 30, 2017, https://news.nd.edu/news/gift-from -jack-kelly-and-gail-weiss-to-support-notre-dame-international-security-center/.

24  Caitlin McLean, "When Was the Internet Invented? What to Know About the Creators of It and More," USA Today, Aug. 28, 2022, https://www.usatoday.com/story /tech/2022/08/28/when-was-internet-created-who-invented-it/10268999002/.

25  DNC v. RNC, US Court of Appeals for the Third Circuit, March 8, 2012, https://elec tionlawblog.org/wp-content/uploads/rncvdnc.pdf.

26  Michael Wines, "Freed by Court Ruling, Republicans Step Up Effort to Patrol Voting," New York Times, May 18, 2020, https://www.nytimes.com/2020/05/18/us /Voting-republicans-trump.html.

27  Ronna McDaniel, "RNC Chairwoman: Democrats Don't Want to Play by the Same Voting Rules," Washington Post, Aug. 27, 2020, https://www.washingtonpost.com /opinions/2020/08/27/ronna-mcdaniel-rnc-chairwoman-poll-watchers-voting/.

28  Jim Rutenberg, "The Attack on Voting in the 2020 Elections," New York Times Magazine, Sept. 30, 2020, https://www.nytimes.com/2020/09/30/magazine/trump-voter -fraud.html.

29  Dennis Wagner, Ryan W. Miller, Nick Penzenstadler, Kevin McCoy, and Donovan Slack, "For These Trump Supporters Primed to Disbelieve Defeat, Challenging the Election Was a Civic Duty," *USA Today*, Dec. 3, 2020, https://www.usatoday.com /in-depth/news/2020/12/03/trump-lawsuits-challenging-election-michigan -arizona-pennsylvania-georgia/6425725002/.

30  Michael Biesecker and Garance Burke, "Trump 'Army' of Poll Watchers Led by Veteran of Fraud Claims," Associated Press, Nov. 2, 2020, https://apnews.com/article /mike-roman-trump-poll-watchers-election-e110e6c9e62c9c8520f4a1a2040 d8cfc.

31  Mike Roman (@mikeroman), "Absentee Ballot processing starts in Detroit today. 134 tables operating for 10 hours. Trump Challengers will have 100% coverage (unless crooked officials try to kick them out)!," Twitter, Nov. 2, 2020, https://twitter .com/mikeroman/status/1323283823241633794.

32  Tresa Baldas, Kristen Jordan Shamus, Niraj Warikoo, M. L. Elrick, Joe Guillen, and Evan Petzold, "'Get to TCF': What Really Happened Inside Detroit's Ballot Counting Center," *Detroit Free Press*, Nov. 6, 2020, https://www.freep.com/story/news /local/michigan/detroit/2020/11/06/tcf-center-detroit-ballot-counting/61735 77002/.

33  Detroit Public TV, One Detroit, "Election 20/20: Detroit to D.C.," YouTube, Jan. 6, 2022, https://www.youtube.com/watch?v=MHjKJi2zGWk.

34  Tim Alberta, "The Inside Story of Michigan's Fake Voter Fraud Scandal," *Politico Magazine*, Nov. 24, 2020, https://www.politico.com/news/magazine/2020/11/24 /michigan-election-trump-voter-fraud-democracy-440475.

35  Brooke Destra, "Philly Sports Fans Agree: 'Bad Things Happen in Philadelphia,'" NBC Sports, Sept. 29, 2020, https://www.nbcsports.com/philadelphia/eagles/donald -trump-philly-sports-fans-agree-bad-things-happen-philadelphia.

36  Ryan J. Reilly, "'Karen Was Upset': Trump Suit Cites Mostly Trivial Complaints by GOP Poll-Watchers," *HuffPost*, Nov. 11, 2020, https://www.huffpost.com/entry /trump-voter-fraud-michigan-lawsuit_n_5fabef94c5b68707d1fb0166.

37  Trump v. Benson, Exhibit 1 (affidavits), November 2020, https://www.document cloud.org/documents/20404085-3-exhibit-1-affidavits-compressed.

38  Tresa Baldas, Kristen Jordan Shamus, Niraj Warikoo, and Evan Petzold, "Chaos Erupts at TCF Center as Republican Vote Challengers Cry Foul in Detroit," *Detroit Free Press*, Nov. 4, 2020, https://www.freep.com/story/news/politics/elections/2020 /11/04/tcf-center-challengers-detroit-michigan/6164715002/.

39  Detroit Public TV, "Election 20/20: Detroit to D.C."

40  Ryan J. Reilly (@ryanjreilly), "Here's a sampling of the insane Facebook content that Michigan right-wingers were consuming," Twitter, Nov. 17, 2020, https://twit ter.com/ryanjreilly/status/1328869084604346369.

41  Mark Phelan, "Detroit Auto Show Canceled; Facility to Be Converted to Coronavirus Field Hospital," *Detroit Free Press*, March 28, 2020, https://www.usatoday.com /story/money/cars/2020/03/28/coronavirus-detroit-auto-show-canceled-tcf-center -hospital/2934839001/.

42  Julia Carrie Wong, "The Year of Karen: How a Meme Changed the Way Americans Talked About Racism," *Guardian*, Dec. 27, 2020, https://www.theguardian.com /world/2020/dec/27/karen-race-white-women-black-americans-racism.

43  Henry Goldblatt, "A Brief History of 'Karen,'" *New York Times*, July 31, 2020, https://www.nytimes.com/2020/07/31/style/karen-name-meme-history.html.

44  Tyler Hayden, "Santa Ynez Official Identified Among January 6 Mob," *Santa Barbara Independent*, Nov. 9, 2022, https://www.independent.com/2022/11/09/santa-ynez-official-identified-among-january-6-mob/.

45  100 Percent FED Up, Facebook Live, Nov. 5, 2020, https://www.facebook.com/100PercentFEDUp/videos/658610644837902.

46  Ryan J. Reilly (@ryanjreilly), "Here's a GOP challenger bragging about how she purposefully walked through a crowd of Detroit elections workers without a mask out of spite. Her friend had a Trump mask on," Twitter, Nov. 18, 2020, https://twitter.com/ryanjreilly/status/1329056772426969090.

47  Facebook video saved by author (filmed Nov. 4, 2020, saved Nov. 12, 2020).

48  Transcribed Interview of William Barr, Select January 6th Committee Final Report and Supporting Materials Collection, June 2, 2022, https://www.govinfo.gov/collection/january-6th-committee-final-report?path=/gpo/January%206th%20Committee%20Final%20Report%20and%20Supporting%20Materials%20Collection/Supporting%20Materials%20-%20Documents%20on%20File%20with%20the%20Select%20Committee.

49  Sarah Cwiek, "Once Again, Trump Supporters Swarm TCF Center in Protest," Michigan Radio, Nov. 6, 2020, https://www.michiganradio.org/politics-government/2020-11-06/once-again-trump-supporters-swarm-tcf-center-in-protest.

50  George Hunter and Jordyn Grzelewski, "Vote Protesters, Counter-demonstrators Face Off at Detroit's TCF Center," *Detroit News*, Nov. 6, 2020, https://www.detroitnews.com/story/news/politics/2020/11/06/group-protests-election-count-detroit-tcf-center-friday/6187278002/.

51  Davey Alba, "There's a Simple Reason Workers Covered Windows at a Detroit Vote-Counting Site," *New York Times*, Nov. 5, 2020, https://www.nytimes.com/2020/11/05/technology/michigan-election-ballot-counting.html.

52  Andy Barr, "51% of GOP Voters: Obama Foreign," Politico, Feb. 15, 2011, https://www.politico.com/story/2011/02/51-of-gop-voters-obama-foreign-049554.

53  Josh Clinton and Carrie Roush, "Poll: Persistent Partisan Divide over 'Birther' Question," NBC News, Aug. 10, 2016, https://www.nbcnews.com/politics/2016-election/poll-persistent-partisan-divide-over-birther-question-n627446.

54  Maggie Haberman, *Confidence Man* (New York: Penguin Press, 2022).

55  Dan Pfeiffer, "President Obama's Long Form Birth Certificate," White House, April 27, 2011, https://obamawhitehouse.archives.gov/blog/2011/04/27/president-obamas-long-form-birth-certificate.

56  Ryan J. Reilly, "Paul Ryan Has News for the Birthers: He's Not One of Them," Talking Points Memo (TPM), Oct. 5, 2012, https://talkingpointsmemo.com/muckraker/paul-ryan-has-news-for-the-birthers-he-s-not-one-of-them.

57  Ryan J. Reilly, "WND's Joseph Farah May Sue Esquire over Birther Parody," TPM, May 18, 2011, https://talkingpointsmemo.com/muckraker/wnd-s-joseph-farah-may-sue-em-esquire-em-over-birther-parody.

58  Ryan J. Reilly, "After-Birthers: How Conspiracy Theorists Reacted to Obama's Long-Form Birth Certificate," TPM, April 28, 2011, https://talkingpointsmemo.com/muckraker/after-birthers-how-conspiracy-theorists-reacted-to-obama-s-long-form-birth-certificate.

59  Ryan J. Reilly, "Orly Taitz: Obama's Long-Form Birth Certificate Should Say 'Negro' Not 'African,'" TPM, April 27, 2011, https://talkingpointsmemo.com/muckraker/orly-taitz-obama-s-long-form-birth-certificate-should-say-negro-not-african.

60   Maggie Haberman, "The Donald: I'm Proud of Myself!" Politico, April 27, 2011, https://www.politico.com/story/2011/04/the-donald-im-proud-of-myself-053801.

61   "Donald Trump Remarks on President Obama's Birth Certificate," C-SPAN, April 27, 2011, https://www.c-span.org/video/?299230-1/donald-trump-remarks-president-obamas-birth-certificate.

62   Virginia Chamlee, "Barack Obama's 'Excruciating' 2011 Roast of Donald Trump Left Him Furious, Chris Christie Says," People, Nov. 17, 2021, https://people.com/politics/chris-christie-on-obamas-2011-roast-of-donald-trump/.

63   Michael Barbaro, "After Roasting, Trump Reacts in Character," New York Times, May 1, 2011, https://www.nytimes.com/2011/05/02/nyregion/after-roasting-trump-reacts-in-character.html.

64   Roxanne Roberts, "I Sat Next to Donald Trump at the Infamous 2011 White House Correspondents' Dinner," Washington Post, April 28, 2016, https://www.washingtonpost.com/lifestyle/style/i-sat-next-to-donald-trump-at-the-infamous-2011-white-house-correspondents-dinner/2016/04/27/5cf46b74-0bea-11e6-8ab8-9ad050f76d7d_story.html.

65   Garrett M. Graff, "'I'd Never Been Involved in Anything as Secret as This,'" Politico, April 21, 2021, https://www.politico.com/news/magazine/2021/04/30/osama-bin-laden-death-white-house-oral-history-484793.

66   "Road to the White House 2012: Donald Trump Presidential Endorsement Announcement," C-SPAN, February 2, 2012, https://www.c-span.org/video/?304163-1/donald-trump-presidential-endorsement-announcement.

67   Donald Trump (@realDonaldTrump), "We can't let this happen. We should march on Washington and stop this travesty. Our nation is totally divided!" Twitter, Nov. 6, 2012, https://twitter.com/realDonaldTrump/status/266034630820507648.

68   Donald Trump (@readDonaldTrump), "This election is a total sham and a travesty. We are not a democracy!" Twitter, Nov. 6, 2012, https://twitter.com/realdonaldtrump/status/266035509162303492?lang=en.

69   Patrice Taddonio, "Watch: For Trump, Romney's Loss Fueled a Tweetstorm. Then, a Trademark," Frontline, Jan. 18, 2017, https://www.pbs.org/wgbh/frontline/article/watch-for-trump-romneys-loss-fueled-a-tweetstorm-then-a-trademark/.

70   Donald Trump (@realDonaldTrump), "Lets fight like hell and stop this great and disgusting injustice! The world is laughing at us," Twitter, Nov. 6, 2012, https://twitter.com/realDonaldTrump/status/266034957875544064.

71   Ryan J. Reilly, "Trump Supporters Have Been Primed for His Bogus Voter Fraud Claims for Years," HuffPost, Oct. 18, 2016, https://www.huffpost.com/entry/trump-voter-fraud_n_58062ef6e4b0b994d4c16848.

72   Jonathan Martin and Alexander Burns, "Officials Fight Donald Trump's Claims of a Rigged Vote," New York Times, Oct. 16, 2016, https://www.nytimes.com/2016/10/17/us/politics/donald-trump-election-rigging.html.

73   Amy Tennery, "Trump Accuses Cruz of Stealing Iowa Caucuses Through 'Fraud,'" Reuters, Feb. 3, 2016, https://www.reuters.com/article/us-usa-election-trump-cruz/trump-accuses-cruz-of-stealing-iowa-caucuses-through-fraud-idUSKCN0VC1Z6.

74   Sam Levine, "Trump Suddenly Ends Voter Fraud Panel," HuffPost, Jan. 3, 2018, https://www.huffpost.com/entry/trump-voter-fraud-panel_n_5a4d6ca1e4b0b0e5a7aad716.

75   Sam Levine and Ryan J. Reilly, "Legal U.S. Immigrants Face Prison and Even Deportation for Voting," HuffPost, Aug. 29, 2018, https://www.huffpost.com/entry/voter-fraud-immigrants-doj-trump_n_5b859e78e4b0162f471d233c.

76  Michael Graff and Nick Ochsner, *The Vote Collectors* (Chapel Hill: University of North Carolina Press, 2021).

77  Frank Dale, "Trump Encourages North Carolina Republican at Center of Possible Election Fraud to 'Stand and Fight,'" ThinkProgress, Jan. 8, 2019, https://archive .thinkprogress.org/donald-trump-mark-harris-north-carolina-leslie-mccrae -dowless-election-fraud-house-congress-455702089660/.

78  Rashaan Ayesh, "Trump Declines to Say He'll Accept Results of Election: 'I Have to See,'" Axios, July 19, 2020, https://www.axios.com/2020/07/19/trump-election-rigged.

79  Michael Crowley, "Trump Won't Commit to 'Peaceful' Post-Election Transfer of Power," *New York Times*, Sept. 23, 2020, https://www.nytimes.com/2020/09/23/us /politics/trump-power-transfer-2020-election.html.

80  Devlin Barrett, *October Surprise: How the FBI Tried to Save Itself and Crashed an Election* (New York: PublicAffairs, 2020).

81  "Review of Four FISA Applications and Other Aspects of the FBI's Crossfire Hurri-cane Investigation," DOJ Office of the Inspector General, December 2019, https:// www.justice.gov/storage/120919-examination.pdf.

82  Nate Silver, "The Comey Letter Probably Cost Clinton the Election," *FiveThirty-Eight*, May 3, 2017, https://fivethirtyeight.com/features/the-comey-letter-probably -cost-clinton-the-election/.

83  Ryan J. Reilly, "James Comey Says Idea FBI Swayed Election Outcome Makes Him 'Mildly Nauseous,'" *HuffPost*, May 3, 2017, https://www.huffpost.com/archive/au /entry/james-comey-hillary-clinton-emails_au_5cd34bc9e4b0acea94fe12db.

84  James Comey, *A Higher Loyalty* (New York: Flatiron Books, 2018).

85  Ryan J. Reilly, "The 'Law and Order' Party Has Unleashed an Extraordinary Attack on Law Enforcement," *HuffPost*, Jan. 27, 2018, https://www.huffpost.com/entry/fbi -attack-trump-mueller-republicans_n_5a6b3fe7e4b0ddb658c5c3fd.

86  Ryan J. Reilly, "'Viva Le Resistance': FBI Anti-Trump Messages Give President More 'Deep State' Fodder," *HuffPost*, June 18, 2018, https://www.huffpost.com /entry/fbi-texts-anti-trump-ig-report_n_5b22ac25e4b0a0a527799ebc.

87  FBI, "FBI Director Christopher Wray's Remarks at Press Conference on Election Security," press release, Oct. 21, 2020, https://www.fbi.gov/news/press-releases/fbi -director-christopher-wrays-remarks-at-press-conference-on-election-security.

88  Ryan J. Reilly, "Lawyers for Trump 'Super Fan' Cesar Sayoc Say Trump's Rhetoric Inspired Terrorist Plot on Democrats," *HuffPost*, July 22, 2019, https://www.huff post.com/entry/cesar-sayoc-trump-bombing_n_5d3635b1e4b004b6adb48758.

89  Andy Campbell, Sebastian Murdock, and Ryan J. Reilly, "Bomb Mailing Suspect Cesar Sayoc Was a Big Trump Fan with a Criminal History," *HuffPost*, Oct. 26, 2018, https://www.huffpost.com/entry/mail-pipe-bomb-cesar-sayoc_n_5bd32dcee4b055 bc948b4c90.

90  "Six Arrested on Federal Charge of Conspiracy to Kidnap the Governor of Michi-gan," Justice Department, Oct. 8, 2020, https://www.justice.gov/opa/pr/six-arrested -federal-charge-conspiracy-kidnap-governor-michigan.

91  US v. Fox, Government's Sentencing Memorandum, Dec. 5, 2022, https://storage .courtlistener.com/recap/gov.uscourts.miwd.99930/gov.uscourts.miwd.99930 .781.0.pdf.

92  Ben Collins and Brandy Zadrozny, "Proud Boys Celebrate After Trump's Debate Callout," NBC News, Sept. 29, 2020, https://www.nbcnews.com/tech/tech-news /proud-boys-celebrate-after-trump-s-debate-call-out-n1241512.

93  "September 29, 2020 Debate Transcript," Commission on Presidential Debates, Sept. 29, 2020, https://www.debates.org/voter-education/debate-transcripts/september-29-2020-debate-transcript/.

94  Jennifer Moore, interview by Select Committee to Investigate the Jan. 6th Attack on the US Capitol, July 26, 2022, https://www.govinfo.gov/content/pkg/GPO-J6-TRANSCRIPT-CTRL0000916069/pdf/GPO-J6-TRANSCRIPT-CTRL0000916069.pdf.

95  "Office of the Director of National Intelligence Progress Report—WMD Commission Recommendations, 2006," Federation of American Scientists, Intelligence Resource Program, https://irp.fas.org/dni/prog072706.pdf.

96  John Mintz, "Homeland Security Employs Imagination," *Washington Post,* June 18, 2004, https://www.washingtonpost.com/archive/politics/2004/06/18/homeland-security-employs-imagination/deeb045f-f3ce-4f0d-ab20-317cb58ffb46/.

97  Kelly Etz, "CBS's *Criminal Minds* Spin-Off Should Punishable by Law," *Michigan Daily,* Feb. 11, 2011, https://www.michigandaily.com/uncategorized/criminal-minds-suspect-behavior/.

98  Adam Goldman and Alan Feuer, "Bias and Human Error Played Parts in F.B.I.'s Jan. 6 Failure, Documents Suggest," *New York Times,* Feb. 1, 2023, https://www.nytimes.com/2023/02/01/us/politics/trump-jan-6-fbi.html.

99  FBI Behavioral Analysis Unit, *Lone Offender: A Study of Lone Offender Terrorism in the United States (1972–2015),* November 2019, https://www.fbi.gov/file-repository/lone-offender-terrorism-report-111319.pdf.

100  Rosa Brooks, "What's the Worst That Could Happen? The Election Will Likely Spark Violence—and a Constitutional Crisis," *Washington Post,* Sept. 3, 2020, https://www.washingtonpost.com/outlook/2020/09/03/trump-stay-in-office/.

101  Transition Integrity Project, "Preventing a Disrupted Presidential Election and Transition," Aug. 3, 2020, https://www.documentcloud.org/documents/7013152-Preventing-a-Disrupted-Presidential-Election-and.html#document/p1.

102  Barton Gellman, "The Election That Could Break America," *Atlantic,* Sept. 23, 2020, https://www.theatlantic.com/magazine/archive/2020/11/what-if-trump-refuses-concede/616424/.

103  Michael C. Bender, 2021 *"Frankly, We Did Win This Election": The Inside Story of How Trump Lost* (New York: Twelve, Hachette, 2021).

104  Donald Trump (@realDonaldTrump), "Last night I was leading, often solidly, in many key States, in almost all instances Democrat run & controlled. Then, one by one, they started to magically disappear as surprise ballot dumps were counted. VERY STRANGE, and the 'pollsters' got it completely & historically wrong!" Twitter, Nov. 4, 2020, https://twitter.com/realDonaldTrump/status/1324004491612618752.

105  Ryan J. Reilly, "Former GOP-Appointed Federal Prosecutors Blast Trump's 'Reckless' Vote Fraud Comments," *HuffPost,* Nov. 5, 2020, https://www.huffpost.com/entry/trump-voter-fraud-gop-us-attorneys_n_5fa4a1d3c5b623bfac4db023.

106  Ryan J. Reilly, "Former GOP Prosecutor and Voter Fraud 'True Believer' Says Trump 'Smells of Desperation,'" *HuffPost,* Nov. 5, 2020, https://www.huffpost.com/entry/trump-voter-fraud-doj-david-iglesias_n_5fa4590ac5b64c88d3fe7417.

107  Donald Trump, "Remarks by President Trump on the Election," White House, Nov. 5, 2020, https://trumpwhitehouse.archives.gov/briefings-statements/remarks-president-trump-election/.

108  US v. Nordean et al., Third Superseding Indictment, Feb. 14, 2022, https://www
    .justice.gov/usao-dc/press-release/file/1510791/download.

109  US v. Rhodes et al., Indictment, Jan. 8, 2021, https://www.justice.gov/usao-dc/press
    -release/file/1462346/download.

110  US v. Milstreed, Facebook Business Record, May 31, 2022, https://storage.courtlis
    tener.com/recap/gov.uscourts.dcd.243449/gov.uscourts.dcd.243449.8.1.pdf.

111  FBI, Statement of Facts, Jan. 20, 2023, https://www.justice.gov/usao-dc/press
    -release/file/1566536/download.

112  America First Bruins at UCLA (@afbruins), "Still think literacy tests are a bad
    idea? Didn't think so," Twitter, Nov. 4, 2020, https://web.archive.org/web/2020
    1104203338/https://twitter.com/afbruins/status/1324087342550843392.

113  Rachel Weiner, "Far-right UCLA Student Who Sat in Pence's Senate Seat on Jan. 6
    Sentenced," *Washington Post*, Oct. 19, 2022, https://www.washingtonpost.com/dc
    -md-va/2022/10/19/christian-secor-ucla-jan6-sentence/.

114  "You About to Lose Yo Job," Know Your Meme, June 9, 2022, https://knowyour
    meme.com/memes/you-about-to-lose-yo-job.

115  Oona Goodin-Smith, "Philly's Four Seasons Total Landscaping Dishes the Dirt on
    the News Conference Heard 'Round the World: 'It Was Nothing We Anticipated,"
    *Philadelphia Inquirer*, Dec. 1, 2020, https://www.inquirer.com/news/philadelphia
    /trump-four-seasons-total-landscaping-what-happened-20201201.html.

116  Amber Jamieson and Julia Reinstein, "Biden Supporters Played Beyoncé to Drown
    Out Trump Campaign Officials in Philly," BuzzFeed News, Nov. 5, 2020, https://
    www.buzzfeednews.com/article/amberjamieson/philladelphia-beyonce-protest
    -biden-trump-election.

117  Alice Park, Charlie Smart, Rumsey Taylor, and Miles Watkins, "An Extremely
    Detailed Map of the 2020 Election," *New York Times*, February 2021, https://www
    .nytimes.com/interactive/2021/upshot/2020-election-map.html.

118  Matt Friedman, "Man Featured at Giuliani Press Conference Is a Convicted Sex
    Offender," Politico, Nov. 9, 2020, https://www.politico.com/states/new-jersey/story
    /2020/11/09/man-featured-at-giuliani-press-conference-is-a-sex-offender-1335241.

119  Olivia Nuzzi, "The Full(est Possible) Story of the Four Seasons Total Landscaping
    Press Conference," *New York*, Dec. 21, 2020, https://nymag.com/intelligencer
    /2020/12/four-seasons-total-landscaping-the-full-est-possible-story.html.

120  US v. Lemke, Affidavit, Jan. 14, 2021, https://www.justice.gov/usao-sdny/press
    -release/file/1360511/download.

121  US v. Lamond, Indictment, May 18, 2023, https://www.documentcloud.org/docu
    ments/23816946-lamonds-indictment.

122  US v. Tarrio, Enrique Tarrio defense exhibit, April 6, 2023.

123  Alan Feuer, "Group Chat Linked to Roger Stone Shows Ties Among Jan. 6 Fig-
    ures," *New York Times*, May 20, 2022, https://www.nytimes.com/2022/05/20/us
    /politics/roger-stone-jan-6.html.

124  Ryan J. Reilly, "Patriot Leader Says Bybee Rejected Him for Clerkship," MainJus
    tice.com, Feb. 25, 2010, https://web.archive.org/web/20100304200256/http://
    www.mainjustice.com/2010/02/25/patriot-leader-says-bybee-rejected-him-for
    -clerkship/.

125  Ryan J. Reilly, "Oath Keepers Founder Says 'Undercover' Poll Watching Effort
    Won't Intimidate Voters," *HuffPost*, Oct. 27, 2016, https://www.huffpost.com/entry
    /oath-keepers-poll-watching_n_58122566e4b0990edc2f8178.

126 Eric Kleefeld, "Right-Wing 'Wagon' Conspiracy Theory Goes Bust in Michigan Vote Count," *Media Matters*, Nov. 5, 2020, https://www.mediamatters.org/voter -fraud-and-suppression/right-wing-wagon-conspiracy-theory-goes-bust-michigan -vote-count.

127 Ryan J. Reilly, "Oath Keeper Charged in Jan. 6 Attack Texted with Andrew Giuliani About Election," NBC News, Sept. 26, 2022, https://www.nbcnews.com /politics/justice-department/oath-keeper-charged-jan-6-attack-texted-andrew -giuliani-election-rcna49483.

128 Ryan J. Reilly, "Oath Keepers Lawyer Says Stewart Rhodes Wanted Her Trump Contacts Before Jan. 6 Capitol Attack," NBC News, July 11, 2020, https://www.nbc news.com/politics/congress/oath-keepers-lawyer-says-stewart-rhodes-wanted -trump-contacts-jan-6-ca-rcna37267.

129 Ross Jones, "Video Falsely Claims Possible Voter Fraud in Detroit. It Actually Shows a WXYZ Photographer Loading Camera Gear," WXYZ, Nov. 5, 2020, https://www.wxyz.com/news/video-claiming-to-show-possible-voter-fraud-in -detroit-is-actually-a-wxyz-photographer-loading-camera-gear.

130 Will Bunch, "How a Blue-Collar Kensington Kid Became Trump's Field General of Voter Suppression," *Philadelphia Inquirer*, Oct. 13, 2020, https://www.inquirer.com /columnists/attytood/mike-roman-voter-suppression-election-2020-trump-gop -20201013.html.

131 Mike Roman, interview by Select Committee to Investigate the Jan. 6th Attack on the US Capitol, Aug. 10, 2022, https://www.govinfo.gov/content/pkg/GPO-J6 -TRANSCRIPT-CTRL0000916109/pdf/GPO-J6-TRANSCRIPT-CTRL 0000916109.pdf.

132 Danny Hakim and Nick Corasaniti, "Trump Campaign Draws Rebuke for Surveil-ling Philadelphia Voters," *New York Times*, Oct. 22, 2020, https://www.nytimes .com/2020/10/22/us/politics/trump-campaign-voter-surveillance.html.

133 Ryan J. Reilly, "His Poll-Watching Video Became a GOP Obsession. Now He Wants Trump to Rein In 'Rigged' Election Talk," *HuffPost*, Oct. 19, 2016, https://www .huffpost.com/entry/new-black-panthers-donald-trump_n_5806805be4b0dd54ce3 5edc8.

134 Ryan J. Reilly, "A GOP Lawyer Tried to Block Hospitalized Voters. But This Philly Med Student Fought Back (and Won)," *HuffPost*, Nov. 8, 2016, https://www.huff post.com/entry/philadelphia-hospital-patients-vote_n_58226fa6e4b0aac 624881348.

135 Ryan Briggs, "Lawyer Behind Trump's Legal Campaign to Win Pa. Has a Long History in Philly," WHYY, Nov. 13, 2020, https://whyy.org/articles/lawyer-behind -trumps-legal-campaign-to-win-pa-has-a-long-history-in-philly/.

136 "Linda Kerns: A Hero Attorney Battles It Out in Philadelphia's Election Integrity War Zone," *Who's Counting with Cleta Mitchell*, YouTube, April 28, 2022, https:// www.youtube.com/watch?v=3cEXYs24Zek.

137 Ryan J. Reilly, "How Bill Fulton Infiltrated Alaska's Right Wing as an FBI Infor-mant," *HuffPost*, Jan. 11, 2013, https://www.huffpost.com/entry/bill-fulton-alaska -fbi-informant_n_2456883.

138 Aaron Blake, "'What's the Downside for Humoring Him?': A GOP Official's Unin-tentionally Revealing Quote About the Trump Era," *Washington Post*, Nov. 10, 2020, https://www.washingtonpost.com/politics/2020/11/10/whats-downside-humo ring-him-gop-officials-unintentionally-revealing-quote-about-trump-era/.

139  Ryan J. Reilly, "Donald Trump's Mass Voter Fraud Conspiracies Could Get Some-body Killed," *HuffPost*, Nov. 10, 2020, https://www.huffpost.com/entry/trump-voter-fraud-conspiracy-theory-violence_n_5fa9b2f6c5b64c88d403e909.

# Chapter 4

1  Emails posted online by former FBI employee Kyle Seraphin, Scribd, Feb. 7, 2023.

2  Jennifer L. Moore, "Captain's Log—Quarantine Day 181," FBI document, Sept. 13, 2020.

3  Ryan J. Reilly and Daniel Barnes, "Oath Keeper Testifies He Was Ready to Die on Jan. 6 to Keep Trump in Office," NBC News, Oct. 18, 2022, https://www.nbcnews.com/politics/justice-department/oath-keeper-testifies-was-ready-die-jan-6-keep-trump-office-rcna52838.

4  US v. Rhodes, Jason Dolan, testimony during Oath Keepers trial, Oct. 18, 2022.

5  Patrick Maks, "Calling the 2020 Presidential Race State by State," Associated Press, Nov. 8, 2020, https://blog.ap.org/behind-the-news/calling-the-2020-presidential-race-state-by-state.

6  Annie Karni and Maggie Haberman, "Fox's Arizona Call for Biden Flipped the Mood at Trump Headquarters," *New York Times*, Nov. 4, 2020, https://www.nytimes.com/2020/11/04/us/politics/trump-fox-news-arizona.html.

7  US v. Rhodes et al., Exhibit 5308, Oct. 18, 2022, https://www.documentcloud.org/documents/23815827-5308.

8  Jane Lytvynenko, "'I Found Election Interference and No One Cared': One US Vet-eran's Fight to Protect His Compatriots Online," *BuzzFeed*, Dec. 30, 2019, https://www.buzzfeednews.com/article/janelytvynenko/kristofer-goldsmith-veteran-disinformation.

9  Kathrin Havrilla-Sanchez, "Alumnus and Military Veteran Fights Spread of Politi-cal Misinformation," Columbia University, Oct. 30, 2020, https://www.gs.columbia.edu/news/alumnus-and-military-veteran-fights-spread-political-misinformation-lead-up-election-day.

10  Jane Lytvynenko, "The White Extremist Group Patriot Front Is Preparing for a World After Donald Trump," *BuzzFeed*, Oct. 27, 2020, https://www.buzzfeednews.com/article/janelytvynenko/patriot-front-preparing-after-trump.

11  News2Share, "III% Security Force Militia Leader Chris Hill—Interview on December 5, 2020," YouTube, Aug. 16, 2021, https://www.youtube.com/watch?v=1tw71c1TArA.

12  US v. Carrillo, Criminal Complaint, June 16, 2020, https://storage.courtlistener.com/recap/gov.uscourts.cand.361522/gov.uscourts.cand.361522.1.0.pdf.

13  Nick Robins-Early, "Far-Right 'Boogaloo' Supporter Charged with Murder in Deaths of Officers," *HuffPost*, June 16, 2020, https://www.huffpost.com/entry/boogaloo-carrillo-oakland-murder-dave-underwood_n_5ee9378cc5b6fdae7db88f2a?cd.

14  Leah Sottile, "Inside the Boogaloo: America's Extremely Online Extremists," *New York Times*, Aug. 19, 2020, https://www.nytimes.com/interactive/2020/08/19/magazine/boogaloo.html.

15  Jared Thompson, "Examining Extremism: The Boogaloo Movement," Center for Strategic and International Studies, June 30, 2021, https://www.csis.org/blogs/examining-extremism/examining-extremism-boogaloo-movement.

16  US v. Carrillo, Criminal Complaint, June 16, 2020.

17  Report obtained by author.

18  Atlantic Council's DFRLab, "#StopTheSteal: Timeline of Social Media and Extremist Activities Leading to 1/6 Insurrection," Just Security, Feb. 10, 2021, https://www.justsecurity.org/74622/stopthesteal-timeline-of-social-media-and-extremist-activities-leading-to-1-6-insurrection/.

19  Michael Edison Hayden, "Far Right Resurrects Roger Stone's #StopTheSteal During Vote Count," Southern Poverty Law Center, Nov. 6, 2020, https://www.splcenter.org/hatewatch/2020/11/06/far-right-resurrects-roger-stones-stopthesteal-during-vote-count.

20  Sheera Frenkel, "The Rise and Fall of the 'Stop the Steal' Facebook Group," *New York Times*, Nov. 5, 2020, https://www.nytimes.com/2020/11/05/technology/stop-the-steal-facebook-group.html.

21  EJ Dickson, "How a Pro-Trump 'Stop the Steal' Group Became 'Gay Communists for Socialism,'" *Rolling Stone*, Nov. 6, 2020, https://www.rollingstone.com/culture/culture-news/stop-the-steal-facebook-group-gay-communists-for-socialism-1086967/.

22  Jonah E. Bromwich, "Whatever It Is, It's Probably Not Hair Dye," *New York Times*, Nov. 19, 2020, https://www.nytimes.com/2020/11/19/style/rudy-giuliani-hair.html.

23  Ryan J. Reilly, "Rudy Giuliani's 'Disgraceful' Arguments to Disenfranchise Pa. Voters Didn't Go So Well," *HuffPost*, Nov. 17, 2020, https://www.huffpost.com/entry/trump-pennsylvania-lawsuit-giuliani_n_5fb3e6cac5b6aad41f73600b.

24  Ryan J. Reilly (@ryanjreilly), "Here it is: Rudy Giuliani—an attorney for the president of the United States and a former U.S. Attorney—botches a basic question about 'strict scrutiny.' 'The normal one.' [edited out an interruption in the middle]," Twitter, Nov. 19, 2020, https://twitter.com/ryanjreilly/status/1329458638197559296 (brackets in the original).

25  Ryan J. Reilly, "Judge Brutally Dismisses Rudy Giuliani's Suit to 'Disenfranchise' Pa. Voters," *HuffPost*, Nov. 21, 2020, https://www.huffpost.com/entry/trump-pennsylvania-giuliani-lawsuit_n_5fb9a3dec5b66bb88c5e7fe9?kp9.

26  Mark Maremont and Corinne Ramey, "How Jenna Ellis Rose from Traffic Court to Trump's Legal Team," *Wall Street Journal*, Dec. 3, 2020, https://www.wsj.com/articles/how-jenna-ellis-rose-from-traffic-court-to-trumps-legal-team-11607038900.

27  Jeremy W. Peters and Alan Feuer, "How Is Trump's Lawyer Jenna Ellis 'Elite Strike Force' Material?," *New York Times*, Dec. 3, 2020, https://www.nytimes.com/2020/12/03/us/politics/jenna-ellis-trump.html.

28  Ryan J. Reilly, "Even 'Birther' Orly Taitz Isn't Sure About Trump Lawyer's Claim He 'Won in a Landslide,'" *HuffPost*, Nov. 24, 2020, https://www.huffpost.com/entry/orly-taitz-rudy-giuliani-jenna-ellis-trump-voter-fraud_n_5fbd18b3c5b63d1b7706ce52.

29  Dan Brooks, "How President Trump Ruined Political Comedy," *New York Times*, Oct. 7, 2020, https://www.nytimes.com/2020/10/07/magazine/trump-liberal-comedy-tv.html.

30  Ben Zimmer, "Truthiness," *New York Times*, Oct. 13, 2010, https://www.nytimes.com/2010/10/17/magazine/17FOB-onlanguage-t.html.

31  Stephen Colbert, "Truthiness," *Colbert Report*, Oct. 17, 2005, https://www.cc.com/video/63ite2/the-colbert-report-the-word-truthiness.

32  "Truthiness Voted 2005 Word of the Year," American Dialect Society, Jan. 6, 2006, https://www.americandialect.org/truthiness_voted_2005_word_of_the_year.

33  Sabrina Tavernise and Brian Stelter, "At Rally, Thousands—Billions?—Respond," *New York Times*, Oct. 30, 2010, https://www.nytimes.com/2010/10/31/us/politics/31rally.html.

34  "Truthiness," *Merriam-Webster* (added April 2020), https://www.merriam-webster.com/dictionary/truthiness.

35  Jane C. Timm, "Fact Check: Echoing Trump, Barr Misleads on Voter Fraud to Attack Expanded Vote-by-Mail," NBC News, Sept. 19, 2020, https://www.nbcnews.com/politics/2020-election/fact-check-echoing-trump-barr-misleads-voter-fraud-attack-expanded-n1240144.

36  Barr, *One Damn Thing After Another.*

37  Bill Carter, "As Citigroup Chief Totters, CNBC Reporter Is Having a Great Year," *New York Times*, Nov. 5, 2007, https://www.nytimes.com/2007/11/05/business/media/05bartiromo.html.

38  Associated Press, "CNBC's Bartiromo Leaving Floor of NYSE," NBC News, May 12, 2004, https://www.nbcnews.com/id/wbna4963290.

39  Maria Bartiromo, "Why Maria Bartiromo Is Grateful to the Men Who Tried to Intimidate Her," *Forbes*, March 8, 2018, https://www.forbes.com/sites/forbes-summit-talks/2018/03/08/why-maria-bartiromo-is-grateful-to-the-men-who-tried-to-intimidate-her/?sh=79222ece67e8.

40  William D. Cohan, "Maria Bartiromo Was a Generational Icon for Financial Television. What Happened?" *Institutional Investor*, Jan. 19, 2019, https://www.institutionalinvestor.com/article/b1cq2nzw56k40k/Maria-Bartiromo-Was-a-Generational-Icon-for-Financial-Television-What-Happened.

41  Fox Business, "Maria Bartiromo on Roger Ailes' Impact on the Media, Her Career," YouTube, May 19, 2017, https://www.youtube.com/watch?v=XUBMNMwfqzU.

42  Brian Steinberg, "Maria Bartiromo Defends Roger Ailes Against Gretchen Carlson Harassment Claims (Exclusive)," *Variety*, July 9, 2016, https://variety.com/2016/tv/news/maria-bartiromo-roger-ailes-gretchen-carlson-1201811345/.

43  John Koblin, Emily Steel, and Jim Rutenberg, "Roger Ailes Leaves Fox News, and Rupert Murdoch Steps In," *New York Times*, July 21, 2016, https://www.nytimes.com/2016/07/22/business/media/roger-ailes-fox-news.html.

44  Clyde Haberman, "Roger Ailes, Who Built Fox News into an Empire, Dies at 77," *New York Times*, May 18, 2017, https://www.nytimes.com/2017/05/18/business/media/roger-ailes-dead.html.

45  Sarah Ellison, "What Happened to Maria Bartiromo?," *Washington Post*, Dec. 23, 2020, https://www.washingtonpost.com/lifestyle/media/maria-bartiromo-fox-news-trump/2020/12/22/35520a90-3fb1-11eb-8db8-395dedaaa036_story.html.

46  US Dominion Inc. v. Fox News Network, Dominion's Brief in Support of Its Motion for Summary Judgment on Liability of Fox News Network, LLC and Fox Corporation, Feb. 16, 2023, https://www.documentcloud.org/documents/23684956-2023-02-16-redacted-dominion-opening-sj-brief-18.

47  Jeremy Barr, "With Fox's Maria Bartiromo as His First Post-Election Interviewer, Trump Found a Sympathetic Ear and Few Questions," *Washington Post*, Nov. 29, 2020, https://www.washingtonpost.com/media/2020/11/29/trump-fox-bartiromo-interview/.

48  William Barr, interview by Select Committee to Investigate the Jan. 6th Attack on the US Capitol, June 2, 2022, https://www.govinfo.gov/content/pkg/GPO-J6

-TRANSCRIPT-CTRL0000083860/pdf/GPO-J6-TRANSCRIPT-CTRL 0000083860.pdf.

49 Dana Milbank, "Have You Heard About How Bill Barr Saved Democracy? Let Bill Barr Tell You," *Washington Post*, March 2, 2022, https://www.washingtonpost.com /opinions/2022/03/04/bill-barr-book-trump-clown-show/.

50 Ryan J. Reilly, "William Barr Kept Silent for Weeks as Trump's Voter Fraud Conspiracy Theories Spread," *HuffPost*, Dec. 1, 2020, https://www.huffpost.com/entry /william-barr-voter-fraud-conspiracy-theories_n_5fc6a25cc5b61d04bfaddb98.

51 FOIA, Justice Department, June 10, 2022, https://www.justice.gov/oip/foia-library /foia-processed/general_topics/meetings_%26communications_barr_06_10_22 /download.

52 David Bauder, "The Story Behind AP Report That Caused Trump to Throw Lunch," Associated Press, June 28, 2022, https://apnews.com/article/capitol-siege-politics -elections-donald-trump-william-barr-a7781d455b044964eb01057534dbb010.

53 Michael Balsamo, "Disputing Trump, Barr Says No Widespread Election Fraud," AP News, June 28, 2022, https://apnews.com/article/barr-no-widespread-election -fraud-b1f1488796c9a98c4b1a9061a6c7f49d.

54 Hearing on the January 6th Investigation, 117th Cong., June 28, 2022, https://www .govinfo.gov/app/details/GPO-J6-HEARING-VIDEO-CTRL0000931167/summary.

55 Craig Mauger, "Dominion Tells Mellissa Carone to Cease 'Defamatory Claims,'" *Detroit News*, Dec. 28, 2020, https://www.detroitnews.com/story/news/politics/2020 /12/28/dominion-tells-mellissa-carone-cease-defamatory-claims/4059316001/.

56 Ryan J. Reilly, "Rudy Giuliani's 'Voter Fraud' Witness Accused of Framing Woman for Stealing Sex Videos," *HuffPost*, Dec. 5, 2020, https://www.huffpost.com/entry /mellissa-carone-wright-rudy-giuliani-trump-voter-fraud-detroit-michigan_n_5fc b94e9c5b63a1534523506.

57 Chad Selweski, "Selweski: Mellissa Carone Turns Overnight Fame into a Drive to Win a Seat, Any Seat," Deadline Detroit, April 25, 2022, https://www.deadlinede troit.com/articles/30390/selweski_mellissa_carone_turns_overnight_fame_into _a_drive_to_win_a_seat_any_seat.

58 Ryan J. Reilly (@ryanjreilly), "Holy smokes the sequel is even better! Rudy Giuliani tries to shush her to calm her down and the Republican even tries to reel her in! She treats this Republican* like he's a Chick-fil-A employee and the milkshake machine went down," Twitter, Dec. 2, 2020, https://twitter.com/ryanjreilly/status/13343 11448340795396.

59 Ryan J. Reilly (@ryanjreilly), "I present this clip of Rudy Giuliani testifying without editing or commentary. (Watch for the 👀)," Twitter, Dec. 2, 2020, https://twit ter.com/ryanjreilly/status/1334344279305695232.

60 Jake Lahut, "Michigan State Representative Confirms Rudy Giuliani Farted During an Election Hearing," *Insider*, Dec. 4, 2020, https://www.businessinsider.com /rudy-giuliani-fart-on-camera-confirmed-by-michigan-state-representative-2020-12.

61 FOIA, Justice Department, June 3, 2022, https://www.justice.gov/oip/foia-library /foia-processed/general_topics/2020_presidential_election_06_03_22_part_1 /download.

62 Donald Trump (@realDonaldTrump), "Melissa is great! https://twitter.com/kyle nabecker/status/1333849243287687172," Twitter, Dec. 6, 2020, https://twitter.com /realDonaldTrump/status/1335464641301008384.

63  Ryan J. Reilly, "Feds Arrest Five Members of 'B Squad' Militia Allegedly Run by Former GOP House Candidate in Jan. 6 Case," NBC News, Aug. 24, 2020, https://www.nbcnews.com/politics/justice-department/feds-charge-members-b-squad-militia-allegedly-run-former-gop-house-can-rcna44621.

64  FBI, Statement of Facts, Aug. 16, 2022, https://www.justice.gov/usao-dc/case-multi-defendant/file/1529756/download.

65  "Yigal Carmon—President and Founder of the Middle East Media Research Institute (MEMRI)," https://docs.house.gov/meetings/FA/FA00/20160706/105161/HHRG-114-FA00-Bio-CarmonY-20160706.pdf.

66  US v. Bies, Affidavit, Aug. 12, 2022, https://www.justice.gov/opa/press-release/file/1526576/download.

67  MEMRI, "Neo-Nazis, White Supremacists on Facebook, Telegram, 4Chan, Instagram, Gab, React to U.S. Elections, Anticipate Coming 'Civil War,'" Nov. 6, 2020, https://www.memri.org/dttm/neo-nazis-white-supremacists-facebook-telegram-4chan-instagram-gab-react-us-elections.

68  Michael Sherwin, interview by Select Committee to Investigate the Jan. 6th Attack on the US Capitol, April 19, 2022, https://www.govinfo.gov/content/pkg/GPO-J6-TRANSCRIPT-CTRL0000071085/pdf/GPO-J6-TRANSCRIPT-CTRL0000071085.pdf.

69  US Department of Homeland Security, "Rightwing Extremism: Current Economic and Political Climate Fueling Resurgence in Radicalization and Recruitment," April 7, 2009, https://irp.fas.org/eprint/rightwing.pdf.

70  Sahil Kapur, "'This Isn't the Final Chapter': Analyst Warns, Again, About Rise of Right-Wing Extremists," NBC News, Jan. 13, 2021, https://www.nbcnews.com/politics/politics-news/isn-t-final-chapter-analyst-warns-again-about-rise-right-n1253950.

71  Roger Hedgecok, "Disagree with Obama? Gov't Has Eyes on You," World Net Daily, April 13, 2009, https://www.wnd.com/2009/04/94799/.

72  Stephen Gordon, "Homeland Security Document Targets Most Conservatives and Libertarians in the Country," Liberty Papers, April 12, 2009, http://www.thelibertypapers.org/2009/04/12/homeland-security-document-targets-most-conservatives-and-libertarians-in-the-country/.

73  Teddy Davis and Ferdous Al-Faruque, "Napolitano Facing Republican Calls for Her Ouster," ABC News, April 23, 2009, https://abcnews.go.com/Politics/story?id=7412992&page=1.

74  Audrey Hudson, "GOP Probes 'Extremism' Report's Origins," *Washington Times*, May 7, 2009, https://www.washingtontimes.com/news/2009/may/07/gop-seeks-explanation-of-extremism-report/.

75  Spencer Ackerman, "DHS Crushed This Analyst for Warning About Far-Right Terror," *Wired*, Aug. 2, 2012, https://www.wired.com/2012/08/dhs/.

76  Ryan J. Reilly, "One Official Tried to Warn Us About Attacks Like Portland. He Was Pushed Out," *HuffPost*, May 31, 2017, https://www.huffpost.com/entry/portland-attack-domestic-terrorism_n_592d92bee4b0df57cbfd8b31.

77  Moore, interview.

78  Aaron Martin, "*11 Investigates* Takes You Inside the FBI's National Threat Operations Center," Nov. 24, 2020, https://www.wpxi.com/news/investigates/11-investigates-takes-you-inside-fbis-national-threat-operations-center/PTTFC5UQ25HX3B2GU63AOPNUQM/.

79  US v. Hostetter, Indictment, June 9, 2021, https://storage.courtlistener.com/recap /gov.uscourts.dcd.232197/gov.uscourts.dcd.232197.1.0.pdf.

80  Ryan J. Reilly, "'Traitors Need to Be Executed': 'Stop the Steal' Organizer Indicted in Jan. 6 Conspiracy Case," *HuffPost*, June 10, 2021, https://www.huffpost.com /entry/alan-hostetter-capitol-attack-arrest-fbi_n_60c27653e4b0e6bab7a5a0be.

81  US v. Rhodes, Testimony, Nov. 11, 2022.

82  US Department of Homeland Security, "Report on DHS Administrative Review into I&A Open Source Collection and Dissemination Activities During Civil Unrest," Jan. 6, 2021, https://www.wyden.senate.gov/imo/media/doc/I&A%20 and%20OGC%20Portland%20Reports.pdf.

83  Benjamin Wittes, "What If J. Edgar Hoover Had Been a Moron?," Lawfare, Aug. 4, 2020, https://www.lawfareblog.com/what-if-j-edgar-hoover-had-been-moron.

84  Shane Harris, "DHS Compiled 'Intelligence Reports' on Journalists Who Published Leaked Documents," *Washington Post*, July 30, 2020, https://www.washingtonpost .com/national-security/dhs-compiled-intelligence-reports-on-journalists-who-pub lished-leaked-documents/2020/07/30/5be5ec9e-d25b-11ea-9038-af089b63ac21 _story.html.

85  Stephanie Dobitsch, interview by Select Committee to Investigate the Jan. 6th Attack on the US Capitol, May 5, 2022, https://www.govinfo.gov/content/pkg /GPO-J6-TRANSCRIPT-CTRL0000082307/pdf/GPO-J6-TRANSCRIPT -CTRL0000082307.pdf.

86  Government Accountability Office, "DHS Employee Morale: Some Improvements Made, but Additional Actions Needed to Strengthen Employee Engagement," Jan. 12, 2021, https://www.gao.gov/products/gao-21-204.

87  Department of Homeland Security Inspector General, "I&A Identified Threats Prior to January 6, 2021, but Did Not Issue Any Intelligence Products Before the U.S. Capitol Breach," March 4, 2022, https://www.oig.dhs.gov/sites/default/files /assets/2022-04/OIG-22-29-Mar22-Redacted.pdf.

88  Department of Homeland Security Inspector General, "I&A Identified Threats."

89  Government Accountability Office, "Federal Agencies' Use of Open Source Data and Related Threat Products Prior to January 6, 2021," May 2022, https://www.gao .gov/assets/gao-22-105963.pdf.

90  Paul Blumenthal and Ryan J. Reilly, "Supreme Court Rejects Trump's Authoritar- ian Bid to Hijack the Presidency," *HuffPost*, Dec. 11, 2020, https://www.huffpost .com/entry/trump-supreme-court-voter-fraud_n_5fd39185c5b66a7584147c7a.

91  Nick Corasaniti and Jim Rutenberg, "Electoral College Vote Officially Affirms Biden's Victory," *New York Times*, Dec. 14, 2020, https://www.nytimes.com/2020 /12/14/us/politics/biden-electoral-college.html.

# Chapter 5

1  Jonathan Swan and Zachary Basu, "Inside the Craziest Meeting of the Trump Presi- dency," Axios, Feb. 2, 2021, https://www.axios.com/2021/02/02/trump-oval-office -meeting-sidney-powell.

2  Eric Herschmann, interview by Select Committee to Investigate the Jan. 6th Attack on the US Capitol, April 6, 2022, https://www.govinfo.gov/content/pkg/GPO-J6 -TRANSCRIPT-CTRL0000062567/pdf/GPO-J6-TRANSCRIPT-CTRL 0000062567.pdf.

3   Donald Trump (@realDonaldTrump), "Peter Navarro releases 36-page report alleging election fraud 'more than sufficient' to swing victory to Trump https://washex.am/3nwaBCe. A great report by Peter. Statistically impossible to have lost the 2020 Election. Big protest in D.C. on January 6th. Be there, will be wild!" Twitter, Dec. 19, 2020, https://twitter.com/realDonaldTrump/status/13401857732205 15840.

4   Donnell Harvin, interview by Select Committee to Investigate the Jan. 6th Attack on the US Capitol, Jan. 24, 2022, https://www.govinfo.gov/content/pkg/GPO -J6-TRANSCRIPT-CTRL0000038866/pdf/GPO-J6-TRANSCRIPT-CTRL 0000038866.pdf.

5   "Trump Tweet: Daddy Says Be in DC on Jan. 6th," TheDonald.win, archived on Dec. 22, 2020, https://archive.is/h4nSI.

6   Robert Evans, "How the Insurgent and MAGA Right Are Being Welded Together on the Streets of Washington D.C.," Bellingcat, Jan. 5, 2021, https://www.belling cat.com/news/americas/2021/01/05/how-the-insurgent-and-maga-right-are-being -welded-together-on-the-streets-of-washington-d-c/.

7   Ryan J. Reilly and Daniel Barnes, "'The Donald' Forum User Convicted of Assault- ing Officers on Jan. 6," NBC News, May 3, 2023, https://www.nbcnews.com/poli tics/justice-department/-donald-forum-user-convicted-assaulting-officers-jan -6-rcna82675.

8   Ryan J. Reilly, "Informant Warned FBI Weeks Before Jan. 6 That the Far Right Saw Trump Tweet as 'a Call to Arms,'" NBC News, Dec. 21, 2020, https://www.nbcnews .com/politics/justice-department/informant-warned-fbi-weeks-jan-6-far-right -saw-trump-tweet-call-arms-rcna62683.

9   US v. Rodriguez, Daniel Rodriguez Interview Transcript, March 31, 2021, https:// www.documentcloud.org/documents/21108067-danny-rodriguez-transcript.

10  US v. Rodriguez, Statement of Offense, Feb. 14, 2023, https://storage.courtlistener .com/recap/gov.uscourts.dcd.229256/gov.uscourts.dcd.229256.160.0_1.pdf.

11  US v. Cua, Statement of Facts, Feb. 13, 2023, https://storage.courtlistener.com /recap/gov.uscourts.dcd.227610/gov.uscourts.dcd.227610.281.0.pdf.

12  Chris Wray, "Remembering Investigative Specialist Saul Tocker," FBI, Dec. 20, 2020, https://www.fbi.gov/news/speeches/remembering-investigative-specialist-saul -tocker.

13  David E. Sanger, Nichole Perlroth, and Eric Schmitt, "Scope of Russian Hacking Becomes Clear: Multiple U.S. Agencies Were Hit," *New York Times*, Dec. 14, 2020, https://www.nytimes.com/2020/12/14/us/politics/russia-hack-nsa-homeland-secu rity-pentagon.html.

14  Ben Fox, "Hack Against US Is 'Grave' Threat, Cybersecurity Agency Says," Associ- ated Press, Dec. 17, 2020, https://apnews.com/article/technology-malware-hacking -russia-software-b3f993fb7bc9390302f0df26ecb6c10e.

15  Ryan J. Reilly, "Teen Testifies About 'Surreal' Experience Tipping Off FBI About His Dad Before Jan. 6 Riot," NBC News, March 3, 2022, https://www.nbcnews.com /politics/justice-department/teen-testifies-surreal-experience-tipping-fbi-dad -jan-6-riot-rcna18526.

16  Donald Trump (@realDonaldTrump), "VOTER FRAUD IS NOT A CONSPIRACY THEORY, IT IS A FACT!!!" Twitter, Dec. 24, 2020, https://twitter.com/realDonald Trump/status/1342212651447967744.

17 Donald Trump (@realDonaldTrump), "MERRY CHRISTMAS!" Twitter, Dec. 25, 2020, https://twitter.com/realDonaldTrump/status/1342463237841948672.

18 Steve Cavendish, Neil MacFarquhar, Jamie McGee, and Adam Goldman, "Behind the Nashville Bombing, a Conspiracy Theorist Stewing About the Government," *New York Times*, Feb. 24, 2021, https://www.nytimes.com/2021/02/24/us/anthony -warner-nashville-bombing.html.

19 Natalie Allison and Adam Tamburin, "Retracing the Key Moments After the Christmas Morning Bombing in Nashville," *Tennessean*, Jan. 10, 2021, https://www .tennessean.com/in-depth/news/local/2021/01/10/timeline-christmas-morning -bombing-nashville/6578915002/.

20 FBI Memphis, "FBI Releases Report on Nashville Bombing," March 15, 2021, https://www.fbi.gov/contact-us/field-offices/memphis/news/press-releases/fbi -releases-report-on-nashville-bombing.

21 Donald Trump (@realDonaldTrump), "The 'Justice' Department and the FBI have done nothing about the 2020 Presidential Election Voter Fraud, the biggest SCAM in our nation's history, despite overwhelming evidence. They should be ashamed. History will remember. Never give up. See everyone in D.C. on January 6th," Twitter, Dec. 26, 2020, https://twitter.com/realDonaldTrump/status/1342821189077622792.

22 "Fact Check: Debunking Conspiracy Links Between Nashville Explosion, Dominion and AT&T," Reuters, Dec. 29, 2020, https://www.reuters.com/article/uk-fact check-att-dominion-nashville-bomb/fact-check-debunking-conspiracy-links -between-nashville-explosion-dominion-and-att-idUSKBN2931BI.

23 Richard Peter Donoghue, interview by Select Committee to Investigate the Jan. 6th Attack on the US Capitol, Oct. 1, 2021, https://www.govinfo.gov/content/pkg/GPO -J6-TRANSCRIPT-CTRL0000034600/pdf/GPO-J6-TRANSCRIPT -CTRL0000034600.pdf.

24 Donoghue, interview by Select Committee.

25 Jeffrey A. Rosen, interview by Select Committee to Investigate the Jan. 6th Attack on the US Capitol, Oct. 13, 2021, https://www.govinfo.gov/content/pkg/GPO-J6 -TRANSCRIPT-CTRL0000034616/pdf/GPO-J6-TRANSCRIPT-CTRL 0000034616.pdf.

26 Donoghue, interview by Select Committee.

27 Donoghue, interview by Select Committee.

28 FBI, FOIA, "United States Capitol Violence and related Events of January 6, 2021," Dec. 31, 2020.

29 FBI, FOIA, "United States Capitol Violence and Related Events of January 6, 2021, Part 23," Dec. 29, 2020, https://vault.fbi.gov/united-states-capitol-violence-and -related-events-of-january-6-2021/united-states-capitol-violence-and-related-events -of-january-6-2021-part-23/view.

30 Zack Budryk, "Ex-Pence Aide Turned Trump Critic 'Very Concerned' About Jan. 6 Violence," *Hill*, Dec. 29, 2020, https://thehill.com/homenews/administration /531923-ex-pence-aide-turned-trump-critic-very-concerned-about-jan-6-violence/.

31 General Services Administration, "FBI Social Media Alerting," Dec. 30, 2020, https://sam.gov/opp/e11b3320747b4ac4a1e8ef02bf055f29/view.

32 Ken Dilanian, "Why Did the FBI Miss the Threats About Jan. 6 on Social Media?," NBC News, March 8, 2021, https://www.nbcnews.com/politics/justice-department /fbi-official-told-congress-bureau-can-t-monitor-americans-social-n1259769.

33  Bowdich, interview.

34  FBI, FOIA, "United States Capitol Violence and Related Events of January 6, 2021, Part 23," Dec. 31, 2020, https://vault.fbi.gov/united-states-capitol-violence-and -related-events-of-january-6-2021/united-states-capitol-violence-and-related-events -of-january-6-2021-part-23/view.

35  Donald Trump (@realDonaldTrump), "The BIG Protest Rally in Washington, D.C., will take place at 11.00 A.M. on January 6th. Locational details to follow. Stop- TheSteal!" Twitter, Jan. 1, 2021, https://twitter.com/realDonaldTrump/status /1345095714687377418.

36  Donald Trump (@realDonaldTrump), "Massive amounts of evidence will be pre- sented on the 6th. We won, BIG!" Twitter, Jan. 1, 2021, https://twitter.com/realDon aldTrump/status/1345100089505755139.

37  Donald Trump (@realDonaldTrump), "January 6th. See you in D.C.," Twitter, Jan. 1, 2021, https://twitter.com/realDonaldTrump/status/1345152408591204352.

38  Committee on Oversight and Reform, "President Trump Pressure Campaign on Department of Justice," June 15, 2021, https://archive.org/details/cor-selected -dojdocuments-2021-6-15-final.

39  Donoghue, interview by Select Committee.

40  Donald Trump (@realDonaldTrump), "An attempt to steal a landslide win. Can't let it happen!" Twitter, Jan. 2, 2021, https://twitter.com/realDonaldTrump/status /1345508977031974918.

41  FBI, Statement of Facts, Jan. 17, 2021, https://www.justice.gov/opa/page/file/1357 391/download.

42  FBI, Statement of Facts, Oct. 26, 2022, https://www.justice.gov/usao-dc/press -release/file/1547816/download.

43  US v. Wood, Statement of Offenses, May 27, 2022, https://www.justice.gov/usao-dc /press-release/file/1509041/download.

44  US v. Rhodes et al., Eighth Superseding Indictment, June 22, 2022, https://www .justice.gov/usao-dc/press-release/file/1514871/download.

45  US v. Donohue, Statement of Offense, April 4, 2022, https://www.justice.gov/usao -dc/press-release/file/1492881/download.

46  US v. Rhodes et al., Exhibit 9727, Feb. 14, 2023.

47  Donald Trump (@realDonaldTrump), "I spoke to Secretary of State Brad Raffens- perger yesterday about Fulton County and voter fraud in Georgia. He was unwill- ing, or unable, to answer questions such as the 'ballots under table' scam, ballot destruction, out of state 'voters', dead voters, and more. He has no clue!" Twitter, Jan. 3, 2021, https://twitter.com/realDonaldTrump/status/1345731043861659650.

48  Donald Trump (@realDonaldTrump), "I will be there. Historic day!" Twitter, Jan. 3, 2021, https://twitter.com/realDonaldTrump/status/1345753534168506370.

49  Donoghue, interview by Select Committee.

50  Rosen, interview by Select Committee.

51  Ryan J. Reilly, "Who Is Jeffrey Clark? Jan. 6 Panel Seeks to Make Trump's Man at DOJ Famous," NBC News, June 23, 2022, https://www.nbcnews.com/politics/jus tice-department/jeffrey-clark-jan-6-panel-seeks-make-trumps-man-doj-famous -rcna34521.

52  Jeffrey Rosen, interview by US Senate Committee on the Judiciary, Aug. 7, 2021, https://www.judiciary.senate.gov/imo/media/doc/Rosen%20Transcript.pdf.

53  Donoghue, interview by Select Committee.

54 Document obtained by Jan. 6 committee, released Dec. 2022, https://www.govinfo .gov/content/pkg/GPO-J6-DOC-CTRL0000930224/pdf/GPO-J6-DOC-CTRL 0000930224.pdf.

55 Richard Donoghue, interview by US Senate Committee on the Judiciary, Aug. 6, 2021, https://www.judiciary.senate.gov/imo/media/doc/Donoghue%20Transcript.pdf.

56 Donald Trump (@realDonaldTrump), "The Swing States did not even come close to following the dictates of their State Legislatures. These States 'election laws' were made up by local judges & politicians, not by their Legislatures, & are therefore, before even getting to irregularities & fraud, UNCONSTITUTIONAL!" Twitter, Jan. 3, 2021, https://twitter.com/realDonaldTrump/status/1345798202650460162.

57 Donald Trump (@realDonaldTrump), "Sorry, but the number of votes in the Swing States that we are talking about is VERY LARGE and totally OUTCOME DETER- MINATIVE! Only the Democrats and some RINO'S would dare dispute this—even though they know it is true!" Twitter, Jan. 3, 2021, https://twitter.com/realDonald Trump/status/1345803569438597121.

58 Dan Scavino, photo only, Facebook, Jan. 1, 2021, https://www.facebook.com/DanS cavino/posts/pfbid02uugf564QgatgSZbVbLUm7Ux2LewWsYHCGnig32TG2vw 5CY3T7G2QSkGcABJz2mJTl.

59 Ciara O'Rourke, "'1776 Flag Flying over White House!' in January 2021. 'Revolution Signal,'" Politifact, Jan. 5, 2021, https://www.politifact.com/factchecks/2021/jan /05/blog-posting/photo-flag-over-white-house-2019/.

60 US v. Tarrio, Evidence, Feb. 9, 2023.

61 Rosen, interview by Select Committee.

62 Rosen, interview by Committee on the Judiciary.

63 Rosen, interview by Committee on the Judiciary.

64 Donoghue, interview by Committee on the Judiciary.

65 Donoghue, interview by Select Committee.

66 Donoghue, interview by Committee on the Judiciary.

67 Amy Gardner, "'I Just Want to Find 11,780 Votes': In Extraordinary Hour-Long Call, Trump Pressures Georgia Secretary of State to Recalculate the Vote in His Favor," *Washington Post*, Jan. 3, 2021, https://www.washingtonpost.com/politics/trump -raffensperger-call-georgia-vote/2021/01/03/d45acb92-4dc4-11eb-bda4-615aae fd0555_story.html.

68 "Here's the Full Transcript and Audio of the Call Between Trump and Raffens- perger," *Washington Post*, Jan. 5, 2021, https://www.washingtonpost.com/politics /trump-raffensperger-call-transcript-georgia-vote/2021/01/03/2768e0cc-4ddd -11eb-83e3-322644d82356_story.html.

69 Brad Hamilton, "Roger Stone Attends Naples Rally Demanding Florida Senators to Object Electoral Votes," NBC 2, Jan. 4, 2021, https://nbc-2.com/news/poli tics/2021/01/03/roger-stone-attends-naples-rally-demanding-florida-senators-to -object-electoral-votes/.

70 Limei (@xlmsnow), "Stop the Steal rally in Naples Florida today. We the people know President Trump won 2020 election landslide. We require the senators stand for our constitution and our justices. Roger Stone speaks the truth & facts," Twitter, Jan. 3, 2021, https://twitter.com/xlmsnow/status/1345875010225909760?lang=en.

71 US v. Scott, Statement of Offense, Feb. 10, 2023, https://storage.courtlistener.com /recap/gov.uscourts.dcd.243944/gov.uscourts.dcd.243944.209.0.pdf.

72 FBI, FOIA, "United States Capitol Violence and related Events of January 6, 2021.

73  Justice Department, FOIA, Jan. 26, 2023, https://www.justice.gov/oip/foia-library /foia-processed/general_topics/white_house_meetings_01_26_23/download.

74  Garrett M. Graff, *Watergate: A New History* (New York: Avid Reader, 2022).

75  Ryan J. Reilly (@ryanjreilly), "This portrait of Elliot Richardson, who stood up to Nixon, was in AG's conference room. Sessions had it replaced," Twitter, March 2, 2017, https://twitter.com/ryanjreilly/status/837425116367765517.

76  Justice Department, FOIA.

77  FOIA, US Secret Service, Jan. 6 documents, https://www.citizensforethics.org/wp -content/uploads/2022/08/July-25-2022_USSS.pdf.

78  US v. Brock, Government Sentencing Memo, March 10, 2023, https://storage .courtlistener.com/recap/gov.uscourts.dcd.227748/gov.uscourts.dcd.227748.88.0.pdf.

79  US v. Brock, Character Letters, March 11, 2023, https://storage.courtlistener.com /recap/gov.uscourts.dcd.227748/gov.uscourts.dcd.227748.90.1.pdf.

80  Donald Trump (@realDonaldTrump), "The 'Surrender Caucus' within the Republican Party will go down in infamy as weak and ineffective 'guardians' of our Nation, who were willing to accept the certification of fraudulent presidential numbers!" Twitter, Jan. 4, 2021, https://twitter.com/realDonaldTrump/status/13461206456 13150208.

81  Rosen, interview by Select Committee.

82  Document reviewed by author, 2022.

83  FOIA, US Secret Service, Jan. 6 documents, https://www.citizensforethics.org/wp -content/uploads/2022/08/July-25-2022_USSS.pdf.

84  Andrew Beaujon, "5 Reasons to Avoid the January 6 Pro-Trump Marches in DC," *Washingtonian*, Jan. 4, 2021, https://www.washingtonian.com/2021/01/04/5-rea sons-to-avoid-the-january-6-pro-trump-marches-in-dc/.

85  David Bowdich, interview by Select Committee to Investigate the Jan. 6th Attack on the US Capitol, Dec. 16, 2021, https://www.govinfo.gov/content/pkg/GPO-J6 -TRANSCRIPT-CTRL0000034627/pdf/GPO-J6-TRANSCRIPT-CTRL 0000034627.pdf.

86  Moore, interview.

87  Micah Lee, "How Northern California's Police Intelligence Center Tracked Protests," *Intercept*, Aug. 17, 2020, https://theintercept.com/2020/08/17/blueleaks-cali fornia-ncric-black-lives-matter-protesters/.

88  Nathan Bernard, "Maine Spy Agency Spread Far-Right Rumors of BLM Protest Violence," *Mainer*, July 7, 2020, https://mainernews.com/maine-spy-agency-spread -far-right-rumors-of-blm-protest-violence/.

89  Sean Carlin, "Bricks Were Placed for Construction, Not to Incite Protesters," Fact Check, June 5, 2020, https://www.factcheck.org/2020/06/bricks-were-placed-for -construction-not-to-incite-protesters/.

90  Jessica Lee, "Were Pallets of Bricks Strategically Placed at US Protest Sites?" Snopes, June 4, 2020, https://www.snopes.com/fact-check/pallets-of-bricks-protest -sites/.

91  Beatrice Dupuy and Arijeta Lajka, "Bricks Become Fodder for False Claims Around Protests," AP News, June 5, 2020, https://apnews.com/article/death-of-george-floyd -george-floyd-2217e8c8af5dcdc1d47622822294ce2e.

92  Ken Dilanian and Julia Ainsley, "Worried About Free Speech, FBI Never Issued Intelligence Bulletin About Possible Capitol Violence," NBC News Jan. 12, 2021, https://www.nbcnews.com/news/us-news/part-due-free-speech-worries-fbi-never -issued-intel-bulletin-n1253951.

93   FBI, FOIA, "United States Capitol Violence and Related Events of January 6, 2021, Part 23," May 5, 2023, https://vault.fbi.gov/united-states-capitol-violence-and -related-events-of-january-6-2021/united-states-capitol-violence-and-related -events-of-january-6-2021-part-23/view.

94   FBI, "Leadership Changes at the Washington Field Office Announced," Aug. 30, 2021, https://www.fbi.gov/news/press-releases/leadership-changes-at-the-washing ton-field-office-announced.

95   Donald Trump (@realDonaldTrump), "See you in D.C.," Twitter, Jan. 5, 2021, https://twitter.com/realDonaldTrump/status/1346478482105069568.

96   Donald Trump (@realDonaldTrump), "The Vice President has the power to reject fraudulently chosen electors," Twitter, Jan. 5, 2021, https://twitter.com/realDon aldTrump/status/1346488314157797389.

97   Document released by Select Committee to Investigate the Jan. 6th Attack on the US Capitol, 2022, https://www.govinfo.gov/content/pkg/GPO-J6-DOC-CTRL00 00930224/pdf/GPO-J6-DOC-CTRL0000930224.pdf.

98   Christopher Mathias and Ryan J. Reilly, "He Attacked Cops at the Capitol. The FBI Interviewed Him. Then He Rejoined the Army," *HuffPost*, Oct. 14, 2021, https:// www.huffpost.com/entry/james-mault-capitol-attack-trump-army-military_n_61 68753de4b065a54971c1d6?hms.

99   US v. Mault, Defendant's Memorandum in Aid of Sentencing, July 4, 2022, https:// storage.courtlistener.com/recap/gov.uscourts.dcd.237239/gov.uscourts.dcd .237239.63.0.pdf.

100  US v. Mault, Government Sentencing Memo, June 30, 2022, https://storage.court listener.com/recap/gov.uscourts.dcd.237238/gov.uscourts.dcd.237238.60.0.pdf.

101  Justice Department, FOIA, "Chat with [Redacted]," Jan. 5, 2021, https://www.jus tice.gov/oip/foia-library/foia-processed/general_topics/comms_doj_congressio nal_state_09_23_22/download.

102  FBI, FOIA, "United States Capitol Violence and Related Events of January 6, 2021," Part 20, Feb. 7, 2023.

103  FBI Norfolk Division Situational Information Report, "Potential for Violence in Washington, D.C. Area in Connection with Planned 'StopTheSteal' Protest on 6 January 2021," Jan. 5, 2021 (released by Jan. 6 committee in 2022), https://www .govinfo.gov/content/pkg/GPO-J6-DOC-CTRL0000001532.0001/pdf/GPO-J6 -DOC-CTRL0000001532.0001.pdf.

104  Department of Homeland Security Inspector General, "I&A Identified Threats Prior to January 6, 2021, but Did Not Issue Any Intelligence Products Before the U.S. Capitol Breach," March 4, 2022, https://www.oig.dhs.gov/sites/default/files /assets/2022-04/OIG-22-29-Mar22-Redacted.pdf.

105  Will Carless, "Proud Boys Leader Arrested on Charges Related to Burning of Black Lives Matter Banner, Police Say," *USA Today*, Jan. 4, 2021, https://www.usatoday .com/story/news/nation/2021/01/04/proud-boys-leader-arrested-ahead-dc -protests-support-trump/4135703001/.

106  Christopher Mathias and Ryan J. Reilly, "Judge Orders Proud Boys Leader to Stay Out of Washington Ahead of Trump Showdown," *HuffPost*, Jan. 5, 2021, https:// www.huffpost.com/entry/proud-boys-enrique-tarrio-case_n_5ff4c9cdc5b65a92291 234b3.

107  Donald Trump (@realDonaldTrump), "Washington is being inundated with peo- ple who don't want to see an election victory stolen by emboldened Radical Left Democrats. Our Country has had enough, they won't take it anymore! We hear you

(and love you) from the Oval Office. MAKE AMERICA GREAT AGAIN!" Twitter, Jan. 5, 2021, https://twitter.com/realDonaldTrump/status/1346578706437963777.

108  Donald Trump (@realDonaldTrump), "I hope the Democrats, and even more importantly, the weak and ineffective RINO section of the Republican Party, are looking at the thousands of people pouring into D.C. They won't stand for a land-slide election victory to be stolen. @senatemajldr @JohnCornyn @SenJohn Thune," Twitter, Jan. 5, 2021, https://twitter.com/realDonaldTrump/status/134 6580318745206785.

109  Donald Trump (@realDonaldTrump), "Antifa is a Terrorist Organization, stay out of Washington. Law enforcement is watching you very closely! @DeptofDefense @TheJusticeDept @DHSgov @DHS_Wolf @SecBernhardt @SecretService @ FBI," Twitter, Jan. 5, 2021, https://twitter.com/realDonaldTrump/status/1346 583537256976385.

110  Donald Trump (@realDonaldTrump), "I will be speaking at the SAVE AMERICA RALLY tomorrow on the Ellipse at 11AM Eastern. Arrive early—doors open at 7AM Eastern. BIG CROWDS!" Twitter, Jan. 5, 2021, https://twitter.com/realDon aldTrump/status/1346588064026685443.

111  Elisha Fieldstadt, Ken Dilanian, Tim Stelloh, and Ryan J. Reilly, "Armed Man Who Was at Capitol on Jan. 6 Is Fatally Shot After Firing into an FBI Field Office in Cincinnati," NBC News, Aug. 11, 2022, https://www.nbcnews.com/news/us -news/armed-man-shoots-fbi-cincinnati-building-nail-gun-flees-leading-inters -rcna42669.

112  Aaron Mak, "On Eve of Congress' Certification, Pro-Trump Protesters Fought with D.C. Police—and One Another," *Slate*, Jan. 6, 2021, https://slate.com/news-and -politics/2021/01/proud-boys-and-pro-trump-protesters-clash-in-d-c-on-tuesday .html.

113  Mary Papenfuss, "Proud Boys' Hotel Hangout Shutting Down on Jan. 6 'Wild Pro-test' Day," *HuffPost*, Dec. 29, 2020, https://www.huffpost.com/entry/hotel-harring ton-closed-proud-boys-january-6_n_5feaac94c5b66809cb33a510.

114  Delia Goncalves, "'Your Silence Is Violence': Black Lives Matter DC Calling on Nearby Hotels to Shut Down for Upcoming Protests," WUSA9, Dec. 30, 2020, https://www.wusa9.com/article/news/community/equality-matters/black-lives -matter-dc-demands-hotels-shut-down-and-condem-pro-trump-protestors/65 -5a5a4747-8f4e-4648-980c-febccd0167be.

115  Will Sommer (@willsommer), "DC's Hotel Harrington, a popular spot for Proud Boys, says it'll be closed on Jan 6—the same day as a last-ditch MAGA rally to stop Congress from counting electoral votes. Angry Trump supporters who lost room reservations are fuming on The Donald," Twitter, Dec. 28, 2020, https://twitter .com/willsommer/status/1343752428609556483.

116  Antichud (@antichud), "Florida Threeper Jeremy Liggett spoke at Freedom Plaza in DC today and mentioned how his Facebook was taken down. Here is the new one: https://facebook.com/jeremy.shae.71 At least one of his crew is visibly armed. So much for DC no gun rule. @wehearttrash @RWParlerWatch @AntiFashGordon," Twitter, Jan. 5, 2021, https://twitter.com/antichud/status/1346572090434981888 /photo/1.

117  US Secret Service, "(U//FOUO) Disruptions to DC Metro Area 01/06/2021 (Online Tip)," Dec. 27, 2020, https://www.govinfo.gov/content/pkg/GPO-J6-DOC-CTRL 0000236995/pdf/GPO-J6-DOC-CTRL0000236995.pdf.

118   Donoghue, interview by Select Committee.

119   Today Show (@TODAYshow), "'It's important that every last person who entered that Capitol be found and charged.' Watch @SavannahGuthrie's full interview with Former FBI Director James @Comey on the U.S. Capitol siege and new inauguration threats," Twitter, Jan. 12, 2021, https://twitter.com/TODAYshow/status /1348972851404156928.

120   Moore, interview.

121   Michael Sherwin, interview by Select Committee to Investigate the Jan. 6th Attack on the US Capitol, April 19, 2022, https://www.govinfo.gov/content/pkg/GPO -J6-TRANSCRIPT-CTRL0000071085/pdf/GPO-J6-TRANSCRIPT-CTRL 0000071085.pdf.

122   Hunton and Williams, *Independent Review of the 2017 Protest Events in Charlottesville, Virginia*, Nov. 24, 2017, https://www.policinginstitute.org/wp-content/uploads /2017/12/Charlottesville-Critical-Incident-Review-2017.pdf.

123   Ken Dilanian and Ryan J. Reilly, "Top Jan. 6 Investigator Says FBI, Other Agencies Could Have Done More to Repel Capitol Mob Had They Acted on Intel," NBC News, Jan. 31, 2023, https://www.nbcnews.com/politics/congress/fbi-stopped-jan -6-capitol-mob-acted-intelligence-rcna68155.

124   Dilanian and Reilly, "Top Jan. 6 Investigator."

125   US v. Badalian, trial exhibits, 2023.

# Chapter 6

1   Bulletin Intelligence (website), Cision Insights Solution, accessed May 19, 2023, https://www.bulletinintelligence.com/.

2   Craig Timberg and Drew Harwell, "Pro-Trump Forums Erupt with Violent Threats Ahead of Wednesday's Rally Against the 2020 Election," *Washington Post*, Jan. 5, 2021, https://www.washingtonpost.com/technology/2021/01/05/parler-telegram -violence-dc-protests/.

3   Donald Trump (@realDonaldTrump), "Get smart Republicans. FIGHT!" Twitter, Jan. 6, 2021, https://twitter.com/realDonaldTrump/status/1346693906990305280.

4   Donald Trump (@realDonaldTrump), "If Vice President @Mike_Pence comes through for us, we will win the Presidency. Many States want to decertify the mistake they made in certifying incorrect & even fraudulent numbers in a process NOT approved by their State Legislatures (which it must be). Mike can send it back!" Twitter, Jan. 6, 2021, https://twitter.com/realDonaldTrump/status/134669821730 4584192.

5   FBI, "No Average Call: A Look Inside the FBI's National Threat Operations Center," Nov. 7, 2019, https://www.fbi.gov/news/stories/inside-the-national-threat-opera tions-center-110719.

6   Jan. 6 committee documents, https://www.govinfo.gov/content/pkg/GPO-J6-DOC -CTRL0000930224/pdf/GPO-J6-DOC-CTRL0000930224.pdf.

7   Michael Sherwin, Jan. 6 investigation press call, Jan. 7, 2021.

8   Christopher Mathias and Ryan J. Reilly, "Judge Orders Proud Boys Leader to Stay Out of Washington Ahead of Trump Showdown," *HuffPost*, Jan. 5, 2021, https:// www.huffpost.com/entry/proud-boys-enrique-tarrio-case_n_5ff4c9cdc5b65a92291 234b3.

9    Nick Quested, interview by Select Committee to Investigate the Jan. 6th Attack on the US Capitol, April 5, 2022, https://www.govinfo.gov/content/pkg/GPO-J6 -TRANSCRIPT-CTRL0000061473/pdf/GPO-J6-TRANSCRIPT-CTRL 0000061473.pdf.

10   Nick Quested, interview.

11   Ryan J. Reilly, "Filmmaker Who Followed Proud Boys During Capitol Siege Wants Jan. 6 Committee to 'Find the Truth,'" NBC News, June 10, 2022, https://www.nbc news.com/politics/congress/filmmaker-followed-proud-boys-capitol-siege-wants -jan-6-committee-find-rcna32641.

12   "Video Investigation: Proud Boys Were Key Instigators in Capitol Riot," *Wall Street Journal*, Jan. 26, 2021, https://www.wsj.com/video/series/in-depth-features/video -investigation-proud-boys-were-key-instigators-in-capitol-riot/37B883B6-9B19 -400F-8036-15DE4EA8A015.

13   Mike Braun, "Senator Braun Statement on Electoral College Vote," Braun's website, Dec. 14, 2020, https://www.braun.senate.gov/senator-braun-statement-electoral -college-vote.

14   "Joint Statement from Senators Cruz, Johnson, Lankford, Daines, Kennedy, Black-burn, Braun, Senators-Elect Lummis, Marshall, Hagerty, Tuberville," Ted Cruz's website, Jan. 2, 2021, https://www.cruz.senate.gov/newsroom/press-releases/joint -statement-from-senators-cruz-johnson-lankford-daines-kennedy-blackburn-braun -senators-elect-lummis-marshall-hagerty-tuberville.

15   Facebook video, Jan. 6, 2021.

16   Mike Braun (@SenatorBraun), "Speaking and listening to Hoosier @realDon-aldTrump supporters who came to DC from Indiana about why I will object today and support an emergency audit into irregularities in the 2020 election," Twitter, Jan. 6, 2021, https://twitter.com/SenatorBraun/status/1346869233314009094.

17   Justice Department, "Tenney III, George Amos," Capitol Breach Cases, sentenced Dec. 5, 2022, https://www.justice.gov/usao-dc/defendants/tenney-iii-george-amos.

18   US v. Dillard, Statement of Facts, Aug. 19, 2022, https://www.justice.gov/usao-dc /press-release/file/1528226/download.

19   Dorothy Gilliam, "The U.S. Attorney and the Barry Case," *Washington Post*, April 2, 1990, https://www.washingtonpost.com/archive/local/1990/04/02/the-us-attorney -and-the-barry-case/ec4e38fd-dc1f-4342-9465-051449c91658/.

20   Nat Hentoff, "Of Course It Was Entrapment," *Washington Post*, Jan. 24, 1990, https:// www.washingtonpost.com/archive/opinions/1990/01/24/of-course-it-was-entrap ment/9691a310-74e7-42cf-a362-9179823c87e1/.

21   Christina Cauterucci, "Marion Barry, the Butt of a Thousand Televised Crack Jokes," *Washington City Paper*, Nov. 24, 2014, https://washingtoncitypaper.com /article/404239/marion-barry-the-butt-of-a-thousand-televised-crack-jokes/.

22   Maureen Dowd, "Liberties; Dressing Down Rudy," *New York Times*, Nov. 29, 1997, https://www.nytimes.com/1997/11/29/opinion/liberties-dressing-down-rudy.html.

23   Miriam Elder, "At Least the Married Lawyers Who Helped Bring Us the Impeach-ment Scandal Are Having Fun," BuzzFeed News, Dec. 11, 2019, https://www .buzzfeednews.com/article/miriamelder/victoria-toensing-joe-digenova-ukraine -giuliani.

24   Donald Trump (@realDonaldTrump), "I look forward to Mayor Giuliani spear-heading the legal effort to defend OUR RIGHT to FREE and FAIR ELECTIONS! Rudy Giuliani, Joseph diGenova, Victoria Toensing, Sidney Powell, and Jenna Ellis,

a truly great team, added to our other wonderful lawyers and representatives!" Twitter, Nov. 14, 2020, https://twitter.com/realDonaldTrump/status/1327811527123 103746.

25  David E. Sanger and Nicole Perlroth, "Trump Fires Christopher Krebs, Official Who Disputed Election Fraud Claims," *New York Times*, Nov. 17, 2020, https://www.nytimes.com/2020/11/17/us/politics/trump-fires-christopher-krebs.html.

26  Elahe Izadi, "Joseph diGenova Resigns from Gridiron Club After Saying Fired Cybersecurity Official Should Be Shot," *Washington Post*, Dec. 2, 2020, https://www.washingtonpost.com/media/2020/12/02/joseph-digenova-gridiron-club/.

27  Justice Department, "Attorney General William P. Barr on the Nomination of Justin E. Herdman to Serve as U.S. Attorney for the District of Columbia and the Designation of Timothy J. Shea to Serve as Acting Administrator for the DEA," press release, May 18, 2020, https://www.justice.gov/opa/pr/attorney-general-william-p-barr-nomination-justin-e-herdman-serve-us-attorney-district.

28  Katie Benner and Adam Goldman, "D.C. Prosecutors' Tensions with Justice Dept. Began Long Before Stone Sentencing," *New York Times*, April 29, 2020, https://www.nytimes.com/2020/02/23/us/politics/justice-department-dc-prosecutors.html.

29  Ryan J. Reilly and Sebastian Murdock, "Roger Stone Guilty on Charges of Lying to Congress, Witness Intimidation," *HuffPost*, Nov. 15, 2019, https://www.huffpost.com/entry/roger-stone-guilty-trump_n_5dcc42f8e4b03a7e0293cda8.

30  Ryan J. Reilly, "DOJ Whistleblowers Testify That William Barr Has Politicized Justice," *HuffPost*, June 24, 2020, https://www.huffpost.com/entry/william-barr-justice-department-politicization_n_5ef37e56c5b66c312680f393.

31  Justice Department, Executive Grant of Clemency, July 10, 2020, https://www.justice.gov/pardon/page/file/1293796/download.

32  Fenit Nirappil, Julie Zauzmer Weil, and Rachel Chason, "'Black Lives Matter': In Giant Yellow Letters, D.C. Mayor Sends Message to Trump," *Washington Post*, June 5, 2020, https://www.washingtonpost.com/local/dc-politics/bowser-black-lives-matter-street/2020/06/05/eb44ff4a-a733-11ea-bb20-ebf0921f3bbd_story.html.

33  Jason Zengerle, "Donald Trump Jr. Is Ready. But for What, Exactly?," *New York Times Magazine*, Aug. 24, 2020, https://www.nytimes.com/2020/08/24/magazine/donald-trump-jr.html.

34  Colby Itkowitz and Josh Dawsey, "Pence Under Pressure as the Final Step Nears in Formalizing Biden's Win," *Washington Post*, Dec. 24, 2020, https://www.washingtonpost.com/politics/pence-biden-congress-electoral/2020/12/24/48f48da8-4604-11eb-a277-49a6d1f9dff1_story.html.

35  Claudia Grisales, "In Georgia, Pence Walks Line Between Trump's Election Falsehoods and GOP Senate Fight," NPR, Nov. 21, 2020, https://www.npr.org/2020/11/21/937167707/in-georgia-pence-walks-line-between-trumps-election-falsehoods-and-gop-senate-fi.

36  Mike Pence, *So Help Me God* (New York: Simon & Schuster, 2022).

37  Donald Trump (@realDonaldTrump), "States want to correct their votes, which they now know were based on irregularities and fraud, plus corrupt process never received legislative approval. All Mike Pence has to do is send them back to the States, AND WE WIN. Do it Mike, this is a time for extreme courage!" Twitter, Jan. 6, 2021, https://twitter.com/realDonaldTrump/status/1346808075626426371.

38  Donald Trump (@realDonaldTrump), "THE REPUBLICAN PARTY AND, MORE IMPORTANTLY, OUR COUNTRY, NEEDS THE PRESIDENCY MORE THAN

EVER BEFORE—THE POWER OF THE VETO. STAY STRONG!" Twitter, Jan. 6, 2021, https://twitter.com/realDonaldTrump/status/1346809349214248962.

39 Donald Trump (@realDonaldTrump), "The States want to redo their votes. They found out they voted on a FRAUD. Legislatures never approved. Let them do it. BE STRONG!" Twitter, Jan. 6, 2021, https://twitter.com/realdonaldtrump/status/1346 822610957561858.

40 Maggie Haberman, "In New Book, Pence Reflects on Trump and Jan. 6," *New York Times*, Nov. 9, 2022, https://www.nytimes.com/2022/11/09/us/politics/pence-trump -jan-6.html.

41 Radley Balko, "More on the Time Rudy Giuliani Helped Incite a Riot of Racist Cops," *Washington Post*, Nov. 16, 2016, https://www.washingtonpost.com/news/the -watch/wp/2016/11/16/more-on-the-time-rudy-giuliani-helped-incite-a-riot-of -racist-cops/?outputType=amp.

42 Laura Nahmias, "White Riot in 1992, Thousands of Furious, Drunken Cops Descended on City Hall—and Changed New York history," *New York*, Oct. 4, 2021, https://nymag.com/intelligencer/2021/10/the-forgotten-city-hall-riot.html.

43 Nat Hentoff, "Rudy's Racist Rants: An NYPD History Lesson," Cato Institute, July 14, 2016, https://www.cato.org/commentary/rudys-racist-rants-nypd-history-lesson.

44 Ryan J. Reilly, "Jan. 6 Defendant Who Stole Liquor, Coat Rack Says He Was 'Follow- ing Presidential Orders,'" NBC News, April 13, 2022, https://www.nbcnews.com /politics/donald-trump/jan-6-defendant-stole-liquor-coat-rack-says-was-presiden tial-orders-rcna24311.

45 Ryan J. Reilly, "Donald Trump Endorses Police Brutality in Speech to Cops," *Huff- Post*, July 28, 2017, https://www.huffpost.com/entry/trump-police-brutality_n_597 b840fe4b02a8434b6575a.

46 Donald Trump, "Read Trump's Jan. 6 Speech, a Key Part of Impeachment Trial," NPR, Feb. 10, 2021, https://www.npr.org/2021/02/10/966396848/read-trumps -jan-6-speech-a-key-part-of-impeachment-trial.

47 Ryan J. Reilly (@ryanjreilly), "A sampling of Jan. 6 defendant Ryan Nichols' politi- cal discourse," Twitter, Feb. 4, 2022, https://twitter.com/ryanjreilly/status/148 9747752020066307.

48 US v. Nichols, Video Exhibits, Dec. 20, 2021.

49 US v. Samsel, Government's Memorandum in Opposition to Defendant's Motion to Revoke Detention Order, June 2, 2021, https://www.documentcloud.org/documents /20795652-samsel.

50 US v. Genco, Government's Sentencing Memorandum, Sept. 12, 2022, https://stor age.courtlistener.com/recap/gov.uscourts.dcd.240613/gov.uscourts.dcd.240613 .69.0.pdf.

51 FBI, "Steven M. D'Antuono Named Assistant Director in Charge of the Washington Field Office," Oct. 13, 2020, https://www.fbi.gov/news/press-releases/steven-m -dantuono-named-assistant-director-in-charge-of-the-washington-field-office.

52 Michael Sherwin, interview by Select Committee to Investigate the Jan. 6th Attack on the US Capitol, April 19, 2022, https://www.govinfo.gov/content/pkg/GPO -J6-TRANSCRIPT-CTRL0000071085/pdf/GPO-J6-TRANSCRIPT-CTRL 0000071085.pdf.

53 Richard Peter Donoghue, interview by Select Committee to Investigate the Jan. 6th Attack on the US Capitol, Oct. 1, 2021, https://www.govinfo.gov/content/pkg/GPO

-J6-TRANSCRIPT-CTRL0000034600/pdf/GPO-J6-TRANSCRIPT-CTRL
0000034600.pdf.

54  US v. Rhodes et al., Trial Testimony, Oct. 2022.

55  Goodwin v. District of Columbia, Memorandum Opinion, Jan. 13, 2022, https://
casetext.com/case/goodwin-v-dist-of-columbia-1.

56  Ryan J. Reilly, "D.C. Residents Open Their Doors to Save Protesters from the
Police," *HuffPost*, June 2, 2020, https://www.huffpost.com/entry/dc-protest-policing
_n_5ed5c1a3c5b651b2b317508e.

57  News2Share, "Inspector Robert Glover Versus Protesters Outside DC Mayor Bowser's House," YouTube, Sept. 3, 2020, https://www.youtube.com/watch?v=wJTz79
pyOa8.

58  Justice Department, "Man Charged in Federal Court for Attempting to Tear Down
Statue of Andrew Jackson in Lafayette Square amid Protests," July 2, 2020, https://
www.justice.gov/usao-dc/pr/man-charged-federal-court-attempting-tear-down
-statue-andrew-jackson-lafayette-square.

59  Ryan J. Reilly (@ryanjreilly), "The National Park Service prepares to haul away
confederate officer Albert Pike," Twitter, Jan. 20, 2020, https://twitter.com/ryan
jreilly/status/1274347280477413376.

60  Council of DC (@councilofdc), "Ever since 1992, members of the DC Council have
been calling on the federal gov't to remove the statue of Confederate Albert Pike (a
federal memorial on federal land). We unanimously renewed our call to Congress to
remove it in 2017," Twitter, June 19, 2020, https://twitter.com/councilofdc/status
/1274187263736532998.

61  Robert Glover, interview by Select Committee to Investigate the Jan. 6th Attack on
the US Capitol, May 2, 2022, https://www.govinfo.gov/content/pkg/GPO-J6
-TRANSCRIPT-CTRL0000082305/pdf/GPO-J6-TRANSCRIPT-CTRL
0000082305.pdf.

62  National Park Service, FOIA, emails, https://www.nps.gov/aboutus/foia/upload
/Jan6_emails_JohnStanwich_REDACTED_Part1.pdf.

63  Mike Spies and Jake Pearson, "Text Messages Show Top Trump Campaign Fundraiser's Key Role Planning the Rally That Preceded the Siege," ProPublica, Jan. 30,
2021, https://www.propublica.org/article/trump-campaign-fundraiser-ellipse-rally.

64  MPD Cadet Corps, Government of the District of Columbia (website), accessed
February 15, 2023, https://joinmpd.dc.gov/metropolitan-police/cadet.

65  Robin Stein, Haley Willis, Danielle Miller, and Michael S. Schmidt, "'We've Lost
the Line!': Radio Traffic Reveals Police Under Siege at Capitol," *New York Times*,
March 21, 2021, https://www.nytimes.com/video/us/100000007655234/weve-lost
-the-line-radio-traffic-reveals-police-under-siege-at-capitol.html.

66  US v. Fitzsimons, Testimony, Aug. 17, 2022.

67  Eric Cortellessa, "Lawmaker Helping Lead Jan. 6 Hearing First Fled Vietnam,
Then Capitol Rioters," *Time*, July 11, 2022, https://time.com/6195712/stephanie
-murphy-jan-6-committee-interview/.

68  Sarah Ferris, "Diamond-Studded Thorns: 2 House Dem Centrists Speak Up on
Their Way Out," Politico, Dec. 26, 2022, https://www.politico.com/news/2022/12
/26/stephanie-murphy-kathleen-rice-centrists-00075509.

69  Alex Gangitano, "Members Bond over 'Miserable' Workout," *Roll Call*, March 28,
2018, https://rollcall.com/2018/03/28/members-bond-over-miserable-workout/.

70  Mariel Padilla, "'We Had to Run Faster': Rep. Kathleen Rice of New York Recalls the Capitol Riot," *19th*, March 16, 2021, https://19thnews.org/2021/03/kathleen -rice-capitol-riot/.

71  Marjorie Hunter, "Coveted on Capitol Hill: The Hideaway," *New York Times*, March 7, 1983, https://www.nytimes.com/1983/03/07/us/coveted-on-capitol-hill-the-hide away.html.

72  "Inside a Senate Hideaway," Politico, Jan. 5, 2015, https://www.politico.com/maga zine/gallery/2015/01/inside-a-senate-hideaway-000120/?slide=0.

73  Jordy Yager, "Haunted House—and Senate," *Hill*, March 3, 2009, https://thehill .com/capital-living/20943-haunted-house-and-senate/.

74  Robert Parker, *Capitol Hill in Black and White* (New York: Dodd, Mead & Co., 1986).

75  Padilla, "'We Had to Run Faster.'"

76  Pete Reinwald, "U.S. Rep. Murphy Reveals She Was in 'Center of the Storm' Jan. 6," Spectrum News, July 27, 2021, https://www.mynews13.com/fl/orlando/news/2021 /07/27/u-s--rep--murphy-reveals-she-was-in--center-of-the-storm--jan--6.

77  Brendan Gutenschwager (@BGOnTheScene@BGOnTheScene), "Alex Jones, Ali Alexander and Nick Fuentes are preparing to 'Storm the Capitol' here in Atlanta #Atlanta #Georgia #StopTheSteal," Twitter, Nov. 18, 2020, https://twitter.com /BGOnTheScene/status/1329116419037024273.

78  Brendan Gutenschwager (@BGOnTheScene), "Alex Jones is leading the crowd of Trump supporters into the Georgia State Capitol now #Atlanta #Georgia #Elec tions2020," Twitter, Nov. 18, 2020, https://twitter.com/BGOnTheScene/status /1329124686941335553.

79  US v. Beckley, Statement of Facts for a Stipulated Trial, Feb. 23, 2023.

80  Matt Fuller, "The Real Tragedy of Jan. 6 Is That It's Still Not Over," Daily Beast, Jan. 5, 2022, https://www.thedailybeast.com/the-real-tragedy-of-jan-6-is-that-its -still-not-over.

81  Richie McGinniss (@RichieMcGinniss), "THREAD: I accompanied protesters all the way to the Senate Gallery doors," Twitter, Jan. 6, 2021, https://twitter.com/rich iemcginniss/status/1346948567911587840.

82  Ryan J. Reilly (@ryanjreilly), "'Dude, I don't fucking know . . . You want to talk about getting caught with you pants down?'—Capitol Police officer on Jan. 6. 'Wow, if only we had some warning that this was gonna happen!' said another officer (I think MPD), dripping with sarcasm," Twitter, Feb. 18, 2023, https://twitter.com /ryanjreilly/status/1627129975114547201 (ellipsis in the original).

83  Ayman M. Mohyeldin and Preeti Varathan, "Rosanne Boyland Was Outside the U.S. Capitol Last January 6. How—and Why—Did She Die?," *Vanity Fair*, Jan. 5, 2022, https://www.vanityfair.com/news/2022/01/capitol-insurrection-rosanne-boyland -how-and-why-did-she-die.

84  Marc Lester, "A Day in the Life of a United States Senator: Lisa Murkowski," *Anchorage Daily News*, Aug. 18, 2019, https://www.adn.com/politics/2019/08/18/a -day-in-the-life-of-a-united-states-senator-lisa-murkowski/.

85  Liz Ruskin, "How Murkowski Escaped Mob Violence at the Capitol," *Alaska Public Media*, Jan. 7, 2021, https://alaskapublic.org/2021/01/07/how-murkowski-escaped -mob-violence-at-the-capitol/.

86  Emily Cochrane, Luke Broadwater, and Ellen Barry, "'It's Always Going to Haunt Me': How the Capitol Riot Changed Lives," *New York Times*, Sept. 16, 2021, https:// www.nytimes.com/interactive/2021/09/16/us/politics/capitol-riot.html.

87  US v. Mattice, Evidence, Oct. 25, 2021, https://www.youtube.com/watch?v =VHRHSc0pDAM.

88  Russ Buettner, Susanne Craig, and David Barstow, "11 Takeaways from the *Times*'s Investigation into Trump's Wealth," *New York Times*, Oct. 2, 2018, https://www .nytimes.com/2018/10/02/us/politics/donald-trump-wealth-fred-trump.html.

89  US v. Belger, Statement of Facts, Sept. 16, 2022, https://storage.courtlistener.com /recap/gov.uscourts.dcd.247414/gov.uscourts.dcd.247414.1.1.pdf.

90  US v. Castro, Government's Sentencing Memorandum, Feb. 11, 2022, https:// extremism.gwu.edu/sites/g/files/zaxdzs2191/f/Mariposa%20Castro%20Govern ment%20Sentencing%20Memorandum.pdf.

91  US v. Castro, Defendants' Sentencing Memorandum, Feb. 11, 2022, https://storage .courtlistener.com/recap/gov.uscourts.dcd.230071/gov.uscourts.dcd.230071 .39.0.pdf.

92  US v. Betancur, Government's Sentencing Memorandum, July 28, 2022, https:// storage.courtlistener.com/recap/gov.uscourts.dcd.226838/gov.uscourts.dcd.226838 .39.0.pdf.

93  *Unblocked*, podcast video, Jan. 6, 2021, https://twitter.com/jeremyreporter/status /1349585893401776128.

94  US v. Brockhoff, Plea Agreement, Oct. 27, 2022, https://storage.courtlistener.com /recap/gov.uscourts.dcd.234607/gov.uscourts.dcd.234607.45.0.pdf.

95  US v. Brockhoff, Detention Memo, Feb. 8, 2022, https://storage.courtlistener.com /recap/gov.uscourts.dcd.234607/gov.uscourts.dcd.234607.32.0.pdf.

96  Emily Yehle and Alison McSherry, "Every Senator Due to Get a Capitol Hideaway," *Roll Call*, Jan. 15, 2010, https://rollcall.com/2010/01/15/every-senator-due-to-get -a-capitol-hideaway/.

97  Fox News, "Sen. James Risch: We Can Do a Lot Better Than Donald Trump," You-Tube, March 8, 2016, https://www.youtube.com/watch?v=6JphvLTdkUM.

98  Paul Kane and Scott Clement, "Just 27 Congressional Republicans Acknowledge Biden's Win, *Washington Post* Survey Finds," *Washington Post*, Dec. 5, 2020, https:// www.washingtonpost.com/politics/survey-who-won-election-republicans-congress /2020/12/04/1a1011f6-3650-11eb-8d38-6aea1adb3839_story.html.

99  Donald Trump (@realDonaldTrump), "25, wow! I am surprised there are so many. We have just begun to fight. Please send me a list of the 25 RINOS. I read the Fake News Washington Post as little as possible!" Twitter, Dec. 5, 2020, https://twitter .com/realDonaldTrump/status/1335281344302104577.

100  "Where Republicans in Congress Stand on Trump's False Claim of Winning the Election," *Washington Post*, Dec. 5, 2020, https://www.washingtonpost.com/graphics /2020/politics/congress-republicans-trump-election-claims/.

101  Bill Buley, "Still Fighting for Trump," *Bonner County Daily Bee*, Jan. 6, 2021, https://bonnercountydailybee.com/news/2021/jan/06/still-fighting-president-snp/.

102  Jim Risch (@SenatorRisch), "This nonsense and violence needs to stop now," Twitter, Jan. 6, 2021, https://twitter.com/SenatorRisch/status/1346930384953618438.

103  US v. Brockhoff, Statement of Offense, Oct. 20, 2022, https://www.justice.gov/usao -dc/case-multi-defendant/file/1547601/download.

104  Jordan Green, "North Carolina Extremists Pledge to Escalate Beyond DC Insurrection," *Triad City Beat*, Jan. 9, 2021, https://triad-city-beat.com/nc-extremists -pledge-escalate-beyond-dc-insurrection/.

105  US v. Pope, Body Worn Camera footage, Exhibit 82-9, Feb. 21, 2023.

106 Ryan J. Reilly, "Gun-toting Jan. 6 Defendant Says 'Instinct Took Over' When He Charged Police Line with Wooden Pallet," NBC News, April 17, 2023, https://www.nbcnews.com/politics/justice-department/gun-toting-jan-6-defendant-says-instinct-took-charged-police-line-wood-rcna79755.

107 Ryan J. Reilly, "Jan. 6 Rioter Who Brought Two Guns to Capitol Sentenced to Five Years in Prison," NBC News, Oct. 21, 2022, https://www.nbcnews.com/politics/justice-department/jan-6-rioter-brought-two-guns-capitol-sentenced-five-years-prison-rcna53454.

108 Paul Gosar (@DrPaulGosar), "This has all the hallmarks of Antifa provocation," Twitter, Jan. 6, 2021, https://twitter.com/DrPaulGosar/status/13469408165148 13953.

109 Tate Ryan-Mosley and Abby Ohlheiser, "How an Internet Lie About the Capitol Invasion Turned into an Instant Conspiracy Theory," *MIT Technology Review*, Jan. 7, 2021, https://www.technologyreview.com/2021/01/07/1015858/capitol-invasion-antifa-conspiracy-lie/.

110 Jordan Williams, "*Washington Times* Removes Article Claiming Facial Recognition Company Identified Some Capitol Rioters as Antifa," *Hill*, Jan. 7, 2021, https://thehill.com/homenews/media/533168-washington-times-retracts-article-claiming-facial-recognition-company/.

111 "Here's Every Word from the Sixth Jan. 6 Committee Hearing on Its Investigation," NPR, June 28, 2022, https://www.npr.org/2022/06/28/1108396692/jan-6-committee-hearing-transcript.

112 US House of Representatives, *Final Report of the Select Committee to Investigate the January 6th Attack on the United States Capitol* (Washington, DC: US Government Publishing Office, 2022), https://www.govinfo.gov/content/pkg/GPO-J6-REPORT/pdf/GPO-J6-REPORT.pdf.

113 US v. Rhodes et al., Signal chat, Exhibit 9904, Feb. 23, 2023.

114 Ryan J. Reilly, "The Feds Say They're in for the Long Haul in the Jan. 6 Investigation. There Is a Time Limit," NBC News, Jan. 6, 2023, https://www.nbcnews.com/politics/justice-department/feds-say-long-haul-jan-6-investigation-rcna61728.

115 Mo Brooks (@RepMoBrooks), "Please, don't be like #FakeNewsMedia, don't rush to judgment on assault on Capitol. Wait for investigation. All may not be (and likely is not) what appears. Evidence growing that fascist ANTIFA orchestrated Capitol attack with clever mob control tactics. Evidence follows," Twitter, Jan. 7, 2021, https://twitter.com/RepMoBrooks/status/1347171347043115008.

116 US v. Zoyganeles, Statement of Facts, Feb. 18, 2022, https://www.justice.gov/usao-dc/case-multi-defendant/file/1475906/download.

117 Andrew Marantz, "When Reporting Becomes a Defense for Rioting," *New Yorker*, Feb. 3, 2021, https://www.newyorker.com/news/us-journal/when-reporting-becomes-a-defense-for-rioting.

118 US v. Sullivan, Government's Memorandum in Opposition to Defendant's Motion to Release Seizure Order and Forbid Seizures of Other Accounts, May 21, 2021, https://storage.courtlistener.com/recap/gov.uscourts.dcd.227101/gov.uscourts.dcd.227101.29.0.pdf.

119 US v. Sullivan, Memorandum Opinion, Dec. 6, 2021, https://storage.courtlistener.com/recap/gov.uscourts.dcd.227101/gov.uscourts.dcd.227101.60.0_2.pdf.

120 Josh Gerstein, "Judge Refuses to Ban Capitol Riot Suspect from Twitter and Facebook," Politico, Feb. 16, 2021, https://www.politico.com/news/2021/02/16/judge-ban-riot-twitter-and-facebook-469219.

121  US v. Sullivan, Affidavit, Jan. 13, 2021, https://storage.courtlistener.com/recap/gov .uscourts.dcd.227101/gov.uscourts.dcd.227101.1.1_1.pdf.

122  US v. Sullivan, Affidavit.

123  US v. Sullivan, Government's Memorandum.

124  US v. Speed, "United States' Opposition to Defendant's Motion in Limine to Exclude Regarding Personal Ideology and Ongoing Criminal Proceedings," Jan. 17, 2023, https://storage.courtlistener.com/recap/gov.uscourts.dcd.245477/gov.us courts.dcd.245477.42.0.pdf.

125  "Tennessee Man Arrested for Assaulting Law Enforcement Officer During Jan. 6 Capitol Breach," Justice Department, May 5, 2022, https://www.justice.gov/usao -dc/pr/tennessee-man-arrested-assaulting-law-enforcement-officer-during-jan -6-capitol-breach.

126  US v. Kelley, Criminal Complaint, Dec. 15, 2022, https://storage.courtlistener.com /recap/gov.uscourts.tned.107793/gov.uscourts.tned.107793.3.0.pdf.

127  Ryan J. Reilly, "Tucker Carlson's Jan. 6 'Agent Provocateur' Is a Big Tucker Fan and an Amateur Cardinals Mascot," *HuffPost*, Dec. 9, 2021, https://www.huffpost.com /entry/tucker-carlson-capitol-attack-jan-6-rally-runner_n_61afb97de4b01fcf12b 89bd9.

128  Ryan J. Reilly, "Trump Fans Charged in Capitol Attack Didn't Like Antifa Getting Credit for Their Work," *HuffPost*, Feb. 16, 2021, https://www.huffpost.com/entry /trump-capitol-attack-antifa_n_602c1319c5b65259c4e52ee6.

129  Jason Lange, "Half of U.S. Republicans Believe the Left Led Jan. 6 Violence: Reuters/Ipsos Poll," Reuters, June 9, 2022, https://www.reuters.com/world/us /half-us-republicans-believe-left-led-jan-6-violence-reutersipsos-2022-06-09/.

130  "President Trump Video Statement on Capitol Protesters," C-SPAN, Jan. 6, 2021, https://www.c-span.org/video/?507774-1/president-trump-claims-election-stolen -tells-protesters-leave-capitol.

131  Annie Karni and Maggie Haberman, "Trump Openly Condones Supporters Who Violently Stormed the Capitol, Prompting Twitter to Lock His Account," *New York Times*, Jan. 6, 2021, https://www.nytimes.com/2021/01/06/us/politics/trump-pro testers.html.

132  Former Feds Group Freedom Foundation, "Officer Craig Body Cam," Rumble, Jan. 6, 2021, https://rumble.com/v26z4mg-january-6-2021-officer-craig-body-cam.html.

133  James Hohmann, "On Jan. 6, an Immigrant Capitol Police Officer Showed Trump and Insurrectionists What Patriotism Is," *Washington Post*, July 28, 2021, https:// www.washingtonpost.com/opinions/2021/07/28/aquilino-gonell-capitol-police -insurrection/.

134  Scaffolding Solutions, "The Morning After the US Capitol Riot: Scaffold Inspection," YouTube, Jan. 13, 2021, https://www.youtube.com/watch?v=PCWtYIP72n4.

135  Lisa Murkowski, "Murkowski Votes to Convict President Donald J. Trump," Feb. 14, 2021, Murkowski's website, https://www.murkowski.senate.gov/press/release /murkowski-votes-to-convict-president-donald-j-trump.

136  James E. Risch, "Risch on U.S. Capitol Riots and Vote to Certify the Results of the 2020 Presidential Election," Risch's website, Jan. 6, 2021, https://www.risch.senate .gov/public/index.cfm/pressreleases?ID=890E0FC2-9192-4C6C-90EF-12D14 136A8DA.

137  James E. Risch, "Risch Votes Against Conviction in Senate Impeachment Trial," Risch's website, Feb. 13, 2021, https://www.risch.senate.gov/public/index.cfm /pressreleases?ID=B4229F68-E161-4F5C-9CD7-B470319CB91A.

138 Ryan J. Reilly and Sahil Kapur, "Jan. 6 Rioters Trashed a GOP Senator's Office, and He Hasn't Acknowledged It," NBC News, March 9, 2023, https://www.nbcnews.com/politics/congress/jan-6-rioters-trashed-republican-senator-jim-risch-office-rcna73913.

139 Mike Braun (@SenatorBraun), "Today's events changed things drastically. Though I will continue to push for a thorough investigation into the election irregularities many Hoosiers are concerned with as my objection was intended, I have withdrawn that objection and will vote to get this ugly day behind us," Twitter, Jan. 6, 2021, https://twitter.com/SenatorBraun/status/1347038109070979077.

140 Mike Braun, "Braun Statement on Impeachment Vote," Braun's website, Feb. 13, 2021, https://www.braun.senate.gov/braun-statement-impeachment-vote.

# Chapter 7

1 Liz Brown-Kaiser, Ryan J. Reilly, and Zoë Richards, "Jury Convicts QAnon Believer Who Thought He Was Storming the White House During the Capitol Riot," Sept. 23, 2022, NBC News, https://www.nbcnews.com/politics/justice-department/jury-deliberates-fate-qanon-believer-thought-was-storming-white-house-rcna49205.

2 US v. Jensen, Doug Jensen Interview Transcript, Jan. 8, 2021, https://storage.courtlistener.com/recap/gov.uscourts.dcd.225865/gov.uscourts.dcd.225865.69.1_2.pdf.

3 Mitch McConnell, "McConnell Remarks on the Electoral College Count," Jan. 6, 2021, McConnell's website, https://www.republicanleader.senate.gov/newsroom/remarks/mcconnell-remarks-on-the-electoral-college-count-.

4 Rebecca Tan, "A Black Officer Faced Down a Mostly White Mob at the Capitol. Meet Eugene Goodman," *Washington Post*, Jan. 14, 2021, https://www.washingtonpost.com/local/public-safety/goodman-capitol-police-video/2021/01/13/08ab3eb6-546b-11eb-a931-5b162d0d033d_story.html.

5 US v. Seefried, Government's Sentencing Memorandum, Feb. 2, 2023, https://www.documentcloud.org/documents/23596743-135-usa-sentencing-memo-re-kevin.

6 Igor Bobic (@igorbobic), "Here's the scary moment when protesters initially got into the building from the first floor and made their way outside Senate chamber," Twitter, Jan. 6, 2021, https://twitter.com/igorbobic/status/1346911809274478594.

7 US v. Jensen, Sentencing Memo, Dec. 6, 2022, https://storage.courtlistener.com/recap/gov.uscourts.dcd.225865/gov.uscourts.dcd.225865.107.0.pdf.

8 Ashley Parker, Carol D. Leonnig, Paul Kane, and Emma Brown, "How the Rioters Who Stormed the Capitol Came Dangerously Close to Pence," *Washington Post*, Jan. 15, 2021, https://www.washingtonpost.com/politics/pence-rioters-capitol-attack/2021/01/15/ab62e434-567c-11eb-a08b-f1381ef3d207_story.html.

9 Paul Kane, "House, Senate Strike Deal to Give Congressional Gold Medal to All Officers Who Responded on Jan. 6, Not Just Goodman," *Washington Post*, June 15, 2021, https://www.washingtonpost.com/powerpost/congressional-gold-medal-goodman-capitol-police/2021/06/15/15aa2896-cd53-11eb-a7f1-52b8870bef7c_story.html.

10 Todd Magel, "Iowa Man Seen Inside U.S. Capitol During Wednesday Violence," KCCI, Jan. 7, 2021, https://web.archive.org/web/20210108042002/https://www.kcci.com/article/des-moines-iowa-man-seen-inside-capitol-during-violence/35153695.

11   FBI, "Director Wray's Statement on Violent Activity at the U.S. Capitol Building," Jan. 7, 2021, https://www.fbi.gov/news/press-releases/director-wrays-statement-on-violent-activity-at-the-us-capitol-building-010721.

12   Ted Cruz (@TedCruz), "Devastating. Heidi and I are lifting up in prayer the family of the U.S. Capitol Police officer who tragically lost his life keeping us safe. He was a true hero. Yesterday's terrorist attack was a horrific assault on our democracy. Every terrorist needs to be fully prosecuted," Twitter, Jan. 8, 2021, https://twitter.com/tedcruz/status/1347411930114359296.

13   Charlie Savage, "Was the Jan. 6 Attack on the Capitol an Act of 'Terrorism'?" *New York Times*, Jan. 7, 2022, https://www.nytimes.com/2022/01/07/us/politics/jan-6-terrorism-explainer.html.

14   Jonathan Weisman and Reid J. Epstein, "G.O.P. Declares Jan. 6 Attack 'Legitimate Political Discourse,'" *New York Times*, Feb. 4, 2022, https://www.nytimes.com/2022/02/04/us/politics/republicans-jan-6-cheney-censure.html.

15   Tyler Pager, Josh Gerstein, and Kyle Cheney, "Biden to Tap Merrick Garland for Attorney General," Politico, Jan. 6, 2021, https://www.politico.com/news/2021/01/06/biden-to-tap-merrick-garland-for-attorney-general-455410.

16   Matt Zapotosky and Ann E. Marimow, "How the Oklahoma City Bombing Case Prepared Merrick Garland to Take on Domestic Terrorism," *Washington Post*, Feb. 19, 2021, https://www.washingtonpost.com/national-security/merrick-garland-oklahoma-city-bombing/2021/02/19/a9e6adde-67f2-11eb-8468-21bc48f07fe5_story.html.

17   Philip Kennicott, "The Dystopian Lincoln Memorial Photo Raises a Grim Question: Will They Protect Us, or Will They Shoot Us?" *Washington Post*, June 3, 2020, https://www.washingtonpost.com/lifestyle/style/the-dystopian-lincoln-memorial-photo-raises-a-grim-question-will-they-protect-us-or-will-they-shoot-us/2020/06/03/7a1c52b4-a5b7-11ea-bb20-ebf0921f3bbd_story.html.

18   Adam Edelman, "Biden Slams Capitol Rioters as 'Domestic Terrorists': 'Don't Dare Call Them Protesters,'" NBC News, Jan. 7, 2021, https://www.nbcnews.com/politics/white-house/biden-slams-capitol-rioters-domestic-terrorists-don-t-dare-call-n1253335.

19   US v. Jensen, Government's Opposition to Defendant's Motion to Suppress Evidence, April 8, 2022, https://storage.courtlistener.com/recap/gov.uscourts.dcd.225865/gov.uscourts.dcd.225865.69.0_1.pdf.

20   "*Washington Post* Reporters Momentarily Arrested Outside the Capitol," *Washington Post*, Jan. 6, 2021, https://www.washingtonpost.com/video/politics/washington-post-reporters-momentarily-arrested-outside-the-capitol/2021/01/06/da04479d-cdb7-4870-b354-9333e4712c74_video.html.

21   US Attorney's Office, "Nicholas R. Ochs Arrested for Unlawful Entry into the United States Capitol Building," press release, Jan. 8, 2021, https://www.justice.gov/usao-hi/pr/nicholas-r-ochs-arrested-unlawful-entry-united-states-capitol-building.

22   Chad Hedrick (@WSAZChadHedrick), "#BREAKING WV Delegate Derrick Evans has been taken into federal custody. He's charged after allegedly entering a restricted area of the US Capitol with rioters Wednesday. A woman saying he was his grandmother came out telling us to leave as he was put in a car. #WSAZ," Twitter, Jan. 8, 2021, https://web.archive.org/web/20210108194140/https://twitter.com/WSAZChadHedrick/status/1347629406043443203.

23  Ryan J. Reilly, "'Derrick Evans Is in the Capitol!' He Yelled. Derrick Evans Has Now Been Sentenced for Storming the Capitol," NBC News, June 22, 2022, https://www .nbcnews.com/politics/justice-department/former-west-virginia-lawmaker -derrick-evans-sentenced-storming-capitol-rcna34329.

24  Justice Department, "Three Men Charged in Connection with Events at U.S. Capitol," press release, Jan. 9, 2021, https://www.justice.gov/usao-dc/pr/three-men -charged-connection-events-us-capitol.

25  Ryan J. Reilly, "Capitol Rioter Photographed with Pelosi's Podium on Jan. 6 Sentenced to Prison," NBC News, Feb. 25, 2022, https://www.nbcnews.com/politics /justice-department/capitol-rioter-photographed-pelosis-podium-jan-6-sentenced -prison-rcna17409.

26  US v. Ciarpelli, Statement of Facts, Jan. 12, 2021, https://www.justice.gov/opa/page /file/1353446/download.

27  Jordan Fischer, "Capitol Riot Suspect Suggested Copyrighting 'Bigo Was Here' Note to Speaker Pelosi," WUSA9, Jan. 18, 2023, https://www.wusa9.com/article/news /national/capitol-riots/capitol-riot-suspect-suggested-copyrighting-bigo-was-here -in-jailhouse-call-richard-barnett-pelosi-arkansas-gravette-feet-up-desk/65-a7 b772d6-f81a-49dc-b0e3-d106ed7ef257.

28  Justice Department, "Man Arrested for Illegally Entering Office of Speaker of the House," press release, Jan. 8, 2021, https://www.justice.gov/opa/pr/man-arrested -illegally-entering-office-speaker-house.

29  US v. Barnett, Defendant's Post Verdict Rule 33 Motion for a New Trial, Feb. 5, 2023, https://storage.courtlistener.com/recap/gov.uscourts.dcd.226952/gov.uscourts .dcd.226952.174.0.pdf.

30  Henry Rodgers (@henryrodgersdc), "Alex Jones speaks to Trump supporters outside the Capitol," Twitter, Jan. 6, 2021, https://twitter.com/henryrodgersdc/status /1346899077473198080.

31  US v. Lamond, Indictment, May 18, 2023, https://www.documentcloud.org/docu ments/23816946-lamonds-indictment.

32  US v. Lolos, Criminal Complaint, Jan. 9, 2021, https://www.justice.gov/usao-dc /case-multi-defendant/file/1371436/download.

33  US v. Chan, Criminal Complaint, Sept. 7, 2021, https://www.justice.gov/usao-dc /case-multi-defendant/file/1457606/download.

34  US v. Moncada, Criminal Complaint, Jan. 16, 2021, https://www.justice.gov/opa /page/file/1361471/download.

35  US v. Bauer and Hemenway, Statement of Facts, Jan. 14, 2021, https://www.justice .gov/opa/page/file/1355721/download.

36  US v. Lyons, Statement of Facts, Dec. 16, 2022, https://storage.courtlistener.com /recap/gov.uscourts.dcd.227059/gov.uscourts.dcd.227059.66.1.pdf.

37  Ryan J. Reilly, "Newly Released Video Shows Capitol Rioters Raiding Nancy Pelosi's Office on Jan. 6," NBC News, Nov. 16, 2022, https://www.nbcnews.com/politics/jus tice-department/newly-released-video-shows-capitol-rioters-raiding-nancy-pelosis -offic-rcna57588.

38  Will Sommer, "Glenn Beck Fired Star Podcaster Elijah Schaffer After Sexual Assault Accusation," *Daily Beast*, Oct. 6, 2022, https://www.thedailybeast.com /conservative-pundit-elijah-schaffer-fired-from-the-blaze-after-sexual-assault -accusation.

39  Glenn Beck, "BIG TECH Cracks Down on BlazeTV Reporter Following DC Riots: 'Today's the Day I Feared,'" YouTube, Jan. 8, 2021, https://www.youtube.com/watch?v=GtF7amvsPlo.

40  US v. Rondon, Statement of Offense, Dec. 5, 2022, https://storage.courtlistener.com/recap/gov.uscourts.dcd.238318/gov.uscourts.dcd.238318.50.0_1.pdf.

41  Katharine Q. Seelye, "John Lewis, Towering Figure of Civil Rights Era, Dies at 80," *New York Times*, July 17, 2020, https://www.nytimes.com/2020/07/17/us/john-lewis-dead.html.

42  Architect of the Capitol, "Lincoln Catafalque," accessed May 21, 2023, https://www.aoc.gov/what-we-do/programs-ceremonies/lying-in-state-honor/lincoln-catafalque.

43  Luke Broadwater, "John Lewis, Lying in State, Is Honored as Part of a 'Pantheon of Patriots,'" *New York Times*, July 27, 2020, https://www.nytimes.com/2020/07/27/us/politics/john-lewis-memorial.html.

44  Architect of the Capitol, "The Lincoln Catafalque in the U.S. Capitol," April 15, 2015, https://www.aoc.gov/explore-capitol-campus/blog/lincoln-catafalque-us-capitol.

45  Architect of the Capitol, "The Columbus Doors," accessed May 21, 2023, https://www.aoc.gov/explore-capitol-campus/art/columbus-doors.

46  Speaker Pelosi, "Madam Speaker: A Behind-the-Scenes Look at the U.S. Speaker of the House ('The Year of Return': Ghana)," Medium, Aug. 5, 2019, https://speakerpelosi.medium.com/madam-speaker-a-behind-the-scenes-look-at-the-u-s-29ce04dc481.

47  Nancy Pelosi (@SpeakerPelosi), "Today was deeply transformative. We saw the horrors of slavery & humbly walked through the 'Door of Return' w/a renewed sense of purpose to fight injustice & inequality everywhere. We honored the rich traditions of Ghana & thanked the Paramount Chiefs for our enduring friendship," Twitter, July 30, 2019, https://twitter.com/SpeakerPelosi/status/1156316882221355008.

48  US v. Lyons, Affidavit, Jan. 12, 2021, https://www.justice.gov/opa/page/file/1353451/download.

49  US v. Lyons, Statement of Facts.

50  US v. Brock, Affidavit, Jan. 9, 2021, https://www.justice.gov/usao-dc/press-release/file/1352026/download.

51  US v. Munchel, Affidavit, Jan. 10, 2021, https://www.justice.gov/usao-dc/press-release/file/1352221/download.

52  US v. Baranyi, Affidavit, Jan. 10, 2021, https://www.justice.gov/opa/page/file/1355731/download.

53  US v. Colt, Statement of Facts, Jan. 9, 2021, https://www.justice.gov/opa/page/file/1355481/download.

54  US v. Packer, Affidavit, Jan. 12, 2021, https://www.justice.gov/usao-dc/press-release/file/1353201/download.

55  US v. Seefried, Statement of Facts, Jan. 13, 2021, https://www.justice.gov/usao-dc/press-release/file/1354306/download.

56  Justice Department, conference call with beat reporters, Jan. 26, 2021, https://www.justice.gov/opa/page/file/1361521/download.

57  Ryan J. Reilly, "Old Acquaintances of U.S. Capitol Attackers Are Narcing on Their Worst Facebook Friends," *HuffPost*, Jan. 27, 2021, https://www.huffpost.com/entry/facebook-friends-frenemies-capitol-attack-trump-fbi_n_6010667ec5b6a0814272448a.

58  US v. Stedman, Affidavit, Jan. 20, 2021, https://www.justice.gov/opa/page/file
    /1357721/download.

59  Ryan J. Reilly, "Feds Arrest Man Who Wore His High School Varsity Jacket to the
    Capitol Siege," *HuffPost*, Jan. 25, 2021, https://www.huffpost.com/entry/high
    -school-varsity-jacket-us-capitol-riot_n_600f365ac5b634dc37378746.

60  Albuquerque Head (@HeadAlbuquerque), "He is seen here with several know antifa
    members I was in DC and 'at'the capital and went to all confederate flags I seen
    because I'm son of Confederate veterans never seen this guy," Twitter, Jan. 8, 2021,
    https://twitter.com/HeadAlbuquerque/status/1347585534399569921.

61  Ryan J. Reilly and Daniel Barnes, "Jan. 6 Rioter Who Dragged Michael Fanone into
    Crowd Sentenced to 7.5 Years in Prison," NBC News, Oct. 27, 2022, https://www
    .nbcnews.com/politics/justice-department/jan-6-rioter-dragged-mike-fanone
    -crowd-sentenced-75-years-prison-rcna54314.

62  US v. Stotts, Criminal Complaint, March 16, 2021, https://www.justice.gov/usao-dc
    /case-multi-defendant/file/1377866/download.

63  Joe Heim, "A More Solemn—and Secure—U.S. Capitol Two Days After Riot,"
    *Washington Post*, Jan. 8, 2021, https://www.washingtonpost.com/local/a-more-sol
    emn--and-secure--us-capitol-two-days-after-riot/2021/01/08/9a35e5c2-51da-11eb
    -bda4-615aaefd0555_story.html.

64  US v. Rodriguez, Indictment, March 24, 2021, https://storage.courtlistener.com
    /recap/gov.uscourts.dcd.229256/gov.uscourts.dcd.229256.1.0.pdf.

65  FBI, FOIA, "United States Capitol Violence and Related Events of January 6, 2021,
    Part 21."

66  Ryan J. Reilly, "Andy Kim Is a South Jersey Boy. The GOP Calls Him 'Not One of
    Us,'" *HuffPost*, Oct. 30, 2018, https://www.huffpost.com/entry/andy-kim-tom
    -macarthur-nj-3rd-district_n_5bbcd8c5e4b01470d055d5eb.

67  US House of Representatives, "Members of the 117th Congress Sworn In," Jan. 5,
    2021, https://www.house.gov/feature-stories/2021-1-5-members-of-the-117th-congress
    -sworn-in.

68  Claire Wang, "Behind the Viral Photo of Rep. Andy Kim Cleaning Up at Midnight
    After Riots," NBC News, Jan. 8, 2021, https://www.nbcnews.com/news/asian-amer
    ica/behind-viral-photo-rep-andy-kim-cleaning-midnight-after-riots-n1253519.

69  JC Whitington and Monica Akhtar, "How One Congressman's Act of Service Will
    Be Forever Memorialized," Politico, Jan. 6, 2022, https://www.politico.com/news
    /2022/01/06/andy-kim-reflects-jan-6-526568.

70  Kimmy Yam, "A Year After Viral Photo, Rep. Andy Kim Reflects on Being a 'Care-
    taker of Our Democracy,'" NBC News, Jan. 6, 2022, https://www.nbcnews.com
    /news/asian-america/year-viral-photo-rep-andy-kim-reflects-caretaker-democracy
    -rcna11068.

71  US v. Ball, Statement of Facts, May 2, 2023, https://www.justice.gov/usao-dc/press
    -release/file/1582351/download.

72  Carter Walker, "Carlisle Proud Boy Member Targeted in Search Warrant Tied to
    Jan. 6 Plot," Lancaster Online, March 12, 2022, https://lancasteronline.com/news
    /politics/carlisle-proud-boy-member-targeted-in-search-warrant-tied-to-jan-6-plot
    /article_c2596928-a258-11ec-a6bb-c79ff2e0e8a7.html.

73  US v. Miller, government's sentencing memorandum, Feb. 16, 2023, https://storage
    .courtlistener.com/recap/gov.uscourts.dcd.227582/gov.uscourts.dcd.227582.141.0.pdf.

74 Ryan J. Reilly and Jesselyn Cook, "Revealed: The Star-Spangled Trumper Filmed Attacking Cops at the Capitol," *HuffPost*, March 5, 2021, https://www.huffpost.com /entry/robert-palmer-capitol-attack-sedition-hunters_n_604231acc5b60208555 e0b6a.

75 Ryan J. Reilly and Jesselyn Cook, "Florida Man Charged for Attacking Cops During Capitol Riot While Wearing 'Trump' Flag Jacket," *HuffPost*, March 17, https:// www.huffpost.com/entry/florida-man-charged-attacking-cops-capitol-riot_n_604a 8249c5b6cf72d094f2e8?uhq.

76 Ryan J. Reilly, "Trump Rioter Who Attacked Cops with Fire Extinguisher Gets Longest Jan. 6 Sentence Yet," *HuffPost*, Dec. 17, 2021, https://www.huffpost.com /entry/capitol-riot-sentence-trump-robert-scott-palmer_n_61b94c34e4b06621e42 bfe72.

77 Matthew Cecil, *Branding Hoover's FBI: How the Boss's PR Men Sold the Bureau to America* (Lawrence: University of Kansas, 2016).

78 US Department of Justice Office of the Inspector General, "The Federal Bureau of Investigation's Efforts to Hire, Train, and Retain Intelligence Analysts," May 2005, https://oig.justice.gov/reports/FBI/a0520/final.pdf.

79 Melanie W. Sisson, "The FBI's 2nd-Class Citizens," *Washington Post*, Dec. 31, 2005, https://www.washingtonpost.com/archive/opinions/2005/12/31/the-fbis-2nd -class-citizens/213e7a5a-74dd-4ead-abd4-e6192869b3f3/.

80 FBI FOIA, "United States Capitol Violence and Related Events of January 6, 2021," accessed May 19, 2023, https://vault.fbi.gov/united-states-capitol-violence-and -related-events-of-january-6-2021.

81 US v. Powell, Detention Hearing, Feb. 9, 2021, https://storage.courtlistener.com /recap/gov.uscourts.dcd.228286/gov.uscourts.dcd.228286.69.1.pdf.

82 US v. Rhodes, Testimony, Oct. 14, 2022.

83 Ryan J. Reilly and Ken Dilanian, "FBI Official Was Warned After Jan. 6 That Some in the Bureau Were 'Sympathetic' to the Capitol Rioters," NBC News, Oct. 14, 2022, https://www.nbcnews.com/politics/justice-department/fbi-official-was-warned -jan-6-bureau-sympathetic-capitol-rioters-rcna52144.

84 FBI FOIA, accessed May 19, 2023, https://vault.fbi.gov/@@dvpdffiles/f/4/f4d8fa 57ba274002a3a5548d7742ec15/normal/dump_7.png.

85 Reilly and Dilanian, "FBI Official Was Warned."

86 Committee on the Judiciary Democratic Staff, "GOP Witnesses: What Their Disclo- sures Indicate About the State of the Republican Investigations," March 2, 2023, https://democrats-judiciary.house.gov/uploadedfiles/2023-03-02_gop_witnesses _report.pdf.

87 Ryan J. Reilly, "A Digital Dragnet Is Coming for the U.S. Capitol Insurrectionists," *HuffPost*, Jan. 22, 2021, https://www.huffpost.com/entry/capitol-attack-trump-fbi -surveillance_n_6009c26fc5b697df1a0d28f0.

88 Ryan J. Reilly, "'Unprecedented': Feds Say Hundreds of Trump Rioters Can Expect a Knock on the Door," *HuffPost*, Jan. 12, 2021, https://www.huffpost.com/entry /trump-capitol-rioters-fbi-doj_n_5ffe07c9c5b63642b6ffd7c7.

89 FBI, "FBI Assistant Director in Charge Steven M. D'Antuono's Remarks at Press Brief- ing Regarding Violence at U.S. Capitol," Jan. 12, 2021, https://www.fbi.gov/contact-us /field-offices/washingtondc/news/press-releases/fbi-assistant-director-in-charge-ste ven-m-dantuonos-remarks-at-press-briefing-regarding-violence-at-us-capitol-011221.

90  FBI, "FBI Assistant Director in Charge Steven M. D'Antuono's Remarks on Press Call Regarding Violence at U.S. Capitol," Jan. 15, 2021, https://www.fbi.gov/contact-us/field-offices/washingtondc/news/press-releases/fbi-assistant-director-in-charge-steven-m-dantuonos-remarks-on-press-call-regarding-violence-at-us-capitol-011521.

91  Devlin Barrett and Spencer S. Hsu, "Justice Department, FBI Debate Not Charging Some of the Capitol Rioters," *Washington Post*, Jan. 23, 2021, https://www.washingtonpost.com/national-security/doj-capitol-rioters-charges-debate/2021/01/23/3b0cf112-5d97-11eb-8bcf-3877871c819d_story.html.

92  Ryan J. Reilly, "The Feds Have Made 625+ Capitol Riot Arrests. They Still Have a Long Way to Go," *HuffPost*, Oct. 6, 2021, https://www.huffpost.com/entry/feds-made-capitol-riot-arrests-quarter-way-there_n_615c6fafe4b0f7776310fe37.

93  Justice Department, conference call with beat reporters, Jan. 7, 2021.

94  FBI, "FBI Assistant Director in Charge Steven M. D'Antuono's Remarks on Press Call Regarding Violence at U.S. Capitol," Jan. 26, 2021, https://www.fbi.gov/contact-us/field-offices/washingtondc/news/press-releases/fbi-assistant-director-in-charge-steven-m-dantuonos-remarks-on-press-call-regarding-violence-at-us-capitol-012621.

# Chapter 8

1  Ryan J. Reilly, "'Comically Minimal Ego-Stroking': Inside the Bumble Takedown of a Violent Capitol Rioter," *HuffPost*, July 29, 2021, https://www.huffpost.com/entry/capitol-attack-bumble-takedown-jan-6-andrew-taake_n_6101b17ce4b00fa7af7ea007.

2  Anna Spiegel, "Lots of DC Bars and Restaurants Are Going into Pandemic Winter Hibernation," *Washingtonian*, Oct. 29, 2020, https://www.washingtonian.com/2020/10/29/dc-bars-and-restaurants-are-going-into-pandemic-winter-hibernation/.

3  Jon Porter, "Bumble Disables Politics Filter After Capitol Rioters Spotted in Dating App," *Verge*, Jan. 15, 2021, https://www.theverge.com/2021/1/15/22232492/bumble-politics-filter-temporarily-disabled-us-capitol-riots.

4  US v. Taake, FBI Affidavit, July 21, 2021, https://storage.courtlistener.com/recap/gov.uscourts.dcd.234024/gov.uscourts.dcd.234024.1.1_1.pdf.

5  US v. Taake, Order of Detention Pending Trial, Aug. 2, 2021, https://storage.courtlistener.com/recap/gov.uscourts.txsd.1837569/gov.uscourts.txsd.1837569.10.0.pdf.

6  Ryan J. Reilly and Paul Blumenthal, "Brent Bozell IV, Son of Prominent Conservative Activist, Charged in Capitol Riot," *HuffPost*, Feb. 16, 2021, https://www.huffpost.com/entry/brent-bozell-capitol-attack_n_602c2e29c5b65ff1f6034cea.

7  Timothy Noah, "The Rise and Fall of the L. Brent Bozells," *New Republic*, Feb. 19, 2021, https://newrepublic.com/article/161431/brent-bozell-trump-capitol-riot.

8  Tim Alberta, "The Deep Roots of Trump's War on the Press," Politico, April 26, 2018, https://www.politico.com/magazine/story/2018/04/26/the-deep-roots-trumps-war-on-the-press-218105/.

9  "Conservatives Call on State Legislators to Appoint New Electors, in Accordance with the Constitution," Dec. 10, 2020, https://conservativeactionproject.com/conservatives-call-on-state-legislators-to-appoint-new-electors-in-accordance-with-the-constitution/.

10   Tim Graham, "Brent Bozell Condemns Capitol Riot on Fox Business," MRC, Jan. 6, 2021, https://www.mrctv.org/videos/brent-bozell-condemns-capitol-riot-fox-business.

11   Ryan J. Reilly, "'Sedition Hunters': Meet the Online Sleuths Aiding the FBI's Capitol Manhunt," *HuffPost*, June 30, 2021, https://www.huffpost.com/entry/sedition -hunters-fbi-capitol-attack-manhunt-online-sleuths_n_60479dd7c5b653040034f749.

12   No Nazis for Me Thanks (@No_Nazis_Please), "Here's the photo I was show, appar- ently of the late Officer Sicknick. 3 people I'd love to ID: 1) CAT Sweatshirt (did he use that crutch as a weapon?) 2) Orange Sweatshirt (with helmet and baton) 3) Scal- loped Backpack," Twitter, Jan. 9, 2021, https://twitter.com/No_Nazis_Please/status /1348037973938475010.

13   US Capitol Police, "Loss of USCP Officer Brian D. Sicknick," press release, Jan. 7, 2021, https://www.uscp.gov/media-center/press-releases/loss-uscp-colleague-brian -d-sicknick.

14   Chris Hayes, "Must-See New Video Shows Capitol Riot Was Way Worse Than We Thought," MSNBC, Jan. 8, 2021, https://www.msnbc.com/all-in/watch/chris-9917 8053752.

15   Donell Harvin, interview by Select Committee to Investigate the Jan. 6th Attack on the US Capitol, Jan. 24, 2022, https://www.govinfo.gov/content/pkg/GPO-J6 -TRANSCRIPT-CTRL0000038866/pdf/GPO-J6-TRANSCRIPT-CTRL 0000038866.pdf.

16   Justice Department Inspector General, "Audit of the Federal Bureau of Investiga- tion's Implementation of Its Next Generation Cyber Initiative," July 30, 2015, https://www.oversight.gov/sites/default/files/oig-reports/a1529.pdf.

17   John Schwartz, "Traces of Terror: The F.B.I.; Computer System That Makes Data Secure, but Hard to Find," *New York Times*, June 8, 2002, https://www.nytimes .com/2002/06/08/us/traces-of-terror-the-fbi-computer-system-that-makes-data-secure -but-hard-to-find.html.

18   Jason Ryan, "A Decade and $451M Later, FBI Computers Just Now Working Together," ABC News, Aug. 1, 2012, https://abcnews.go.com/Blotter/sentinel-decade -451m-fbi-computers-now-working/story?id=16904032.

19   Justice Department, "Attorney General Holder Announces Significant Policy Shift Concerning Electronic Recording of Statements," press release, May 22, 20114, https://www.justice.gov/opa/pr/attorney-general-holder-announces-significant -policy-shift-concerning-electronic-recording.

20   Erik Wemple, "*New York Post* Settles 'Bag Men' Defamation Suit," *Washington Post*, Oct. 2, 2014, https://www.washingtonpost.com/blogs/erik-wemple/wp/2014/10/02 /new-york-post-settles-bag-men-defamation-suit/.

21   David Montgomery, Sari Horwitz, and Marc Fisher, "Police, Citizens and Technol- ogy Factor into Boston Bombing Probe," *Washington Post*, April 20, 2013, https:// www.washingtonpost.com/world/national-security/inside-the-investigation-of -the-boston-marathon-bombing/2013/04/20/19d8c322-a8ff-11e2-b029-8fb7e 977ef71_print.html.

22   Jay Caspian Kang, "Should Reddit Be Blamed for the Spreading of a Smear?," *New York Times*, July 25, 2013, https://www.nytimes.com/2013/07/28/magazine/should -reddit-be-blamed-for-the-spreading-of-a-smear.html.

23   "Boston News Conference on Bombing Suspect," C-SPAN, April 19, 2013, https:// www.c-span.org/video/?312281-1/boston-news-conference-bombing-suspect.

24  Ryan J. Reilly, "Former Boston Police K-9 Officer Charged with Attacking Law Enforcement on Jan. 6," NBC News, March 30, 2023, https://www.nbcnews.com /politics/justice-department/former-boston-police-officer-charged-jan-6-assault -capitol-police-rcna77312.

25  Ryan J. Reilly, "Instagram Posts Help FBI Nab Trump-Loving Romance Novel Model Who Beat Capitol Cops," *HuffPost*, Aug. 17, 2021, https://www.huffpost.com /entry/logan-barnhart-trump-capitol-attack-fbi-arrest_n_6086e1b3e4b02e74d21 d4a28.

26  US v. Griffin, Sentencing Memo, June 10, 2022, https://storage.courtlistener.com /recap/gov.uscourts.dcd.227183/gov.uscourts.dcd.227183.112.0.pdf.

27  Tom Jackman and Spencer S. Hsu, "Cowboys for Trump Founder Couy Griffin Sentenced for Trespassing," *Washington Post*, June 17, 2022, https://www.washington post.com/national-security/2022/06/17/couy-griffin-sentenced/.

# Chapter 9

1  Molly Ball, "What Mike Fanone Can't Forget," *Time*, Aug. 5, 2021, https://time .com/6087577/michael-fanone-january-6-interview/.

2  Michael Fanone and John Shiffman, *Hold the Line: The Insurrection and One Cop's Battle for America's Soul* (New York: Atria, 2022).

3  Leonard Wu (@iamleonardwu), Twitter, "One of my dear high school friends, Hsin-Yi Wang, her ex-husband was on the frontlines in DC. His story is too important not to share," Twitter, Jan. 8, 2021, https://web.archive.org/web/20210115195530 /https://twitter.com/iamleonardwu/status/1347808951014887429.

4  Peter Hermann, "'We Got to Hold This Door,'" *Washington Post*, Jan. 14, 2021, https://www.washingtonpost.com/dc-md-va/2021/01/14/dc-police-capitol-riot/.

5  Fanone and Shiffman, *Hold the Line*.

6  Ryan J. Reilly (@ryanjreilly), "'Let's go MPD!' 'Old school CDU.' 'We are not losing the U.S. Capitol today.' 'We are not losing the U.S. Capitol,'" Twitter, Sept. 2, 2022, https://twitter.com/ryanjreilly/status/1565723062338813955.

7  Michael Schaffer, "The Secret Tapes of Michael Fanone," Politico, Oct. 22, 2022, https://www.politico.com/news/magazine/2022/10/07/michael-fanone-jan-6 -riot-cop-00060556.

8  Ryan J. Reilly, "The Feds Caught the Most Infamous, Viral Insurrectionists. Now Comes the Hard Part," *HuffPost*, Jan. 19, 2021, https://www.huffpost.com/entry /viral-insurrectionists-capitol-riot-fbi_n_6001e296c5b6efae62f8764a.

9  Ryan J. Reilly, "Jan. 6 Defendant Who Said She's 'Definitely Not Going to Jail' Sentenced to Prison," *HuffPost*, Nov. 4, 2021, https://www.huffpost.com/entry/jenna -ryan-sentenced-capitol-attack-trump_n_6182bb4fe4b0c8666bd6f913.

10  Jason Rink and Paul Escandon, *Q Sent Me—Part 1*, Nov. 11, 2022, https://www .mymoviesplus.com/videos/q-sent-me-episode-one.

11  Justice Department, "Chansley, Jacob Anthony (aka Jacob Angeli)," Capitol Breach Cases, last updated Feb. 4, 2022, https://www.justice.gov/usao-dc/defendants/chan sley-jacob-anthony.

12  FBI, FOIA, "United States Capitol Violence and Related Events of January 6, 2021."

13  Donald Trump (@RealDonaldTrump), "I hear that the great Agents & others in the FBI are furious at FBI leadership for what they are doing with respect to political

weaponization against a President (me) that always had their backs, and that they like (love!) a lot. They don't like being 'used' by people they do not agree with, or respect. Likewise, they are not exactly thrilled with the leadership at DOJ! Similar to the revolt against Comey when he exonerated Crooked Hillary, but was forced, by them, to withdraw the exoneration!" Truth Social, Aug. 20, 2022, https://truthso cial.com/@realDonaldTrump/posts/108858518234900212.

14  "Don Jr. Gives Fiery Final Debate Analysis Exclusively on *Hannity*," Fox News, Oct. 22, 2020, https://www.youtube.com/watch?v=pidFPxne9o4.

15  Ryan J. Reilly, "The 'Law and Order' Party Has Unleashed an Extraordinary Attack on Law Enforcement," *HuffPost*, Jan. 27, 2018, https://www.huffpost.com/entry/fbi -attack-trump-mueller-republicans_n_5a6b3fe7e4b0ddb658c5c3fd.

16  Spencer Ackerman, "'The FBI Is Trumpland': Anti-Clinton Atmosphere Spurred Leaking, Sources Say," *Guardian*, Nov. 4, 2016, https://www.theguardian.com/us -news/2016/nov/03/fbi-leaks-hillary-clinton-james-comey-donald-trump.

17  Devlin Barrett, *October Surprise: How the FBI Tried to Save Itself and Crashed an Election* (New York: PublicAffairs, 2020).

18  Kyle Seraphin (@KyleSeraphin), "Her signature removed protected whistleblowers from their security clearances (and therefore paychecks), Jenn is famous for these emails. My buddy Phil got the ax for calling the J6 response: "Crossfire Hurricane 2: the Revenge" on a survey. Congress has over 100 pages," Twitter, Nov. 7, 2022, https://twitter.com/KyleSeraphin/status/1589611292348018688."

19  Steve Friend, interview by *Washington Examiner*, YouTube, Feb. 25, 2023, https:// www.youtube.com/watch?v=AL-Ga7tfQoo.

20  US v. Richter, Statement of Facts, Feb. 17, 2023, https://storage.courtlistener.com /recap/gov.uscourts.dcd.252273/gov.uscourts.dcd.252273.1.1.pdf.

21  US v. Irwin, Criminal Complaint, Aug. 12, 2021, https://www.justice.gov/usao-dc /case-multi-defendant/file/1428541/download.

22  "Who Is a Journalist: A Conversation w Stephen Horn," Public Report, YouTube, June 22, 2021, https://www.youtube.com/watch?v=ZFT0Q0jDgE0.

23  Ryan J. Reilly, "Feds Boost Reward to $500K for Information on Capitol Pipe Bomber," NBC News, Jan. 4, 2023, https://www.nbcnews.com/politics/justice-de partment/feds-boost-reward-500k-information-capitol-pipe-bomber-rcna64268.

24  US v. Irwin, Criminal Complaint.

25  Sarah Knapp, "Federal Agents Searched Homer Residence in Connection to the Jan. 6 Capitol Riots," *Homer News*, May 5, 2021, https://www.homernews.com/news /homer-residents-said-home-was-searched-by-fbi-in-connection-to-the-jan-6-capitol -riots/.

26  Ryan J. Reilly, "After Hundreds of Arrests in Sprawling Capitol Hunt, the FBI Just Made a Pretty Big Mistake," *HuffPost*, May 7, 2021, https://www.huffpost.com /entry/fbi-capitol-raid-alaska-trump_n_60930634e4b0c15313fbe48f.

27  US v. Rondon, Statement of Facts, Sept. 2, 2021, https://extremism.gwu.edu/sites/g /files/zaxdzs5746/files/Maryann%20Mooney-Rondon%20and%20Rafael%20 Rondon%20Statement%20of%20Facts.pdf.

28  Joel Davidson, "With Guns Drawn, FBI Raids Homer Couple's Home Looking for Nancy Pelosi's Laptop," *Alaska Watchman*, April 29, 2021, https://alaskawatchman .com/2021/04/29/with-guns-drawn-fbi-raids-homer-couples-home-looking-for -nancy-pelosi-laptop/.

29  Justice Department, "Watertown Man Pleads Guilty to Possession of a Sawed-Off Shotgun," press release, Dec. 16, 2022, https://www.justice.gov/usao-ndny/pr/water town-man-pleads-guilty-possession-sawed-shotgun.

30  Ryan J. Reilly, "FBI Arrests 'Airhead' Mother-Son Team Who Raided Pelosi's Office During Capitol Attack," *HuffPost*, Oct. 1, 2021, https://www.huffpost.com/entry /capitol-attack-arrest-mother-son-pelosi-office_n_6155e986e4b099230d2095e4.

31  US v. Cua, Government's Sentencing Memorandum, May 5, 2023, https://storage .courtlistener.com/recap/gov.uscourts.dcd.227610/gov.uscourts.dcd.227610.328.0.pdf.

32  US v. Cua, "Opposition to Defendant's Emergency Motino for Pre-Trial Release," March 2, 2021, https://storage.courtlistener.com/recap/gov.uscourts.dcd.227610/gov .uscourts.dcd.227610.12.0.pdf.

33  Ryan J. Reilly, "'Embarrassed' Trump Supporter Begs Judge for Son's Freedom After Capitol Riot Arrest," *HuffPost*, Feb. 13, 2021, https://www.huffpost.com/entry /bruno-cua-father-trump-capitol-attack_n_60273806c5b680717ee7d557.

34  Ryan J. Reilly, "MAGA Mom Whose Son Stormed Capitol Feels 'Stupid' for Buying Trump's Voter Fraud Lies," *HuffPost*, March 3, 2021, https://www.huffpost.com /entry/bruno-cua-trump-capitol-attack_n_603fb5f8c5b6d7794ae37145.

35  Ryan J. Reilly, "Jeff Sessions Was Deemed Too Racist to Be a Federal Judge. He'll Now Be Trump's Attorney General," *HuffPost*, Nov. 17, 2016, https://www.huffpost .com/entry/trump-attorney-general-jeff-sessions-racist-remarks_n_582cd73ae4b09 9512f80c0c2?cn5id4ygk5rf9lik9.

36  Ryan J. Reilly, "A Woman Is on Trial for Laughing During a Congressional Hearing," *HuffPost*, May 2, 2017, https://www.huffpost.com/entry/laughing-congressio nal-hearing-jeff-sessions-code-pink_n_59076a93e4b05c3976810a3a.

37  Ryan J. Reilly (@ryanjreilly), "Another protester escorted out of Sessions hearing. Her original offense appeared to be simply laughing," Twitter, Jan. 10, 2017, https:// twitter.com/ryanjreilly/status/818837991217123328.

38  John Gramlich, "Only 2% of Federal Criminal Defendants Go to Trial, and Most Who Do Are Found Guilty," Pew Research Center, June 11, 2019, https://www .pewresearch.org/fact-tank/2019/06/11/only-2-of-federal-criminal-defendants-go -to-trial-and-most-who-do-are-found-guilty/.

39  Ryan J. Reilly, "DOJ Lawyers Insist Laughing at Their Boss Jeff Sessions Can Be a Criminal Offense," *HuffPost*, May 2, 2017, https://www.huffpost.com/entry/jeff-ses sions-laughter_n_5908c55ee4b0bb2d08726a91?kuw.

40  Ryan J. Reilly, "Jury Convicts Woman Who Laughed at Jeff Sessions During Senate Hearing," *HuffPost*, May 3, 2017, https://www.huffpost.com/entry/jeff-sessions -laugh-congressional-hearing_n_590929bbe4b05c39768420ef.

41  Ryan J. Reilly, "Judge Tosses Jury's Conviction of Woman Who Laughed at Jeff Sessions, Orders New Trial," *HuffPost*, July 14, 2017, https://www.huffpost.com/entry /protester-laughed-jeff-sessions-sentenced_n_5967de92e4b0d6341fe7a9e2.

42  Ryan J. Reilly, "Jeff Sessions' DOJ to Put Woman Who Laughed at Jeff Sessions on Trial Yet Again," *HuffPost*, Sept. 1, 2017, https://www.huffpost.com/entry/woman -laughed-jeff-sessions_n_59a8cb8ae4b0b5e530fd7c8b.

43  Ryan J. Reilly, "Jeff Sessions' DOJ Drops Prosecution of Woman Who Laughed at Jeff Sessions," *HuffPost*, Nov. 7, 2017, https://www.huffpost.com/entry/laughing -jeff-sessions-case-dropped_n_5a00f081e4b0368a4e868e0c.

44  Ryan J. Reilly, "'QAnon Shaman' Jacob Chansley, a Capitol Riot 'Flag-Bearer,' Sentenced to Prison," *HuffPost*, Nov. 17, 2021, https://www.huffpost.com/entry/qanon -shaman-sentenced-trump-capitol-riot_n_618d4779e4b04e5bdfccfadc.

45  Josh Gerstein and Kyle Cheney, "Capitol Riot Cases Strain Court System," Politico, March 10, 2021, https://www.politico.com/news/2021/03/10/capitol-riot-court-cases -475081.

46  Josh Gerstein, "Feds Admit Breaking Law with Delay in Case Against Alleged Jan. 6 Rioter," Politico, March 14, 2022, https://www.politico.com/news/2022/03/14/feds -admit-breaking-law-with-delay-in-case-against-alleged-jan-6-rioter-00017003.

47  US v. Reeder, FBI Affidavit, Feb. 24, 2021, https://www.justice.gov/usao-dc/case -multi-defendant/file/1371496/download.

48  Ryan J. Reilly and Christopher Mathias, "Feds Weigh New Charges for Capitol Rioter Allegedly Caught on Video Attacking Cops," *HuffPost*, Aug. 18, 2021, https:// www.huffpost.com/entry/robert-reeder-capitol-rioter-assault-sedition-hunters_n_6 11d5e7ce4b0caf7ce2c5e0d.

49  Josh Gerstein and Kyle Cheney, "Feds Agree to Pay $6.1M to Create Database for Capitol Riot Prosecutions," Politico, July 9, 2021, https://www.politico.com/news /2021/07/09/doj-database-capitol-riot-prosecutions-498911.

50  US v. Nordean et al., "Nordean's Notice of Argument in Support of His Ability to Cross-examine Witness on Credibility and Opposition to the Government's Email Motion to Provide Witness with Impeachment Materials in the Middle of Her Cross-Examination," March 12, 2023, https://storage.courtlistener.com/recap/gov .uscourts.dcd.228299/gov.uscourts.dcd.228299.688.0.pdf.

51  US v. Nordean et al., "Nordean's Notice of Argument in Support of Impeachment of Witness with Hidden Jencks-Related Communications," March 9, 2023, https:// www.documentcloud.org/documents/23699546-nordean-filing-on-fbi-testimony.

52  US v. Nordean et al., "Nordean's Response to the Government's Motion to Preclude Cross-Examination on Topics He Has Identified and Motion to Strike Miller's Testimony," March 12, 2023, https://storage.courtlistener.com/recap/gov.uscourts.dcd .228300/gov.uscourts.dcd.228300.691.0.pdf.

53  US v. Perna, Statement of Offense, Dec. 9, 2021, https://www.justice.gov/usao-dc /press-release/file/1457391/download.

54  "Townhall's Frontline Footage of Capitol Building Riots," Bearing Arms' Cam & Co, YouTube, Jan. 11, 2021, https://www.youtube.com/watch?v=YzxvVi8wkrU&t=983.

55  "Matthew Lawrence Perna | 1984–2022 | Obituary," John Flynn Funeral Home, Feb. 27, 2022, https://web.archive.org/web/20220227224509/https://www.flynnfu neralhome.com/obituary/matthew-perna.

56  US v. Perna, Government's Suggestion of Death and Request for Abatement of Prosecution by USA as to Matthew Perna, March 9, 2022, https://storage.courtlistener .com/recap/gov.uscourts.dcd.228265/gov.uscourts.dcd.228265.49.0_1.pdf.

57  Gary Grumbach and Ryan J. Reilly, "D.C. Officer Daniel Hodges Testifies Against Capitol Rioter Who 'Crushed' Him with Shield," NBC News, Aug. 30, 2022, https:// www.nbcnews.com/politics/justice-department/dc-officer-daniel-hodges-testifies -capitol-rioter-crushed-shield-rcna45533.

58  Arden Dier, "How Did a Red Scarf from Sweden End Up at Capitol Riot?," *Newser*, Jan. 15, 2021, https://www.newser.com/story/301334/red-scarf-connects-swedish

-city-to-a-capitol-riot-mystery.html; Christian Christensen (@ChrChristensen), "1) This pic published in the Daily Mail showing debris left after terrorist attack on US Capitol seems relatively unremarkable. But a man in Sweden noticed something: the red scarf at the top with the name of the northern Swedish city of Skellefteå. Then it got interesting," Twitter, Jan. 13, 2021, https://twitter.com/ChrChristensen/status /1349400117586046976.

59  Ronan Farrow, "A Pennsylvania Mother's Path to Insurrection," *New Yorker*, Feb. 1, 2021, https://www.newyorker.com/news/news-desk/a-pennsylvania-mothers-path -to-insurrection-capitol-riot.

60  US v. Young, Sept. 27, 2022, https://extremism.gwu.edu/sites/g/files/zaxdzs2191/f /Kyle%20Young%20Sentencing%20Hearing%20Transcript.pdf.

61  Ryan J. Reilly and Jesselyn Cook, "Revealed: Meet the Trump Fanatic Who Used a Stun Gun on a Cop at the Capitol Insurrection," *HuffPost*, Feb. 26, 2021, https:// www.huffpost.com/entry/fanone-taser-maga-trump-capitol_n_6037bb23c5b69ac3 d35cdb71.

62  US v. Rodriguez, Government's Opposition to Defendant's Motion to Suppress Statements, Oct. 29. 2021, https://storage.courtlistener.com/recap/gov.uscourts.dcd.229 256/gov.uscourts.dcd.229256.42.0_1.pdf.

63  Fanone and Shiffman, *Hold the Line*.

64  US v. Rodriguez, Response to United States' Omnibus Motion in Limine (ECF no. 134), Feb. 3, 2023, https://storage.courtlistener.com/recap/gov.uscourts.dcd.229256 /gov.uscourts.dcd.229256.150.0.pdf.

65  US v. Roche, Statement of Facts, April 7, 2021, https://www.justice.gov/usao-dc /case-multi-defendant/file/1386736/download.

66  Ryan J. Reilly, "Trumper on FBI Capitol Wanted List Attends Giuliani Event for GOPer Who Pushed Big Lie," *HuffPost*, May 18, 2021, https://www.huffpost.com /entry/samuel-lazar-trump-mastriano-capitol-attack_n_60a31164e4b03e1dd38d718e.

67  Ryan J. Reilly, "Trumper Who Wore 'Back the Blue' Shirt to Rally Arrested for Assaulting Officers on Jan. 6," *HuffPost*, July 27, 2021, https://www.huffpost.com /entry/samuel-lazar-fbi-doug-mastriano-trump-capitol-attack-jan-6_n_60be410 be4b099fb31ca9350.

68  Ryan J. Reilly, "Capitol Rioter Who Palled Around with Pa. Republicans Ordered Jailed Until Trial," *HuffPost*, Aug. 31, 2021, https://www.huffpost.com/entry/capi tol-rioter-samuel-lazar-jail-until-trial_n_612e87ece4b0eab0ad9147d1.

69  Ryan J. Reilly, "'Sedition Hunters': Meet the Online Sleuths Aiding the FBI's Capitol Manhunt," *HuffPost*, June 30, 2021, https://www.huffpost.com/entry/sedition-hunt ers-fbi-capitol-attack-manhunt-online-sleuths_n_60479dd7c5b653040034f749.

70  US v. Rodriguez, Statement of Offense, Feb. 14, 2023, https://storage.courtlistener .com/recap/gov.uscourts.dcd.229256/gov.uscourts.dcd.229256.160.0_1.pdf.

71  Government Accountability Office, "Further Actions Needed to Strengthen FBI and DHS Collaboration to Counter Threats," February 2023, https://www.gao.gov /assets/gao-23-104720.pdf.

72  Ryan J. Reilly, "The Feds Say They're in for the Long Haul in the Jan. 6 Investigation. There Is a Time Limit," NBC News, Jan. 6, 2023, https://www.nbcnews.com /politics/justice-department/feds-say-long-haul-jan-6-investigation-rcna61728.

73  Ryan J. Reilly, "Judge Acquits Federal Defense Contractor on Jan. 6 Charges," NBC News, April 6, 2022, https://www.nbcnews.com/politics/justice-department/judge -acquits-federal-defense-contractor-jan-6-charges-rcna23245.

74  Zoe Tillman, "DOJ Told Court to Expect a Deluge of New Jan. 6 Prosecutions," *Bloomberg*, March 15, 2023, https://www.bloomberg.com/news/articles/2023-03-15/doj-told-court-to-expect-a-deluge-of-new-jan-6-prosecutions.

# Chapter 10

1  Ryan J. Reilly, "A Police Officer Died by Suicide After Jan. 6. Here's What He Went Through at the Capitol," *HuffPost*, Jan. 27, 2022, https://www.huffpost.com/entry/jeffrey-smith-capitol-riot-suicide_n_6172c53de4b06573573a9ba4.

2  Reilly, "A Police Officer Died by Suicide."

3  Andrew Glass, "Reporter Fatally Shoots Ex-lawmaker in U.S. Capitol, Feb. 28, 1890," Politico, Feb. 28, 2018, https://www.politico.com/story/2018/02/28/reporter-fatally-shoots-ex-lawmaker-in-us-capitol-feb-28-1890-423402.

4  Peter Overby, "A Historic Killing in the Capitol Building," NPR, Feb. 19, 2007, https://www.npr.org/2007/02/19/7447550/a-historic-killing-in-the-capitol-building.

5  "Kentucky's Silver-Tongued Taulbee Caught in Flagrante, or Thereabouts, with Brown-Haired Miss Dodge," NPR, Dec. 10, 1887, https://media.npr.org/programs/morning/features/2007/feb/taulbee/lvt_scandal.pdf.

6  Joanne B. Freeman, *The Field of Blood* (New York: Picador Farrar, Straus and Giroux, 2018).

7  US House of Representatives, "The Shooting of Congressman William Taulbee on the Steps of the U.S. Capitol," accessed May 19, 2023, https://history.house.gov/Historical-Highlights/1851-1900/The-death-of-Congressman-William-Taulbee-on-the-steps-of-the-U-S--Capitol/.

8  Shaila Dewan, "He Killed Himself After the Jan. 6 Riot. Did He Die in the Line of Duty?" *New York Times*, July 29, 2021, https://www.nytimes.com/2021/07/29/us/police-suicides-capitol-riot.html.

9  Ryan J. Reilly, "'Sedition Hunters': Meet the Online Sleuths Aiding the FBI's Capitol Manhunt," *HuffPost*, June 30, 2021, https://www.huffpost.com/entry/sedition-hunters-fbi-capitol-attack-manhunt-online-sleuths_n_60479dd7c5b653040034f749.

10  Ryan J. Reilly, "A D.C. Cop at the Jan. 6 Riot Died by Suicide. Sleuths Identified 1 of the Rioters He Battled," *HuffPost*, Aug. 13, 2021, https://www.huffpost.com/entry/jeff-smith-capitol-police-assault-sedition-hunters_n_611687c0e4b07c1403147cbb.

11  Reilly, "Police Officer Died By Suicide."

12  Ryan J. Reilly, "Online Sleuths Identify Second Jan. 6 Rioter Seen Battling DC Cop Who Died by Suicide," *HuffPost*, August 23, 2021, https://www.huffpost.com/entry/sedition-hunters-capitol-attack-jeff-smith-trump-riot_n_6117e45fe4b01da700f646dd.

13  Ryan J. Reilly and Leigh Ann Caldwell, "D.C. Police Officer's Suicide After Jan. 6 Riot Declared Line-of-Duty Death," NBC News, March 9, 2022, https://www.nbcnews.com/politics/politics-news/dc-police-officer-died-suicide-jan-6-riot-declared-line-duty-death-rcna19433.

14  Ryan J. Reilly, "D.C. Chiropractor Who Stormed Capitol Arrested on Jan. 6 Charges," NBC News, June 8, 2022, https://www.nbcnews.com/politics/justice-department/dc-chiropractor-stormed-capitol-arrested-jan-6-charges-rcna32679.

15  "VID 20220706 113019071 of Storming the Crapitol 2.0 7/6/22," Tftx22, YouTube, July 6, 2022, https://www.youtube.com/watch?v=Ban5zDEPDAw.

16  Sadie Gurman, "Justice Department Chiefs Can't Get Enough of the Patron Saint of the Rule of Law," *Wall Street Journal*, July 13, 2019, https://www.wsj.com/articles /justice-department-chiefs-cant-get-enough-of-the-patron-saint-of-the-rule-of-law -11563019202.

17  Robert Jackson, "The Federal Prosecutor" (address delivered at the Second Annual Conference of United States Attorneys, Washington, DC, April 1, 1940), https:// www.justice.gov/sites/default/files/ag/legacy/2011/09/16/04-01-1940.pdf.

18  Justice Department, "Attorney General: Robert Francis Kennedy," accessed May 19, 2023, https://www.justice.gov/ag/bio/kennedy-robert-francis.

19  Nicholas deB. Katzenbach, *Some of It Was Fun* (New York: W. W. Norton, 2008).

20  Ronald Kessler, *In the President's Secret Service* (New York: Crown, 2009).

21  Neil A. Lewis, "Edward H. Levi, Attorney General Credited with Restoring Order After Watergate, Dies at 88," *New York Times*, March 8, 2000, https://www.nytimes .com/2000/03/08/us/edward-h-levi-attorney-general-credited-with-restoring-order -after-watergate.html.

22  Katie Benner, "On First Day, Garland Vows to Restore Justice Dept. Independence," *New York Times*, March 11, 2021, https://www.nytimes.com/2021/03/11/us/politics /merrick-garland-attorney-general.html.

23  Justice Department, "Readout of Justice Department Meeting with Families of Fallen Officers," press release, Aug. 5, 2022, https://www.justice.gov/opa/pr/readout -justice-department-meeting-families-fallen-officers.

24  Zoë Richards and Ryan J. Reilly, "Widow of Officer Who Died by Suicide After Jan. 6 Says White House Didn't Give New Line-of-Duty Law 'the Attention It Deserves,'" NBC News, Aug. 17, 2022, https://www.nbcnews.com/politics/politics-news/widow -officer-died-suicide-jan-6-says-white-house-didnt-give-newly-sig-rcna43599.

25  David P. Weber, faculty information page, Salisbury University, accessed May 19, 2023, https://www.salisbury.edu/faculty-and-staff/dpweber.

26  US v. Walls-Kaufman, Statement of Offense, Jan. 19, 2023, https://storage.courtlis tener.com/recap/gov.uscourts.dcd.244502/gov.uscourts.dcd.244502.24.0.pdf.

27  Ryan J. Reilly, "Judge Questions DOJ's Handling of a Jan. 6 Rioter Who 'Scuffled' with an Officer Who Died by Suicide," NBC News, May 5, 2023, https://www.nbc news.com/politics/justice-department/judge-questions-dojs-handling-jan-6-rioter -scuffled-officer-died-suici-rcna82979.

28  Carol D. Leonnig, Devlin Barrett, Perry Stein, and Aaron C. Davis, "Showdown Before the Raid: FBI Agents and Prosecutors Argued over Trump," *Washington Post*, March 1, 2023, https://www.washingtonpost.com/national-security/2023/03/01 /fbi-dispute-trump-mar-a-lago-raid/.

29  Elisha Fieldstadt, Ken Dilanian, Tim Stelloh, and Ryan J. Reilly, "Armed Man Who Was at Capitol on Jan. 6 Is Fatally Shot After Firing into an FBI Field Office in Cincinnati," NBC News, Aug. 11, 2023, https://www.nbcnews.com/news/us-news /armed-man-shoots-fbi-cincinnati-building-nail-gun-flees-leading-inters-rcna 42669.

30  Ben Collins, Ryan J. Reilly, Jason Abbruzzese, and Jonathan Dienst, "Man Who Fired Nail Gun at FBI Building Called for Violence on Truth Social in Days After Mar-a-Lago Search," NBC News, March 20, 2022, https://www.nbcnews.com/tech /tech-news/man-fired-nail-gun-fbi-building-called-violence-days-mar-lago-search -rcna42749.

31  Stephen M. Friend, Declaration, Sept. 21, 2022, https://www.documentcloud.org /documents/23010763-steve-friend-declaration.

32  Miranda Devine, "FBI Hero Paying the Price for Exposing Unjust 'Persecution' of Conservative Americans," *New York Post*, Sept. 21, 2022, https://nypost.com /2022/09/21/fbi-hero-paying-the-price-for-exposing-unjust-persecution-of-conser vative-americans/.

33  Ryan J. Reilly, "Feds Arrest Five Members of 'B Squad' Militia Allegedly Run by Former GOP House Candidate in Jan. 6 Case," NBC News, Aug. 24, 2022, https:// www.nbcnews.com/politics/justice-department/feds-charge-members-b-squad -militia-allegedly-run-former-gop-house-can-rcna44621.

34  US House of Representatives, "GOP Witnesses: What Their Disclosures Indicate About the State of the Republican Investigations," March 2, 2023, https://demo crats-judiciary.house.gov/uploadedfiles/2023-03-02_gop_witnesses_report.pdf.

35  Tom Dreisbach, "Experts See 'Red Flags' at Nonprofit Raising Big Money for Capitol Riot Defendants," NPR, Jan. 20, 2022, https://www.npr.org/2022/01/20/1073061575 /experts-see-red-flags-at-nonprofit-raising-big-money-for-capitol-riot-defendants.

36  Geri Perna (@GeriPerna), "President Trump is the ONLY person who will give my nephew Matthew Perna a Posthumous Pardon. I know this because he told me him- self," Twitter, May 11, 2023, https://twitter.com/GeriPerna/status/1656648735122 702336.

37  Ryan J. Reilly, "Nazi Sympathizer and Army Reservist Who Stormed the Capitol Sentenced to 4 Years in Jan. 6 Case," NBC News, Sept. 22, 2022, https://www.nbc news.com/politics/justice-department/nazi-sympathizer-jan-6-rioter-claimed -didnt-know-congress-met-capitol-rcna48677.

38  US v. Straka, Statement of Offense, Oct. 6, 2021, https://www.justice.gov/usao-dc /case-multi-defendant/file/1441146/download.

39  "There's Only This (Brando, One Man Show)," Brandon Straka, YouTube, Feb. 20, 2017, https://www.youtube.com/watch?app=desktop&v=-g3aEu2H0js.

40  Tim Fitzsimons, "Meet Brandon Straka, a Gay Former Liberal Encouraging Others to #WalkAway from Democrats," NBC News, Aug. 21, 2018, https://www.nbcnews .com/feature/nbc-out/meet-brandon-straka-gay-former-liberal-encouraging-others -walkaway-democrats-n902316.

41  US v. Straka, Statement of Offense, Oct. 6, 2021, https://www.justice.gov/usao-dc /case-multi-defendant/file/1441146/download.

42  "Texas Man Found Guilty of Felony and Misdemeanor Charges Related to Capitol Breach," Justice Department, March 30, 2023, https://www.justice.gov/usao-dc/pr /texas-man-found-guilty-felony-and-misdemeanor-charges-related-capitol-breach-1.

43  US v. Jenkins, Exhibit 204.21, March 20, 2023.

44  US v. Straka, Transcript of Sentencing Hearing, Jan. 24, 2022, https://storage .courtlistener.com/recap/gov.uscourts.dcd.235647/gov.uscourts.dcd.235647.49.0.pdf.

45  Ryan J. Reilly, "Trump Influencer Brandon Straka Walks Away from Jan. 6 Case with Home Detention," Jan. 24, 2022, https://www.huffpost.com/entry/brandon -straka-walk-away-capitol-riot-trump-january-6_n_61eeb38de4b087281f870529?xwc.

46  Michele Nelson, "Man Finds Success with Forum on KMOG Radio," *Payson Roundup*, Aug. 13, 2021, https://www.paysonroundup.com/news/local/man-finds -success-with-forum-on-kmog-radio/article_62b50160-48ec-56b6-8532-245bb120 e1c7.html.

47  Kyle Seraphin, "*The Kyle Seraphin Show*: The Sec D Experience," UncoverDC, Jan. 27. 2023, https://www.uncoverdc.com/2023/01/27/the-kyle-seraphin-show-the-sec-d-experience/.

48  Heather Schwedel, "The Story Behind 'Let's Go Brandon,' the Secretly Vulgar Chant Suddenly Beloved by Republicans," *Slate*, Oct. 22, 2021, https://slate.com/news-and-politics/2021/10/lets-go-brandon-meaning-nascar-republicans-joe-biden.html.

49  Kyle Seraphin, "Hey FBI, I'm breaking up with you . . ." *The Kyle Seraphin Show*, April 14, 2023, https://rumble.com/v2i4c18-opr-whistleblower.html.

50  "Sonoma Ranch Body Cam Feb9," Kyle Seraphin, YouTube, Oct. 25, 2022, https://www.youtube.com/watch?v=HgLA2OaJxaU.

51  Kyle Seraphin, "Why I Was Suspended—Roll the Tape!," Rumble, October 25, 2022, https://rumble.com/v1pq1bh-why-i-was-suspended-roll-the-tape.html.

52  Kyle Seraphin, "FBI Whistleblower Kyle Seraphin Speaks About the FBI and Why He Was Targeted," Rumble, October 14, 2022, https://rumble.com/v1nwdka-fbi-whistleblower-kyle-seraphin-speaks-about-the-fbi-and-why-he-was-targete.html.

53  Evan Osnos, "Dan Bongino and the Big Business of Returning Trump to Power," *New Yorker*, Dec. 27, 2021, https://www.newyorker.com/magazine/2022/01/03/dan-bongino-and-the-big-business-of-returning-trump-to-power.

54  Kyle Seraphin (@KyleSeraphin), "Busy day, and I'll be talking to @RealRogerStone and @ivoryhecker this afternoon. The FBI saw to it to OFFICIALLY revoke my Top Secret Clearance today. (Below) And got my name wrong on the first page of an investigative report. The FBI is completely full of clowns.—'Bradley,'" Twitter, Feb. 6, 2023, https://twitter.com/KyleSeraphin/status/1622674001419198464/photo/2.

55  Kyle Seraphin (@KyleSeraphin), "'During 2022, you demonstrated inappropriate behavior through your routine use of derogatory, racist, sexist, and/or homophobic language and comments which co-workers found offensive; and made unauthorized releases of sensitive government information.'—FBI So, I'm Trump?" Twitter, Feb. 6, 2023, https://twitter.com/KyleSeraphin/status/1622699105016504321.

56  Kyle Seraphin (@KyleSeraphin), "Just had a friend at FBI Washington Field tell me these emails are making management 'panic.' Read these clowns in their own words!" Truth Social, Feb. 11, 2023, https://truthsocial.com/@kyleseraphin/posts/109846687875317032.

57  US v. Evans, Evidence, "Derrick Evans January 6 Capitol Attack Video," YouTube, Jan. 6, 2023, https://www.youtube.com/watch?v=5hxNU6RTVHE.

58  Christopher Wright, "Brandon Straka Singing National Anthem at WalkAway 1 Year Celebration Event!!! #WalkAway," YouTube, May 20, 2019, https://www.youtube.com/watch?v=U6qwkr7FlMQ.

59  Tess Owen, "The Surreal Spectacle of Marjorie Taylor Greene and the Capitol Riot Rage Cage," *Vice*, Aug. 6, 2022, https://www.vice.com/en/article/y3ppy5/cpac-marjorie-taylor-greene-and-the-capitol-riot-rage-cage.

60  Ryan J. Reilly, "'Suit Macer,' Subject of Jan. 6 Conspiracy Theories, Admits Bear-Spraying Capitol Officers," NBC News, March 14, 2023, https://www.nbcnews.com/politics/justice-department/jan-6-rioter-accused-undercover-cop-pled-guilty-spraying-officers-rcna74844.

61  Alanna Durkin Richer, "Judge Scolds 1/6 Lawyer, but Won't Seek Disciplinary Action," AP News, April 21, 2023, https://apnews.com/article/jan-6-capitol-riot-judge-rebukes-lawyers-711e97b67291b218341e8e5eee92e5c3.

62  Office of the Inspector General, "Federal Bureau of Investigation's Management of Information Technology Investments," December 2002, https://oig.justice.gov /reports/FBI/a0309/intro.htm.

63  Justice Department, *FYs 2022–2026 Strategic Plan*, July 2022, https://www.justice .gov/doj/book/file/1516901/download.

64  Alberto Brandolini (@ziobrando): "The bullshit asimmetry: the amount of energy needed to refute bullshit is an order of magnitude bigger than to produce it," Twitter, Jan. 13, 2013, https://twitter.com/ziobrando/status/289635060758507521

65  "Disaster Girl," Know Your Meme, Dec. 30, 2008, https://knowyourmeme.com /memes/disaster-girl.

66  Ken Dilanian, Ryan J. Reilly, and Jonathan Allen, "Jan. 6 Committee Staffers Told Preliminary Plan for Final Report Would Focus Largely on Trump, Not on Law Enforcement Failures, Sources Say," NBC News, Nov. 11, 2022, https://www.nbc news.com/politics/justice-department/jan-6-committee-staffers-told-preliminary -plan-final-report-focus-larg-rcna56802.

67  Robert Draper and Luke Broadwater, "Inside the Jan. 6 Committee," *New York Times Magazine*, Dec. 23, 2022, https://www.nytimes.com/2022/12/23/magazine /jan-6-committee.html.

68  Select January 6th Committee Final Report and Supporting Materials Collection, US Government Publishing Office, Dec. 22, 2022, https://www.govinfo.gov/collec tion/january-6th-committee-final-report.

69  United States Government Accountability Office (GAO), "Federal Agencies Identi- fied Some Threats, but Did Not Fully Process and Share Information Prior to Janu- ary 6, 2021," February 2023, https://www.gao.gov/assets/gao-23-106625.pdf.

70  Ryan J. Reilly, "Romance Novel Model Pleads Guilty to Dragging Officer Down Capitol Steps on Jan. 6," NBC News, Sept. 28, 2022, https://www.nbcnews.com/poli tics/justice-department/jan-6-cases-romance-novel-cover-model-pleads-guilty -dragging-cop-capit-rcna49340.

71  Ryan J. Reilly, "Romance Novel Cover Model Who Dragged Capitol Officer on Jan. 6 Sentenced to 3 Years in Prison," NBC News, April 13, 2023, https://www.nbcnews .com/politics/justice-department/jan-6-romance-novel-cover-model-attacked -police-sentenced-rcna79422.

72  Ryan J. Reilly and Jesselyn Cook, "The Hidden Hand of Facial Recognition in the Capitol Insurrection Manhunt," *HuffPost*, March 26, 2021, https://www.huffpost .com/entry/facial-recognition-capitol-matthew-beddingfield_n_605cc93fc5b67ad3 871dad4e.

73  US v. Beddingfield, Statement of Offense, Feb. 16, 2023, https://storage.courtlis tener.com/recap/gov.uscourts.dcd.240916/gov.uscourts.dcd.240916.40.0_1.pdf.

74  US v. Beddingfield, Plea Agreement, Feb. 16, 2023, https://storage.courtlistener .com/recap/gov.uscourts.dcd.240916/gov.uscourts.dcd.240916.39.0.pdf.

75  USA v. Search Warrant, Application for Search Warrant, Residence Located in Ket- tering, Ohio, Southern District of Ohio, Aug. 11, 2021.

76  Ryan J. Reilly and Alanna Vagianos, "FBI Arrests 'Tunnel Commander,' an Anti- abortion Extremist Who Fought Cops on Jan. 6," *HuffPost*, Aug. 12, 2021, https:// www.huffpost.com/entry/fbi-arrests-tunnel-commander-anti-abortion-capitol-riot _n_61152eb9e4b07c1403123f7e.

77  Justice Department, "Ohio Man Sentenced on Felony and Misdemeanor Charges Related to Capitol Breach," press release, Feb. 24, 2023, https://www.justice.gov

/usao-dc/pr/ohio-man-sentenced-felony-and-misdemeanor-charges-related-capitol
-breach.

78 Sam Lazar, "Letter 01/03/23," *Jan 6 Defendant Samuel Lazar* (blog), Jan. 3, 2023,
https://web.archive.org/web/20230324140731/https://xenabeauty1208.wixsite
.com/mysite/post/letter-01-03-23.

79 Ryan J. Reilly, "Trump Fan Who Assaulted Officer Fanone on Jan. 6 Sentenced to
More Than 7 Years in Prison," NBC News, Sept. 27, 2022, https://www.nbcnews
.com/politics/justice-department/trump-fan-assaulted-officer-fanone-jan-6-sen
tenced-7-years-prison-rcna49448.

80 "Ashli Babbitt's Mother Speaks Out After Being Arrested on Two-Year Mark of
Daughter's Jan 6 Killing," Newsmax, YouTube, Jan. 6, 2023, https://www.youtube
.com/watch?v=v_EbESkg_C8.

81 Ryan J. Reilly, Ryan Nobles, and Liz Brown-Kaiser, "Kevin McCarthy Met Ashli
Babbitt's Mom Ahead of GOP Visit with Jan. 6 Prisoners," NBC News, March 23,
2023, https://www.nbcnews.com/politics/congress/kevin-mccarthy-met-ashli-bab
bitts-mom-ahead-gop-visit-jan-6-prisoners-rcna76419.

82 Ryan J. Reilly (@ryanjreilly), "This guy has a song about Ashli Babbitt," Twitter,
Sept. 18, 2021, https://twitter.com/ryanjreilly/status/1439258075182084102.

83 Joe Heim, "Official U.S. Capitol Tour Guides Told to Only Mention Jan. 6 If Asked,"
*Washington Post*, Jan. 5, 2023, https://www.washingtonpost.com/dc-md-va/2023/01
/04/january-6-attack-capitol-tour/.

# Epilogue

1 Ryan J. Reilly, "Jan. 6 Rioter Who Said He Was Following Trump's 'Marching
Orders' and Wanted to Arrest Biden and Pelosi Is Found Guilty," NBC News, April 4,
2023, https://www.nbcnews.com/politics/justice-department/jan-6-defendant-said
-followed-trumps-marching-orders-wanted-arrest-bid-rcna77947.

2 Ryan J. Reilly (@ryanjreilly), "Ed Badalian, freshly convicted on Jan. 6 charges,
compares Trump's arrest to the citizen's arrest of Nancy Pelosi and Joe Biden he
hoped 'patriots' and law enforcement would pull off," Twitter, April 4, 2023, https://
twitter.com/ryanjreilly/status/1643314081489625103.

3 Ryan J. Reilly, "Paul Manafort to Serve 7 and a Half Years in Prison in Mueller
Cases," *HuffPost*, March 19, 2019, https://www.huffpost.com/entry/paul-manafort
-sentenced_n_5c882287e4b038892f485569.

4 Executive Grant of Clemency, Paul Manafort, Dec. 23, 2020, https://www.justice
.gov/file/1349071/download.

5 Dan Berman and Katelyn Polantz, "'The American People Cared. And I Care.' Top
Lines from Judge Amy Berman Jackson During the Roger Stone Sentencing," CNN,
Feb. 21, 2020, https://www.cnn.com/2020/02/20/politics/amy-berman-jackson
-quotes/index.html.

6 Executive Grant of Clemency, Roger Jason Stone Jr., Dec. 23, 2020, https://www
.justice.gov/file/1349096/download.

7 Ryan Lucas, "Spotlight Lands on Amy Berman Jackson, Judge in Stone Case, After
a Lengthy Career," NPR, Feb. 25, 2020, https://www.npr.org/2020/02/25/808966785
/after-lengthy-career-spotlight-lands-on-amy-berman-jackson-judge-in-stone
-case.

8    "District Judge Amy Berman Jackson," United States District Court, District of Columbia, accessed May 16, 2023, https://www.dcd.uscourts.gov/content/district -judge-amy-berman-jackson.

9    Emily Heil, "Washington 'Jeopardy' Contestant Matt Jackson Is a Pop-Culture Sensation—and the Son of a Federal Judge," *Washington Post*, Oct. 7, 2015, https://www .washingtonpost.com/news/reliable-source/wp/2015/10/07/washington-jeopardy-con testant-matt-jackson-is-a-pop-culture-sensation-and-the-son-of-a-federal-judge/.

10   Ryan J. Reilly, "Jesse Jackson Jr. Sentenced for Defrauding Campaign," *HuffPost*, Aug. 14, 2013, https://www.huffpost.com/entry/jesse-jackson-jr-sentenced_n_3752476.

11   Deep State Dogs (@1600PennPooch): "THE HANDOFF! Danny Rodriguez #Taser-Prick was indeed handed an electroshock taser in the tunnel by a stranger. That stranger appears to be Kyle Young #AscendDad, who is already charged with attacking Ofc. Michael Fanone! #SeditionHunters," Twitter, Dec. 23, 2021, https:// twitter.com/1600PennPooch/status/1474008133274148864.

12   US v. Young, Sentencing Hearing, Sept. 27, 2022, https://extremism.gwu.edu /sites/g/files/zaxdzs5746/files/Kyle%20Young%20Sentencing%20Hearing%20 Transcript.pdf.

13   Ryan J. Reilly, "Trump Fan Who Assaulted Officer Fanone on Jan. 6 Sentenced to More than 7 Years in Prison," NBC News, Sept. 27, 2022, https://www.nbcnews .com/politics/justice-department/trump-fan-assaulted-officer-fanone-jan-6 -sentenced-7-years-prison-rcna49448.

14   Ryan J. Reilly and Daniel Barnes, "Jan. 6 Rioter Who Dragged Michael Fanone into Crowd Sentenced to 7.5 Years in Prison," NBC News, Oct. 27, 2022, https://www .nbcnews.com/politics/justice-department/jan-6-rioter-dragged-mike-fanone -crowd-sentenced-75-years-prison-rcna54314.

15   Katelyn Polantz, "Judge Amy Berman Jackson Strikes Again," CNN, May 28, 2021, https://www.cnn.com/2021/05/28/politics/judge-amy-berman-jackson/index.html.

16   Steve Peoples, "Republicans Set Opening Presidential Debate for August," AP News, Feb. 23, 2023, https://apnews.com/article/politics-united-states-government -2022-midterm-elections-milwaukee-52b1a9bb8168af7251ddbb17113ff987.

17   Jonah E. Bromwich, William K. Rashbaum, and Kate Christobek, "Dilemma for Judge in Trump Case: Whether to Muzzle the Former President," *New York Times*, April 6, 2023, https://www.nytimes.com/2023/04/06/nyregion/trump-case-judge -juan-merchan.html.

18   Isaac Arnsdorf, Meg Kelly, Rachel Weiner, and Tom Jackman, "Behind Trump's Musical Tribute to Some of the Most Violent Jan. 6 Rioters," *Washington Post*, May 4, 2023, https://www.washingtonpost.com/investigations/interactive/2023/trump-j6 -prison-choir/.

19   Josh Gerstein, "California Salon Owner Charged in Capitol Riot Lingers in L.A. Jail," *Politico*, Feb. 8, 2021, https://www.politico.com/news/2021/02/08/california -salon-owner-capitol-riot-467574.

20   US v. Bisignano, Order Setting Conditions of Release, Feb. 26, 2021, https://storage .courtlistener.com/recap/gov.uscourts.dcd.226937/gov.uscourts.dcd.226937.21.0.pdf.

21   "Karen Goes on Homophobic Rant During L.A. Protest," TMZ, Dec. 1, 2020, https://www.tmz.com/2020/12/01/woman-homophobic-rant-protesting-covid -lockdown-los-angeles/.

22   "Real Patriots of Beverly Hills—Trailer," Real Patriots of Beverly Hills, YouTube, Dec. 6, 2021, https://www.youtube.com/watch?v=TRhjjjeO3uY.

23  US v. Bisignano, Plea Agreement, Aug. 4, 2021, https://storage.courtlistener.com
/recap/gov.uscourts.dcd.226937/gov.uscourts.dcd.226937.39.0_3.pdf.

24  US v. Rodriguez, Indictment, Nov. 17, 2021, https://www.documentcloud.org/docu
ments/21117606-rodriguez-badalian-indictment.

25  Michael Edison Hayden, "'There's Nothing You Can Do': The Legacy of #Pizza-
gate," Southern Poverty Law Center, July 7, 2021, https://www.splcenter.org/hate
watch/2021/07/07/theres-nothing-you-can-do-legacy-pizzagate.

26  Jeremy B. White, "Pelosi Attacker Was Immersed in 2020 Election Conspiracies,"
*Politico*, Oct. 28, 2022, https://www.politico.com/news/2022/10/28/pelosi-attacker
-online-hints-conspiracy-immersion-00064093; US v. DePage, Criminal Complaint,
Oct. 31, 2022, https://www.justice.gov/opa/press-release/file/1548106/download.

27  Carter Young, "Hundreds Clash with Police Following 'Rage' Concert," ABC News,
Aug. 14, 2000, https://abcnews.go.com/Politics/story?id=123137&page=1.

28  Sarah Ferguson, "We Couldn't Move Quickly Enough," *Village Voice*, August 15,
2000, https://www.villagevoice.com/2000/08/15/we-couldnt-move-quickly-enough/.

29  Ken Dilanian and Ryan J. Reilly, "GOP Witnesses Undermined Jan. 6 Cases with
Conspiracy Theories, FBI Says," NBC News, May 18, 2023, https://www.nbcnews
.com/politics/congress/gop-witnesses-undermined-jan-6-cases-conspiracy-theories
-fbi-says-rcna85095.

30  Ryan J. Reilly, "FBI Says Former Agent Arrested Over Jan. 6 Called Officers Nazis
and Encouraged Mob to 'Kill 'Em'," NBC News, May 2, 2023, https://www.nbcnews
.com/politics/justice-department/fbi-says-former-agent-arrested-jan-6-called-officers
-nazis-rcna82567.

31  Ministry of Foreign Affairs of the Russian Federation, "Statement of the Russian
Foreign Ministry in Connection with the Introduction of Personal Sanctions Against
US Citizens," May 19, 2023.

32  Marjorie Taylor Greene, Press Release, "Congresswoman Marjorie Taylor Greene
Introduces Articles of Impeachment Against U.S. Attorney Matthew Graves," May
16, 2023, https://greene.house.gov/news/documentsingle.aspx?DocumentID=437;
Marjorie Taylor Greene, Press Release, "Congresswoman Marjorie Taylor Greene
Introduces Articles of Impeachment Against FBI Director Christopher Wray," May
16, 2023, https://greene.house.gov/news/documentsingle.aspx?DocumentID=438.

33  Alicia Powe, "Obama Banned from Russia, Hundreds Blacklisted for Direct
'Involvement in the Persecution' of J6 'Dissidents'," Gateway Pundit, May 21, 2023,
https://www.thegatewaypundit.com/2023/05/obama-banned-russia-hundreds
-blacklisted-direct-involvement-pesercution/.

34  Peter Baker, "Russia's Latest Sanctions on U.S. Officials Turn to Trump Enemies,"
*New York Times*, May 21, 2023, https://www.nytimes.com/2023/05/21/world
/europe/russia-sanctions-trump.html.

35  Ryan J. Reilly and Daniel Barnes, "A Jovial Stewart Rhodes Tries to Woo Jurors at
Oath Keepers Seditious Conspiracy Trial," NBC News, Nov. 4, 2020, https://www
.nbcnews.com/politics/justice-department/stewart-rhodes-set-testify-oath-keepers
-jan-6-seditious-conspiracy-tri-rcna55531.

36  Ryan J. Reilly and Daniel Barnes, "Stewart Rhodes Wrote Message to Trump After
Jan. 6 Calling on Him to 'Save the Republic' and Arrest Members of Congress," NBC
News, Nov. 2, 2022, https://www.nbcnews.com/politics/justice-department/stewart
-rhodes-wrote-message-trump-jan-6-calling-republic-arrest-membe-rcna55216.

37  Ryan J. Reilly, Daniel Barnes, and Gary Grumbach, "Oath Keepers Founder Sentenced to 18 Years in Jan. 6 Seditious Conspiracy Case," NBC News, May 25, 2023,

https://www.nbcnews.com/politics/justice-department/oath-keepers-founder-sentenced-18-years-jan-6-seditious-conspiracy-cas-rcna85852.

38  Philip Bump, "The White House Tells Media to Ask Kris Kobach to Prove There's Voter Fraud. They Do. He Doesn't," *Washington Post*, Feb. 13, 2017, https://www.washingtonpost.com/news/politics/wp/2017/02/13/the-white-house-tells-media-to-ask-kris-kobach-to-prove-theres-voter-fraud-they-do-he-doesnt/.

39  Melissa Brunner, "KBI Director Tony Mattivi Talks Targeting Fentanyl, Setting Priorities," WIBW, March 20, 2023, https://www.wibw.com/2023/03/20/kbi-director-tony-mattivi-talks-targeting-fentanyl-setting-priorities/.

40  "Jury Convicts Four Leaders of the Proud Boys of Seditious Conspiracy Related to U.S. Capitol Breach," Justice Department, May 4, 2023, https://www.justice.gov/opa/pr/jury-convicts-four-leaders-proud-boys-seditious-conspiracy-related-us-capitol-breach.

41  Ryan J. Reilly, "Philly Proud Boy Testifies at Seditious Conspiracy Trial That He Doesn't 'Recall' If He Used Pepper Spray on Officers on Jan. 6," NBC News, April 19, 2023, https://www.nbcnews.com/politics/justice-department/proud-boy-testifies-jan-6-seditious-conspiracy-trial-rcna79215.

42  "MPD Lieutenant Charged with Obstruction of Justice and False Statements," Justice Department, May 19, 2023, https://www.justice.gov/usao-dc/pr/mpd-lieutenant-charged-obstruction-justice-and-false-statements.

43  Bill Simmons, "13 White Supremacists Acquitted in Arkansas Murder and Sedition Trial," *Washington Post*, April 8, 1988, https://www.washingtonpost.com/archive/politics/1988/04/08/13-white-supremacists-acquitted-in-arkansas-murder-and-sedition-trial/21c30cbe-c120-40ac-8fec-33420d1b0d2e/.

44  Beth Teitell, "Want New Life? Get on a Jury," *Boston Herald*, Oct. 23, 1994.

45  Kathleen Below, *Bring the War Home: The White Power Movement and Paramilitary America* (Cambridge, MA: Harvard University Press, 2018); Southern Poverty Law Center, "Terrorist, '14 Words' Author, Dies in Prison," SPLC Intelligence Report, Oct. 1, 2007, https://www.splcenter.org/fighting-hate/intelligence-report/2007/terrorist-14-words-author-dies-prison.

46  Below, *Bring the War Home*.

47  Heidi L. Beirich, PhD, "The Role of the Proud Boys in the January 6th Capitol Attack and Beyond," Written Statement Before the Congress of the United States Select Committee to Investigate the January 6th Attack on the United States Capitol, March 22, 2022, https://www.justsecurity.org/wp-content/uploads/2023/03/Heidi-Beirich-Expert-Statement.pdf.

48  US v. Lang, Affidavit in Support of Criminal Complaint and Arrest Warrant, Jan. 15, 2021, https://www.justice.gov/opa/page/file/1355866/download.

49  Steve Baker, "From Prison, My Interview with J6 Defendant, Jake Lang," Pragmatic Constitutionalist, Feb. 28, 2023, https://thepragmaticconstitutionalist.podbean.com/e/from-prison-my-interview-with-j6-defendant-jake-lang/.

50  FBI Washington Field (@FBIWFO), "#FBIWFO released photos of this woman who allegedly participated in the U.S. Capitol riots on January 6, 2021. If you recognize her, call 1-800-225-5324 or visit http://tips.fbi.gov to submit a tip. Refer to photo 537 in your tip," Twitter, April 27, 2023, https://twitter.com/FBIWFO/status/1651617941723512832.

51  Ryan J. Reilly, "DOJ Charges 'Pink Beret' Jan. 6 Rioter IDed After an Ex Spotted Her in a Viral FBI Tweet," NBC News, May 8, 2023, https://www.nbcnews.com/politics/justice-department/government-charges-pink-beret-jan-6-rioter-ided-ex-spotted-viral-fbi-t-rcna83339.

52  User comment, "'Pink Beret' Jan. 6 Rioter Charged After Ex Spotted Her in an FBI Tweet," MSNBC, YouTube, May 9, 2023, https://www.youtube.com/watch?v=z0uc OudtHok.

53  Ryan J. Reilly, "Jan. 6 Rioter Who Electroshocked Michael Fanone Shouts 'Trump Won' After He's Sentenced To 12½ Years," NBC News, June 21, 2023, https://www .nbcnews.com/politics/justice-department/jan-6-rioter-electroshocked-dc-officer -michael-fanone-sentenced-125-ye-rcna89388.

## Update, July 2023

1  Justice Department, "30 Months Since the Jan. 6 Attack on the Capitol," July 6, 2023, https://www.justice.gov/usao-dc/30-months-jan-6-attack-capitol.

2  Ryan J. Reilly and Jonathan Dienst, "Long Island Funeral Home Owner Arrested Two Years After Jan. 6 Sleuths ID'd Him," NBC News, June 7, 2023, https://www .nbcnews.com/politics/justice-department/ny-funeral-home-owner-arrested -jan-6-sleuths-ided-rcna88111.

3  Ryan J. Reilly, "DOJ Charges 'Bob's Burgers,' 'Arrested Development' Actor In Jan. 6 Capitol Riot," NBC News, June 7, 2023, https://www.nbcnews.com/politics /justice-department/feds-charge-bobs-burgers-arrested-development-actor-jan -6-capitol-riot-rcna88135.

4  Ryan J. Reilly, "D.C. Chiropractor Who 'Scuffled' With Officers on Jan. 6 is Sentenced to Prison," NBC News, June 14, 2023, https://www.nbcnews.com/politics/justice -department/dc-chiropractor-scuffled-officers-jan-6-sentenced-prison-rcna88931.

5  Ryan J. Reilly, "Judge Orders Jan. 6 Rioter Who Showed Up Outside Obama's Home Detained Until Trial," NBC News, July 12, 2023, https://www.nbcnews.com/politics /justice-department/judge-orders-jan-6-rioter-showed-obamas-home-detained-trial -rcna93411.

6  US v. Sorelle, Order, June 20, 2023, https://storage.courtlistener.com/recap/gov .uscourts.dcd.246841/gov.uscourts.dcd.246841.36.0.pdf.

7  Ryan J. Reilly, Fiona Glisson and Daniel Barnes, "Trump Campaign's Election Day Operations Official Appears Before Jan. 6 Grand Jury," NBC News, June 22, 2023, https://www.nbcnews.com/politics/donald-trump/trump-campaign-official -appears-jan-6-grand-jury-dc-rcna90684.

8  Ryan J. Reilly, "Trump-appointed judge gives a 'break' to Jan. 6 rioter who wants to be a police officer," NBC News, July 7, 2023, https://www.nbcnews.com/politics/justice -department/trump-appointed-judge-gives-break-jan-6-rioter-sentencing-rcna93170.

9  Ryan J. Reilly, "Rioter Who Stormed Capitol While On Bail On An Attempted Murder Charge Gets 3 Years," NBC News, July 11, 2023, https://www.nbcnews .com/politics/justice-department/rioter-stormed-capitol-bail-attempted-murder -charge-gets-3-years-rcna93410.

10  Ryan J. Reilly, "'Idiot' Jan. 6 Rioter Who Stole John Lewis Photo From Nancy Pelosi's Office Gets 4 Years," NBC News, July 14, 2023, https://www.nbcnews.com/politics /justice-department/idiot-jan-6-rioter-stole-john-lewis-photo-nancy-pelosis-office -gets-4-rcna94150.

11  Ryan J. Reilly, "FBI Charges Now-Husband Of 'Pink Beret' Jan. 6 Rioter Who Was Turned In By Her Ex," NBC News, July 17, 2023, https://www.nbcnews.com/politics /justice-department/fbi-charges-now-husband-pink-beret-jan-6-rioter-was-turned -ex-rcna94632.

# Index

Credit: Chase McAlpine

**Ryan J. Reilly** is a justice reporter for NBC News. His unparalleled reporting on the Capitol attack has been cited by the Jan. 6 committee and the FBI. He was also a 2017 Livingston Award finalist for his reporting on jail deaths at *HuffPost*. He regularly appears on MSNBC and NBC News Now, and has been a guest on many other television and radio programs, including *Fresh Air with Terry Gross*. He and his family live in Washington, DC.

@ryanjreilly on Twitter/Instagram